D1448560

PORTLAND PRESS RESEARCH MONOGRAPH

# Plant Cell Division

X

*Other books in this series:*

**Nutrient Regulation of Insulin Secretion**
edited by
P.R. Flatt

**The Annexins**
edited by
S.E. Moss

**Structural and Dynamic Properties of Lipids and Membranes**
edited by
P.J. Quinn and R.J. Cherry

**Hemicellulose and Hemicellulases**
edited by
M.P. Coughlan and G.P. Hazlewood

**Mitochondria: DNA, Proteins and Disease**
edited by
V. Darley-Usmar and A.H.V. Schapira

**Pulmonary Vascular Remodelling**
edited by
J.E. Bishop, J.T. Reeves and G.J. Laurent

**Oxidative Stress, Lipoproteins and Cardiovascular Dysfunction**
edited by
C. Rice-Evans and K.R. Bruckdorfer

**Clinical Pulmonary Hypertension**
edited by
A.H. Morice

**Channelling in Intermediary Metabolism**
edited by
L. Agius and H.S.A. Sherratt

PORTLAND PRESS RESEARCH MONOGRAPH    X

# Plant Cell Division

Edited by
**D. Francis**
**D. Dudits**
**D. Inzé**

Portland Press
London and Miami

571·8442 P

Published by Portland Press Ltd, 59 Portland Place, London W1N 3AJ, U.K.
In North America orders should be sent to Ashgate Publishing Co., Old Post Road,
Brookfield, VT 05036-9704, U.S.A.

Copyright © 1998 Portland Press Ltd, London

ISBN 1 85578 089 5    ISSN 0964-5845

British Library Cataloguing-in-Publication Data
A catalogue record for this book is available from the British Library

Typeset by Portland Press Ltd
Printed in Great Britain by Information Press Ltd, Eynsham, UK

# Contents

# Preface

In October 1995, an EMBO workshop on the plant cell cycle took place at the Institute of Plant Biology in Szeged, Hungary. This meeting drew internationally recognized workers to discuss the cell cycle from the molecular, to the cell, through to the developmental levels [1]. Given this captive audience, we came up with the idea of asking each invited participant to write a substantial review of their research on the plant cell cycle. The majority of the invited speakers responded enthusiastically and the result is what follows. We emphasize that the book is not a proceedings volume; the chapters set out to review and explore each aspect of the plant cell cycle rigorously. We thank all the authors for responding so well.

The current book is the third major volume on the plant cell cycle to appear over the last 11 years. Indeed, it is fascinating to note the growth of the subject, and the turnover of techniques in the current volume compared with the first [2]. This shows how enormously the subject has changed to embrace the techniques of molecular biology and is reflected in our observation that the index of the current volume covers aspects of the cell cycle which were neither recognizable nor measurable 11 years ago [2]. Moreover, the 'new' methodologies were only just starting to appear on the plant cell cycle stage a mere 4 years ago [3]. We hope you will have as much enjoyment reading this book as we had in editing it.

The first eight chapters cover the plant cyclin-dependent protein kinases (Cdks) and the plant cyclins together with cell-cycle-dependent gene expression in plants.

Chapter 1, by L. De Veylder, M. Van Montagu and D. Inzé, reviews current progress on the cell cycle in *Arabidopsis*. Much is now known about the *Arabidopsis* homologues of Cdks. *cdc2At* emerges as the functional homologue of the fission yeast *cdc2* gene, and *cdc2bAt* may well have a more specialized role in regulating cell division. The authors also examine the role of the *cdc2* homologues during plant development by transforming *Arabidopsis* and tobacco with dominant negative mutant forms of *cdc2At*. They specifically try to answer the question of the extent to which developmental programmes depend on cell division.

Chapter 2, by D. Dudits et al., focuses on *cdc2* and cyclin homologues of alfalfa (*Medicago sativa*). In particular, they examine the similarity of the alfalfa *cdc2* to other plant homologues, as well as animal and human cdks. What emerges is that the conserved EGVPSTAIREISLLKE sequence places the alfalfa homologues firmly in a plant grouping which reflects the evolutionary divergence of the plant cdks from their animal and yeast counterparts. A challenging model is also presented which postulates the interaction between $Ca^{2+}$-dependent signal transduction chains, plant growth regulator-dependent signal transduction mechanisms and the expression of plant *cdc* genes.

Chapter 3, by V. Sundaresan and J. Colasanti, reviews cyclin-dependent kinases in relation to the spatial and temporal controls of cell division. The review starts with an overview that 'contrasts' and 'compares' cell cycle regulation. A cautionary conclusion is that at present there is inconclusive evidence for functionally different Cdc2 proteins in higher plants; this theme is taken up at various points throughout this volume. A model is also presented which features Cdc2 kinase in the spatial control of cell division in plants.

Chapter 4, by J.P. Renaudin, A. Savouré, M. Van Montagu, D. Inzé and P. Rouzé, is an in-depth review of plant cyclins. A meticulous analysis of all plant cyclins, from both published sources and from the authors' laboratories, has led to what will hopefully become a universal and logical classification system for the plant cyclins. Phylogenetic analysis of 51 plant cyclins is explained through multiple alignment of conserved amino acid sequences and phylogenetic trees.

Chapter 5, by J.A.H. Murray et al., focuses on the D cyclins which function in late $G_1$ of the cell cycle. By forming heterodimers with Cdks, thereby facilitating kinase activity in the latter, they regulate entry of cells into the cell cycle in animals and, almost certainly, do the same in plant meristems. The authors develop a model which features a Cdk–cyclin D complex responding to auxin- and cytokinin-mediated signal transduction chains in the context of meristem cellular organization and function.

Chapter 6, by H. Hirt, L. Bögre, I. Meskiene, M. Dahl, K. Zwerger and E. Heberle-Bors, explains the advantages and disadvantages of functional analysis of plant cell cycle genes through complementation in yeast. Notably, removal of sequence elements that target proteins for proteolysis in specific alfalfa cyclin genes was necessary to obtain complementation of the function in yeast $G_1$ cyclins. This chapter highlights the usefulness of the complementation approach, but is also careful to emphasize its limitations in understanding the role of plant cell cycle genes *in situ*.

Chapter 7, by B. Plesse, J. Fleck and P. Genschik, examines the ubiquitin-dependent proteolytic pathway and cell cycle control. Until fairly recently, the perpetual puzzle was: how does mitosis end? How are enzymic activities switched off so abruptly? This paper explains how and reviews the elegant ubiquitin-dependent degradation mechanism that exists to bring about the proteolytic destruction of the cyclin partner of the cdk–cyclin complex in anaphase of mitosis. Accumulated evidence is now revealing that ubiquitin-dependent pathways are central to the degradation of cyclins throughout the cell cycle. Exactly how this might be achieved in plant cells is also explored.

Chapter 8, by M. Ito, is a meticulous review of cell-cycle-dependent gene expression patterns in plants. It identifies plant genes that are regulated periodically in the cell cycle and collates information on various genes including proliferating cell nuclear antigen and ribonucleotide reductase. Also, the expression of auxin-dependent genes, which seem to be important for cell cycle progression, is given a thorough airing. The chapter concludes with views on how specific elements within the promoters may mediate the expression of these cell-cycle-dependent genes.

Chapters 9–11 examine cell cycle control in relation to developmental control.

Chapter 9, by D. Francis, considers the extent to which a cell sizer control is an integral component of normal plant development. Each of the major meristematic regions of the plant is reviewed in this context and well-established models of size control are thoroughly discussed. In the absence of native regulators of plant cdks (i.e. no plant homologue to *cdc25* or *wee1*, as yet!), data from experiments which have used the fission yeast *cdc25* to perturb plant development, as well as those emanating from dominant negative mutations of plant *cdc2s*, are collated and presented in a model which tries to explain what the role of a cell sizer may be in plant development.

Chapter 10, by J.H. Doonan, reviews cell proliferation in apical meristems of *Antirrhinum* as they undergo the vegetative to floral transition. RNA *in situ* hybridization experiments show the spatial distribution of gene expression in dividing cells within meristems, and this chapter shows how several cell-cycle-related genes are expressed in a phase-specific manner. In particular, the visualization of cycle genes provides a useful marker of how cell proliferation changes during morphogenesis.

Chapter 11, by A. Fehér, M. Schultze and E. Kondorosi, examines how external signal transduction chains affect endogenous controls of the cell cycle. In particular, the manner in which the rhizobial Nod factor might trigger mitogenic signalling cascades is given close attention. A model is presented which portrays how signal transduction chains may impinge on the cell cycle. Cytokinin/auxin signals, together with metabolic signals and positional information, are all considered in relation to the induction of cell division.

Chapters 12–14 consider DNA replication and histone biochemistry.

Chapter 12, by P.A. Sabelli, S.R. Burgess, A.K. Kush and P.R. Shewry, surveys the so-called licensing factors necessary for DNA replication and revisits Cdc25 phosphatase activity during mitotic activation. The initiation of replication, origins of replication and proteins interacting with replication origins are reviewed. The transition from $G_1$ to S phase is elegantly portrayed through the interaction of various cofactors with origin replication complexes, and a model of how Cdc25 might regulate and activate M-phase-promoting factor is presented.

Chapter 13, by N. Chaubet and C. Gigot, reviews histone gene activity in plants. Through mRNA *in situ* hybridization, expression of histones is examined in developing tissue systems, and, through Northern blotting techniques, histone gene expression during the cell cycle is discussed. Particular attention is given to histone promoter analysis in transgenic plants. Finally, there is a consideration of *cis*-acting elements, thereby highlighting a major difference between animal and plant histone promoters.

Chapter 14, by M. Iwabuchi, T. Nakayama and T. Meshi, reviews transcriptional regulation of plant histone genes. The theme of both *cis*- and *trans*-elements on wheat histone genes is taken up, together with sequence-specific DNA-binding proteins. The review concludes by addressing the probability that various transcription factors are involved in the regulation of plant histone gene expression. The authors also stress the need to study post-translational modification of transcription factors which regulate histone gene expression.

Chapters 15 and 16 cover structural aspects of mitosis and developmental/cell cycle mutants.

Chapter 15, by M. Vantard, V. Stoppin and A.-M. Lambert, reviews plant microtubules in relation to their central role in mitosis. The authors point out molecular mechanisms which regulate microtubule dynamics and microtubule organization. In particular, the microtubular-organizing centres (MTOCs) are given close scrutiny as is how microtubule-nucleating activity is cell cycle regulated. Microtubule-activating (interacting) proteins [MAPs, not to be confused with mitogen-activated protein (MAP) kinases] are also reviewed in relation to their role in microtubule organization. The authors point the way forward in addressing plant MTOC proteins as putative substrates for activated Cdks. Moreover, the preprophase band is given deserved special attention.

Chapter 16, by J. Traas and P. Laufs, is a detailed survey of developmental mutants of higher plants which exhibit abnormal meristematic behaviour. The mutants are discussed in the general framework of meristem function, spatial regulation of mitotic activity and developmental control. This fascinating tour of KNOTTED, SHOOT MERISTEMLESS, ROOTMERISTEMLESS, to name but a few, highlights the potential that exists in studying them. Clearly, a multitude of mutants exist which exhibit normal cell division patterns, but in the wrong order or in the right order, but not in the normal number of layers or even combinations of the two. The authors present a strong argument for basing studies on cell cycle control on specific mutants which, in turn, will further resolve our understanding of cell division in development.

We hope very much that this volume has captured the vigorous and stimulating research which currently is directed at the plant cell cycle. We now have a multitude of homologues to the fission yeast protein kinase, p34cdc2. We also have as many, if not more, homologues to the cyclins, the protein subunit to which the Cdk binds, but without which there is no kinase activity. One major challenge is to understand the extent of the complexity of these homologues, and to develop functional assays that pin-point which catalytic subunit binds with which specific cyclin and, where, in the cell cycle, these cyclin-dependent kinases function. This is a theme that permeates throughout the entire volume. What also emerges as a powerful force is how the expression of plant cdc genes is linked to endogenous developmental programmes. Moreover, the extent to which signal transduction chains impinge on the cell cycle will clearly be a major thrust of future plant cell cycle research. This potentially exciting field, captured in its infancy in this volume, will no doubt carry plant cell cycle research well into the next millenium. Hence, we hope that this book will stimulate and encourage researchers as we enter the next exciting phase of the plant cell cycle story.

Finally, what about the title? Controversial? We think that the assembled chapters cover just about everything that a cell needs, that a cell possesses and that a cell responds to, in order to divide. We cannnot give a definitive answer to what controls mitosis in plants, but we are content that what follows is all to do with 'Plant Cell Division'.

**D. Francis, D. Dudits and D. Inzé,**
*1996*

### References

1.   Francis, D. (1996) Trends Plant Sci. 1, 43–44
2.   Bryant, J.A. and Francis, D. (1985) The Cell Division Cycle in Plants, Cambridge University Press, Cambridge
3.   Ormrod, J.C. and Francis, D. (1993) Molecular and Cell Biology of the Plant Cell Cycle, Kluwer Academic Publishers, Dordrecht

# Abbreviations

## Key

Throughout this volume we have tried to adopt a uniform nomenclature for *cdc* and *cdc*-related genes and their encoded products. Given the perpetual confusion that exists in the cell cycle literature from journal to journal, we cannot guarantee absolute consistency!

*cdc (cdk)* for plant and fission yeast genes
*CDC* for budding yeast gene (and higher eukaryote homologues) (also *CDK*)
Cdc (Cdk) for the protein encoded by that particular *cdc* gene, e.g. Cdc2 denotes the protein kinase encoded by *cdc2*
CDC for the protein encoded by the budding yeast or human homologue

| | |
|---|---|
| **ABA** | Abscisic acid |
| **ACS** | ARS core sequence |
| **ARS** | Autonomously replicating sequences |
| **BAP** | Benzyladenine |
| **CaDPK, CDPK** | Calcium-dependent protein kinase |
| **CAK** | Cdk-activating kinase |
| **CAM** | Calmodulin |
| **CAMKIN** | Calmodulin-dependent protein kinase |
| **CDK** | Cyclin-dependent kinase |
| **CEP** | Carboxyl extension protein |
| **CK** | Casein kinase |
| **CKI** | Cyclin-dependent kinase inhibitor |
| **CLN** | Cyclin(s) of budding yeast |
| **CPIB** | $\alpha$-4-Chlorophenoxy isobutyric acid |
| **CZ** | Central zone |
| **2,4-D** | 2,4-Dichlorophenoxyacetic acid |
| **DAG** | Diacylglycerol |
| **DAPI** | 4-6 Diamidodino-2-phenyl indole |
| **DHFR** | Dihydrofolate reductase |
| **EST** | Expressed sequence tag |
| **FT** | Flow-through |
| **GA** | Gibberellic acid |

| GST | Glutathione S-transferase |
|-----|---------------------------|
| GUS | Glucuronidase |
| HALF-1 | HBP-1-associated leucine-zipper factor-1 |
| HBP | Histone-promoter-binding protein |
| HRGP | Hydroxyproline-rich glycoproteins |
| HSP | Heat shock protein |
| IAA | Indol-3-ylacetic acid |
| $IP_3$ | Inositol 1,4,5-trisphosphate |
| 2-IP | Isopentenyladenine |
| IVT | In vitro transcription–translation |
| LCO | Lipo-chito-oligosaccharide |
| MAP | Microtubule-activating protein |
| MAPK | Mitogen-activated protein kinase |
| MAPKK | Mitogen-activated protein kinase-kinase |
| MBF | MCB-binding factor |
| MCB | MluI cell cycle box |
| MCM | Minichromosome maintenance |
| ME | MPM-2 epitope |
| MI | Mitotic index |
| MPF | Maturation promoting factor |
| MS | Murashige and Skoog culture medium |
| MTOC | Microtubular-organizing centre |
| NAA | Naphthylacetic acid |
| NLS | Nuclear localization site |
| NON | Nonamer |
| NPA | N-(1-Naphthyl)phthalamic acid |
| dNTP | deoxyribonucleoside triphosphate |
| OBRF | Octamer-binding regulatory factor |
| OCT | Octamer |
| ORC | Origin replication complex |
| PCNA | Proliferating cell nuclear antigen |
| PCR | Polymerase chain reaction |
| PGR | Plant growth regulators |
| PKC | Protein kinase C |
| PL | Phospholipase |
| PP1 | Alfalfa phosphatase type 1 |
| PP2 | Alfalfa phosphatase type 2 |
| PP2A | Protein phosphatase 2A |
| PPB | Preprophase band |
| PRM | Pith rib meristem |
| PZ | Peripheral zone |
| R | Restriction point |
| Rb | Retinoblastoma protein |

| | |
|---|---|
| **RNR** | Ribonucleotide reductase |
| **RT** | Reverse transcription (as in RT-PCR) |
| **SAM** | Shoot apical meristem |
| **SBF** | SCB-binding factor |
| **SCB** | Swi4-Swi6 cell cycle box |
| **SEM** | Scanning electron microscopy |
| **ssDBP** | Single-stranded DNA-binding protein |
| **SV40** | Simian virus 40 |
| **TIBA** | 2,3,5-Tri-iodobenzoic acid |
| **TS** | Thymidylate synthase |
| **UBC** | Ubiquitin conjugating |
| **UFO** | Unusual floral organs |
| **UTR** | Untranslated region |
| **WZF** | Wheat zinc-finger protein |

# Cell cycle control in *Arabidopsis*

## Lieven De Veylder*, Marc Van Montagu*† and Dirk Inzé‡

*Laboratorium voor Genetica, Department of Genetics, Flanders Interuniversity Institute for Biotechnology, and ‡Laboratoire Associé de l'Institut National de la Recherche Agronomique (France), Universiteit Gent, K.L. Ledeganckstraat 35, B-9000 Gent, Belgium.

## Introduction

The basic mechanisms controlling the progression through the cell cycle appear to be conserved in all higher eukaryotes, although the temporal and spatial control of cell division can differ largely between organisms. Plants have unique developmental features which are not found in either animals or fungi. First, due to the presence of a rigid cell wall, plant cells cannot move and consequently organogenesis is dependent on cell division and cell expansion at the site of formation of new organs. Secondly, cell divisions are confined to specialized regions, called meristems. These meristems continuously produce new cells which, as they move away from the meristem, become differentiated. The meristem identity itself can change from a vegetative to a reproductive phase, resulting in the formation of flowers. Thirdly, plant development is largely post-embryonic. During embryogenesis, the main developmental event is the establishment of the root–shoot axis. Most plant growth occurs after germination, by iterative development at the meristems. Lastly, as a consequence of the sessile life of plants, development and cell division are, to a large extent, influenced by environmental factors such as light, gravity, wounding, nutrients, and stress conditions.

All these specific features must be somehow reflected in a plant-specific regulation of the factors controlling cell division. In recent years, *Arabidopsis thaliana* has become the model system for genetic and molecular studies. Its small genome size and well-characterized development makes this plant an attractive model for cell cycle research. Fate maps of the root meristem [1] and the shoot apical meristem [2] allow us to trace back differentiated cells to initial divisions in the meristem. Furthermore, many developmental mutants have been isolated and described. In addition, the establishment of a refined genetic map of the genome allows the rapid isolation of novel genes for which only the phenotype is known.

To study how plants exactly regulate cell division, we initiated the characterization of *A. thaliana* genes controlling cell cycle progression and studied their expression pattern during development. This review summarizes our current knowledge and discusses future developments.

†To whom correspondence should be sent at the following address: Laboratorium voor Genetica, Universiteit Gent, K.L. Ledeganckstraat 35, B-9000 Gent, Belgium.

# Cyclin-dependent kinases in *A. thaliana*

## Cloning of *A. thaliana* CDK genes

The control of cell cycle progression is mainly exerted at two transition points: one late in $G_1$, before DNA synthesis, and one at the $G_2/M$ boundary. Progression through these control points is mediated by the sequential activation of heteromeric Ser/Thr-specific protein kinase complexes, known as cyclin-dependent kinases (CDKs). These CDK complexes consist of a catalytic CDK subunit and a regulatory cyclin subunit.

The first described gene encoding a CDK was identified in the yeast *Schizosaccharomyces pombe*, and nominated *cdc2* [3]. Genetic research revealed that the *S. pombe cdc2* gene product, p34[cdc2], is required for both the induction of DNA synthesis (S phase) and the start of mitosis (M phase). A similar function was attributed to the CDC28 gene product, the p34[cdc2] homologue of *Saccharomyces cerevisiae*. Subsequently, *cdc2* homologues were isolated from many organisms including *Homo sapiens*, *Xenopus laevis*, *Gallus gallus* and *Drosophila melanogaster*. Plant CDK genes were identified in *Pisum sativa* (pea), *Medicago sativa* (alfalfa), *Zea mays* (maize), *Oryza sativa* (rice), *Petunia hybrida*, *Glycine max* (soybean), *Antirrhinum majus*, and *A. thaliana* (reviewed by [4]). In contrast to yeasts, higher eukaryotes have evolved to use multiple CDKs to regulate different stages of the cell cycle (reviewed by [5]).

By polymerase chain reaction (PCR) amplification and screening of genomic libraries with the *S. pombe cdc2* gene, two *Arabidopsis* CDK genes were isolated, designated *cdc2aAt* and *cdc2bAt* [6–8]. The encoded proteins share 56% amino acid identity. Cdc2aAt shows the highest identity with the other non-plant eukaryotic Cdc2 homologues (63–67%). The Cdc2bAt protein has a lower identity (50–58%). Cdc2aAt contains a PSTAIRE motif, which is conserved in many eukaryotic CDK proteins, whereas Cdc2bAt contains a PPTALRE motif. These motifs are part of a stretch of 16 amino acids found in all CDK proteins. Mutation studies have shown that this region is essential for cyclin binding [9]. The presence of different motifs in Cdc2aAt and Cdc2bAt suggest that both CDKs might bind specific cyclins belonging to different classes.

In yeasts and mammals, the residues Thr-14, Tyr-15 and Thr-161 (or any equivalent Thr residue) show a cell-cycle-dependent phosphorylation/dephosphorylation pattern, correlating with modulation of the CDK activity [10–15]. These residues are also conserved in Cdc2aAt and Cdc2bAt, but their cell-cycle-dependent phosphorylation *in vivo* has yet to be demonstrated.

The functionality of the Cdc2aAt protein was proven by the complementation of yeast temperature-sensitive cdc2 mutants [6,7]. In mutants expressing the *Arabidopsis cdc2aAt* gene, the elongated shape, characteristic of the *cdc2/CDC28* mutants, was restored to the normal phenotype. Similarly to human, *Drosophila* and other plant CDKs, this complementation was incomplete. Partial complementation is not surprising because the cdc2 protein has to interact with numerous other proteins involved in cell cycle control (e.g. cyclins, p13[suc1], and various substrates). Some of these interactions may be species-specific. Alternatively, it is possible that distinct plant CDKs are active in different cell cycle phases and may only be able to complement mutants affected in specific phases.

The *cdc2bAt* gene product did not complement yeast *cdc2/CDC28* mutants. This result suggests that *cdc2aAt* might be the functional homologue of the yeast *cdc2* gene and that *cdc2bAt* might have a more specialized role in the regulation of plant cell division. One such function could be the activation of the Cdc2aAt protein. In all eukaryotes, full activation of CDK activity requires the phosphorylation of a Thr residue located at position 161 (or the equivalent Thr residue). In humans, this phosphorylation is carried out by the MO15 kinase, representing itself a CDK/cyclin complex [16–19]. Possibly, Cdc2bAt plays a role similar to that of MO15 kinase in plants, leading to the cell-cycle-specific activation of Cdc2aAt. Alternatively, Cdc2aAt and Cdc2bAt could bind different cyclins and phosphorylate different substrates, or could be functional at different time points of the cell cycle (see below).

## Developmental regulation of *cdc2aAt* and *cdc2bAt* expression

In higher eukaryotes the expression of CDK genes is, at least in part, transcriptionally regulated. In mammals, a positive correlation is found between the proliferative state of the tissues and the abundance of *cdc2* mRNA [20,21]. Also in *Arabidopsis*, variable amounts of *cdc2aAt* transcript levels were detected in different tissues as revealed by RNA gel blot analysis [6,22]. *Cdc2aAt* transcript is most abundant in roots, high in the stem and flowers, but not present in mature leaves.

A more detailed picture of the *cdc2aAt* expression pattern was obtained by the use of *in situ* hybridization techniques and the use of transgenic plants containing the *cdc2aAt* promoter fused to the β-glucuronidase (*gus*) reporter gene [22,23]. Early after seed germination, *cdc2aAt* expression is only found in the root apical meristem. Later in development, expression is also found in the shoot apical meristem and at the site of adventitious root formation. This timing of *cdc2aAt* expression corresponds with the order of activation of the meristems. [³H]Thymidine-labelled roots have shown that the early pattern of *cdc2aAt* expression coincides with that of [³H]thymidine incorporation. Later in development, expression of *cdc2aAt* is restricted to the shoot apical meristem and developing leaves. In older leaves, *cdc2aAt* expression declines rapidly. In the roots, *cdc2aAt* is expressed in the root meristem, but also in the parenchyma of the vascular cylinder and in the entire pericycle. The pericycle is a differentiated tissue surrounding the vascular cylinder which retains a potential to divide and is involved in lateral root formation and root thickening. Expression in these pericycle cells was lower in older parts of the root.

The absence of light delays the expression of *cdc2aAt* in the aerial part of the plant. Light-grown seedlings show β-glucuronidase (GUS) activity in the shoot apical meristem 4 days after germination. In contrast, dark-grown seedlings show no GUS activity in this region until they are 7 days old. Interestingly, as demonstrated by the absence of [³H]thymidine incorporation, no active divisions occur in the meristem of 7-day-old etiolated seedlings. Also in wounded leaves there is a high level of *cdc2aAt* expression, although here too no significant cell divisions occurred in the conditions used.

The expression data show that *cdc2aAt* is highly expressed in dividing cells. The correlation between the abundance of the CDK transcript and the proliferative state of the tissue was similarly demonstrated in maize and alfalfa [24,25]. In differentiated *Arabidopsis* cells, the *cdc2aAt* transcript level ceases. *Cdc2aAt* transcription is, however, not completely restricted to dividing cells. Cells with high competence to divide retain *cdc2aAt* transcription (e.g. expression in the pericycle cells). Furthermore, expression in the shoot apical meristem of etiolated seedlings and wounded leaves suggests that *cdc2aAt* expression is a reflection of the competence of these cells to divide. These exciting observations are in contrast to those found in mammals, where no *cdc2* transcript is detected in differentiated cells [20,21]. This difference of spatial expression of *cdc2* between plants and animals might reflect unique features of their developmental programmes. The expression pattern of *cdc2aAt* in plants can be regarded as the sign that cells are on 'standby' to initiate division, waiting for intrinsic signals or favourable environmental conditions. Such a cellular state was defined as 'competence' [23]. In plants, fully differentiated cells are, most often, able to dedifferentiate and undergo division. Differentiated cells do not express *cdc2aAt* and, before division, *cdc2aAt* should be induced. At this moment, a totipotent and differentiated cell becomes a competent cell. A clear example is the activation of *cdc2aAt* in wounded *Arabidopsis* leaves. The expression data also indicate that *cdc2aAt* expression is not sufficient to trigger cell division and that additional components are essential; the most putative candidates are cyclins (see below).

Whole-mount *in situ* hybridizations with a *cdc2bAt*-specific probe revealed an expression pattern almost identical to that for *cdc2aAt* [105]. The presence of *cdc2bAt* transcript in the shoot apical meristem of etiolated seedlings indicates that, as for *cdc2aAt*, *cdc2bAt* expression is linked with competence of division. Interestingly, *cdc2bAt* is additionally expressed in the hook region of etiolated seedlings, where no *cdc2aAt* transcript is detectable. The involvement of phytohormones (auxins and gibberellins) and endo-reduplication in the process of hook formation might trigger the specific transcription of *cdc2bAt* in this region.

Surprisingly, expression of *cdc2bAt* shows a cell-cycle-dependent expression pattern as revealed by *in situ* hybridizations of longitudinal sections of *Arabidopsis* vegetative apices. *Cdc2aAt* transcripts are uniformly distributed in all cells of young leaf primordia, procambium, and meristems. In contrast, *cdc2bAt* transcripts show a patchy expression pattern in the same tissues. These results indicate that *cdc2bAt* transcription is cell cycle regulated. Stained cells most probably reflect cells captured into a specific cell cycle phase. Hybridization of sorted nuclei, and *in situ* hybridizations on root tips treated with specific cell cycle phase inhibitors, indicate that *cdc2bAt* transcription occurs in the S and G$_2$ phases of the cell cycle. Similar experiments revealed that *cdc2aAt* is constitutively expressed through the cell cycle [105].

A cell-cycle-dependent expression pattern of CDKs has also been reported for other CDK plant homologues. In *Antirrhinum*, four different CDK homologues were identified. At least two of these CDKs (*cdk2c* and *cdk2d*) show a fluctuating expression pattern through the cell cycle ([26]; see J.H. Doonan, Chapter 10). The *Antirrhinum* Cdc2c protein contains the PPTALRE motif, which is similar to that found in the *Arabidopsis* Cdc2bAt protein. The *Antirrhinum* Cdc2c and *Arabidopsis* Cdc2bAt proteins share 83% amino acid identity [106]. These data indicate that in plants at least two different classes of CDK genes exist. A first class contains constitutively expressed CDKs with an intact PSTAIRE motif. The second class comprises CDKs showing a cell-cycle-regulated expression pattern and contain the PPTALRE motif. The presence of several CDK classes in plants is analogous to the situation in mammals where different classes of CDK proteins have been shown to bind different cyclins and to be functional at distinct time-points of the cell cycle [5]. The differences between *cdc2aAt* and *cdc2bAt* again suggests specific roles for each of the CDKs in cell cycle regulation, rather than being redundant genes.

## Effects of hormones on *cdc2aAt* expression

Plant hormones, in particular auxins and cytokinins, have a pronounced effect on cell division. How auxins and cytokinins influence the cell cycle is almost completely unknown. To study the effects of plants hormones on *cdc2aAt* expression, roots of transgenic plants containing the *cdc2aAt* promoter fused to the *gus* gene were chosen as a model system [23]. Auxins (indole-3-acetic acid, naphthalene-2-acetic acid and 2,4-dichlorophenoxyacetic acid) have a 2-fold effect on *cdc2aAt* expression. On the one hand, they inhibit *cdc2aAt* expression in a well-defined zone of the root apical meristem and, on the other hand, they stimulate expression in zones distal to the inhibition zone. The relevance of this observation is still unknown. *In situ* hybridizations using a *cdc2aAt*-specific probe gave similar results. [³H]Thymidine labelling of auxin-treated roots indicated that DNA synthesis occurred in the zone in which *cdc2aAt* expression was inhibited.

Cytokinins cause an increase of *cdc2aAt* transcription in the upper part of the main root. [³H]Thymidine incorporation showed that this is correlated with cell division. Treatment with abscisic acid inhibits *cdc2aAt* transcription in the vascular tissue of the entire root. The expression in the main root tip is, however, unaffected. Application of ethylene precursors does not have any significant effect on the expression of *cdc2aAt*.

In all cases it is important to stress that the observed expression patterns are not only due to the exogenous hormones, but are rather caused by the interaction of the applied hormones with internal factors produced by the different parts of the plants. The auxin effect on root tips and the induction of *cdc2aAt* expression by cytokinins were seen only in intact plants, and not in excised root tissue that might be rapidly depleted of endogenous hormones [27,28].

*Cdc2aAt* expression during de-differentiation was examined in leaf mesophyll protoplasts derived from tobacco plants expressing the *gus* gene under the control of the *cdc2aAt* promoter [23]. In the presence of appropriate concentrations of auxin and cytokinin to stimulate division, an induction of GUS activity was detected. Cultivation with either auxin or cytokinin alone was insufficient to induce cell division, but, nevertheless, both hormones induced *cdc2aAt* transcription. This indicates again that the induction of competence to divide, triggered by the hormones, is sufficient to induce *cdc2aAt* expression.

The rapid (within 10 min) induction of *cdc2aAt* by auxins indicates that the induction is probably at the level of transcription [29]. In agreement with this, auxin-responsive elements are located within the *cdc2aAt* promoter [30]. Also an abscisic-acid-responsive element and a number of putative regulatory motifs (Myc and Myb protein-binding sequences) were identified, although binding to these motifs remains to be revealed.

## Cellular distribution of Cdc2aAt during the cell cycle

CDK/cyclin complexes were shown to have a cell cycle phase-dependent subcellular localization. For example, human cyclin A is nuclear from its time of synthesis during S phase and remains predominantly nuclear in the S and $G_2$ phases until its destruction. In contrast, human cyclin B1 accumulates in the cytoplasm of interphase cells and enters the nucleus immediately before mitosis [31]. Even the two closely related cyclins B1 and B2 show a completely different localization pattern during cell cycle progression [32]. The translocation to specific parts of the cell of the CDK units by their binding to particular cyclins is believed to determine the substrate specificity of each CDK/cyclin complex.

Using specific antibodies, the cellular distribution of Cdc2aAt during the cell cycle was studied in *Arabidopsis* root tip cells (H. Stals, unpublished work). During the interphase, Cdc2aAt is predominantly present in the nucleus, excluding the nucleolus. When cells enter mitosis Cdc2aAt associates with the preprophase band (PPB) and later colocalizes with the spindle. During the metaphase, Cdc2aAt disappears from the condensed chromosomes. At the anaphase, it temporally re-colocalizes with the chromosomes, but it disappears again during late anaphase. In the telophase, Cdc2aAt migrates to the newly formed daughter nuclei, excluding the nucleolus.

The Cdc2aAt localization pattern is largely similar to the situation found in mammals [33]. Its different localizations during the cell cycle enables Cdc2aAt to perform distinct roles at various time-points, by phosphorylating different substrates. Putative intranuclear substrates for Cdc2aAt could be the retinoblastoma protein and p53 (both involved in checkpoint transitions) and proteins involved in the process of DNA replication (reviewed in [34]; see P. Sabelli et al., Chapter 12). Chromosome condensation is accompanied by intensive phosphorylation of histone H1, which is an excellent substrate of CDK/cyclin complexes *in vitro* [35]. The association with the spindle and the chromosomes in mitosis suggests that Cdc2aAt-mediated phosphorylation might also be involved in the establishment of the spindle architecture and the chromosome alignment (see V. Sundaresan and J. Colasanti, Chapter 3).

Interestingly, as for a maize CDK [36], colocalization of Cdc2aAt with the PPB and phragmoplast was observed. The PPB is a plant-specific cytoskeletal structure which girdles the cell at the prophase of the cell cycle, precisely at the place of the future cell wall. At the beginning of mitosis, before the appearance of the metaphase spindle, the PPB disappears again [37,38]. The absence of a PPB-like distribution of p34$^{cdc2}$ in cells where PPB formation is inhibited suggests that a major role of the PPB might be to direct the p34$^{cdc2}$ kinase to specific sites located on the cell cortex [36]. p34$^{cdc2}$-mediated phosphorylation of various unknown targets could, therefore, be involved in the imprinting of the place of new cell wall synthesis.

Immunolocalization patterns of the different isolated *Arabidopsis* cyclins will provide us with more clues about the distinct roles of each CDK/cyclin complex and will clarify whether the shunting of the Cdc2aAt protein to and from the chromosomes during the cell cycle is mediated through its binding with specific cyclins.

## Overexpression of dominant mutant *cdc2aAt* genes in plants

Morphogenesis and histogenesis occur throughout plant life due to continuous iterative growth at the meristems. This implies that cell division, elongation, and differentiation occur throughout development and must be highly co-ordinated. To study to what extent these processes are linked to each other, the effects of overexpression of wild type and dominant mutant *cdc2aAt* genes on plant development were studied.

Overexpression of the wild-type *cdc2aAt* gene in *Arabidopsis* caused a 2-fold increase of extractable Cdc2 histone H1 kinase activity, without having any noticeable effect on the plant phenotype or change of cell cycle kinetics [39]. Only occasionally did plants show a tendency to form more shoots. The absence of severe phenotypes is not surprising since full Cdc2aAt activation requires several regulatory proteins that might be limiting. Moreover, this result is in agreement with the observation that *cdc2aAt* expression in plants is not linked with division *per se*.

Overexpression of a *cdc2aAt* allele (Cdc2.A14Y15), carrying mutations of the Thr-14 and Tyr-15 residues to non-phosphorylatable amino acids, did not result in abnormal plant development, with exception of the loss of apical dominance of some plants. In fission yeast, the Tyr-15 residue of p34$^{cdc2}$ becomes phosphorylated during the G$_2$ phase of the cell cycle [10]. A phosphoryl group on this residue, which is located in the ATP-binding domain, prevents correct ATP binding and thereby inhibits kinase activity. At the start of mitosis, the Tyr-15 residue is dephosphorylated by a specific phosphatase (Cdc25), and thereby the p34$^{cdc2}$ kinase is activated. In *S. pombe*, mutation of the Tyr-15 residue to the non-phosphorylatable Phe-15 causes a premature mitosis (the Wee phenotype), suggesting that Tyr-15 phosphorylation is crucial for the timing of p34[cdc2] activation, and that of entry in mitosis. In higher eukaryotes, the Thr-14 residue has also been identified as a major phosphorylation site [12–14]. The phosphorylation of Thr-14 works in conjugation with Tyr-15 phosphorylation, inhibiting p34[cdc2] activation.

Despite the presence of Tyr-15 in essentially all reported plant p34[cdc2] homologues, no rigorous biochemical evidence for the regulatory phosphorylation at this site has been demonstrated in plants yet. The lack of phenotype in *Arabidopsis* plants overproducing Cdc2.A14Y15 suggests that the regulation of Cdc2aAt by Thr-14 and Tyr-15 phosphorylation may not be important for controlling the onset of mitosis, just as described for *S. cerevisiae* [40,41]. In agreement, despite many attempts, no plant gene encoding a Cdc25 phosphatase has been isolated up to now. In favour of a role of Thr-14 and Tyr-15 phosphorylation is the observation that overexpression of the *S. pombe cdc25* gene in tobacco plants causes an altered leaf morphology and changes to the timing of flowering ([42]; see D. Francis, Chapter 9).

The overexpression of a dominant mutant allele of *cdc2aAt* (*cdc2.N146*), however, has drastic effects upon cell division. This mutant *cdc2aAt* gene has a substitution of Asp-146 for Asn-146. The three-dimensional structure of a human CDK protein showed that the Asp-146 residue is essential for the Mg–ATP binding [43]. Mutation of this residue blocks the transfer of the γ-phosphate group of ATP to substrates, eliminating CDK kinase activity. Expression of similar *cdc2* mutants in wild-type cells of *S. cerevisiae* and human caused a complete block of cell division [44,45].

*Arabidopsis* plants overexpressing the *cdc2.N146* gene could not be obtained. From three independent transformations with the CaMV 35S-*cdc2.N146* construct, only three transformants were obtained: two escapes and one containing a non-expressed transgene. Transformation of tobacco plants with the same construct resulted in the isolation of only three plant lines showing low *cdc2.N146* expression. The dominant negative effect of *cdc2.N146* can be explained by assuming that the mutant Cdc2 kinase is still capable of binding to regulatory factors, resulting in a competition between the wild-type and mutant allele. Studies with an analogous mutation in human CDKs showed that cyclin titration is, at least, part of the mechanism of kinase inhibition [45]. The ability to rescue some *cdc2.N146*-overexpressing tobacco plants might result from a less efficient interaction of the *Arabidopsis* Cdc2.N146 mutant protein with regulatory molecules in a heterologous system.

The expression of *cdc2.N146* is correlated with a reduction of CDK activity in plant extracts. The transgenic tobacco seedlings have shorter roots with a reduced number of lateral roots, in comparison with wild-type seedlings. Cotyledons and first leaves have a reduced and more elongated size. Flowering plants show smaller internodes, leaves, flowers, and seed pods. Microscopic analysis revealed that cells in all tissues (with exception of flowers) are 5–10-fold larger than normal, without affecting the anatomical structure of the organs. The shoot apical meristem has a reduced size, but the cells are not larger than in controls. The root apical meristem, normally well organized in a tiered set of initials, is completely disorganized. Most interestingly, the developmental timing of these transgenic plants is unaffected. New leaves were initiated at the same rate as in untransformed controls.

Flow cytometric analysis, using isolated nuclei from mature leaves of 2-month-old plants, showed that the *cdc2.N146*-overexpressing plants have a similar ratio of $G_1/G_2$ nuclei as wild-type plants. These results indicate that all cell cycle phases are equally affected by the overproduction of the mutant Cdc2aAt protein and suggest that Cdc2aAt is involved in the regulation of both the $G_1/S$ and $G_2/M$ checkpoint transitions.

The *cdc2.N146* overexpression data show that a reduction of cell division results in smaller plants without having any effect on cell differentiation or morphology. The plant structure is maintained, with all organs conserving their normal number of cell layers. The problem of the fewer cells is compensated by a polarized increase of cell growth. These results support the lack of a causal relationship between cell division and morphogenesis. Similar conclusions were also obtained by other researchers. Traas et al. [46] demonstrated that *Arabidopsis* mutants deficient in normal cell expansion and alignment of their division planes exhibit normal organogenesis and differentiation (see J. Traas and P. Laufs, Chapter 16). Furthermore, it has recently been shown that the elimination of root initial cells by laser ablation had no severe effect upon root differentiation [47]. Moreover, the analysis of the *shoot meristemless* mutant from *Arabidopsis* suggests that an organized shoot apical meristem might not be essential for leaf development [48]. All these individual observations support the model that the organismal shape of plants develops independently from cell divisions. Extreme examples are found in diverse marine algae. These algae (*Caulerpa*) have a plant body differentiated in various structures, without showing cell partitioning. The lack of a causal relationship between cell division and morphogenesis has also been demonstrated in γ-irradiated wheat seedlings [49,50]. These seedlings showed, until one week after irradiation, normal plant growth without any DNA replication or cell division. In both the Cdc2.N146-overproducing and γ-irradiated seedlings, plant development is solely mediated by cell enlargement, without affecting cell differentiation. The uncoupling of cell division from morphogenesis illustrates the extraordinary developmental flexibility of plants, and suggests plants have developed a redundancy of mechanisms to regulate their morphogenesis.

The same *cdc2a.N146* mutant was also expressed in *Arabidopsis* plants under the control of the 2S2 (seed storage albumin) promoter. This promoter drives specific expression during embryo development [51]. Analysis of 13 independent lines transformed with the *2S2-cdc2a.N146* construct revealed three lines showing a low percentage of abnormal seed germination. In one line, approximately 12% of the seeds did not germinate. In the second line, 5–10% of seedlings had three cotyledons, fused cotyledons, or an abnormal pattern of first leaves. The third line exhibited 5–10% of completely abnormal seedlings, consisting of cotyledon-like structures without root development (A. Hemerly and P. Ferreira, unpublished work).

During early embryo development, most of the divisions are formative and fixed in their orientation. Perturbations in the normal sequence of initial divisions in the embryo give rise to abnormal embryogenesis, as shown for the *gnom* and *monopteros* mutants [52,53]. Therefore, in contrast to post-embryonic development, frequencies and division planes of cell divisions during embryo formation are important. The expression of *cdc2a.N146* driven by the 2S2 promoter in the embryo will most probably have an effect on the normal sequence of cell divisions during embryogenesis, causing drastic consequences on the subsequent development. A detailed microscopic analysis of embryogenesis in these mutants will help reveal the role of established divisions during embryo development.

In yeast many other mutations of *cdc2* have been described [54–58]. In general, four distinct classes of mutants have been isolated. A first class of mutants arrests after DNA replication without displaying chromosome condensation, suggesting a block in $G_2$. A second class displays a premature entry into mitosis. The third class of mutants exhibits an arrest in the M phase, as revealed by the presence of condensed chromosomes and abnormal septa. The last class exhibits an increasing DNA content in the absence of mitosis, suggesting an uncoupling of the S phase from the M phase. It is of considerable interest to construct similar mutations in plant CDKs and study the effect of their expression on cell division and plant development.

## Cyclins of *A. thaliana*

### Isolation and characterization of mitotic *Arabidopsis* cyclins

Monomeric CDK proteins have almost no kinase activity. Full activation of the CDK protein activity requires (at least) the binding of the CDK catalytic subunit to regulatory proteins, called cyclins. Enzymic measurements of kinase activity of monomeric CDKs versus the cyclin-bound forms have shown that cyclin binding leads to a 40,000-fold increase of kinase activity [59]. The three-dimensional structure of the human cyclin A–CDK2 complex revealed that, most probably, this increase of kinase activity is completely due to the conformational changes in the CDK2 protein caused by the cyclin binding [60].

All the cyclins share a highly homologous region of 100 amino acids, termed the cyclin box, which is required for their interaction with the CDK catalytic subunit [60,61]. Since cyclins bind to a region located closely to the CDK substrate recognition domain, cyclin binding probably plays a role in determining the substrate specificity. Besides, it is believed that cyclins regulate the substrate specificity of the CDKs by translocating the complexes to specific locations in the cell.

Cyclins are furthermore characterized by their cell-cycle-dependent timing of appearance. They are classified into different groups according to their sequence and cell cycle phase-specific expression pattern. In every organism studied so far, the complexity of the cyclin gene family is huge. In a relative simple organism such as *S. cerevisiae*, one CDK protein is accompanied by at least nine different cyclins belonging to two different classes. One class of cyclins functions in the $G_1$ phase of the cell cycle, and is called CLN cyclins [62,63]. The second class, termed B-type cyclins, promotes both the onset of S phase (CLB5, CLB6) and M phase progression (CLB1, CLB2, CLB3, and CLB4) [64–68]. In mammals, up to eight different cyclin groups are identified until now, nominated A through H. A-, B- and F-type cyclins are required for entry into mitosis [69–73] and are, therefore, referred to as mitotic cyclins. A-type cyclins also play a role during the S phase [74,75]. Three groups of $G_1$ cyclins, nominated C, D, and E, have been identified in animal cells by their ability to rescue yeast CLN mutants [76]. E-type cyclins are required for progression from $G_1$ to the S phase [77].

The instability of mitotic cyclins is attributed to the presence of a conserved sequence at the N-terminus, termed the destruction box. This sequence is a target for ubiquitin-mediated proteolysis [78,79]. $G_1$-specific cyclins have no destruction box, but contain PEST sequences, which are rich in proline, glutamic acid, serine and threonine residues, accounting for great instability to the cyclin proteins (reviewed in [80]).

In *Arabidopsis*, until now, 11 different cyclins have been reported and classified into four groups. The first *Arabidopsis* cyclin gene (*cyc1At*) was isolated by PCR using primers deduced from conserved regions of the cyclin box [81]. *Cyc1At* shows significant homology to mitotic (M phase)-specific cyclins, including the presence of a mitotic destruction box in the N-terminus. *Cyc1At* mRNA micro-injection could drive *Xenopus* oocyte maturation, thereby demonstrating its functionality. Screening of cDNA and genomic libraries of *Arabidopsis* with PCR fragments helped to identify four additional mitotic cyclin homologues (*cyc2aAt, cyc2bAt, cyc3aAt, and cyc3bAt*) [82]. The deduced amino acid sequences of *cyc2aAt* and *cyc2bAt* are 71% similar, and those of *cyc3aAt* and *cyc3bAt* 72%. *Cyc1At* is only approximately 35% similar to the cyclin classes 2 or 3. The cyclins of both classes 2 and 3 have an N-terminally located, destruction box. In addition, *cyc2aAt* and *cyc2bAt* contain PEST sequences.

Sequence analysis did not allow us to clearly classify the different mitotic *Arabidopsis* cyclins as A- or B-type. All isolated plant cyclins appear to be more related to each other than to any particular group of animal or yeast cyclins. *Arabidopsis* root tips treated with oryzalin, which inhibits microtubule polymer-ization and arrests cells at the metaphase, revealed accumulation of *cyc1At* transcript, as shown by whole-mount *in situ* hybridizations. In contrast, treatment with hydroxyurea, which blocks DNA replication and arrests cells at early S phase, led to an absence of *cyc1At* transcript [82]. This indicates that *cyc1At* transcript is most abundant in the $G_2$ and M phases of the cell cycle, which fits with its higher expression in $G_2$ nuclei [81]. These results suggest that *cyc1At* acts more like a B-type cyclin.

In contrast to what is found for *cyc1At*, oryzalin-treated *Arabidopsis* root tips showed lower expression of *cyc2aAt*, *cyc2bAt*, *cyc3aAt* and *cyc3bAt* than hydroxyurea-treated tips, suggesting that these cyclins are functional in the S and $G_2$ phases of the cell cycle. Hybridizations of nuclei from *Arabidopsis* sorted by flow cytometry on the basis of their DNA content, using the *Arabidopsis* cyclins as probes, confirmed these results.

To obtain a clearer picture of the transcriptional regulation of the *cyc1At* and *cyc3aAt* genes, their transcriptional activation was followed during a complete cell cycle. Since *Arabidopsis* cell suspension cultures are difficult to synchronize, tobacco BY2 cell cultures were chosen for the ease of synchronization [83]. These cultures were stably transformed with chimeric constructs containing the *cyc1At* or *cyc3aAt* promoter fused to the *gus* reporter gene. The progression through the cell cycle was followed by measurement of histone H4 transcription (as a marker for S phase) and mitotic index determination (as a marker for M phase). The transcriptional activity of the *cyc1At* and *cyc3aAt* promoters was determined by the measurement of *gus* steady-state mRNA levels on RNA gel blots.

The rate of transcription of the *cyc1At* promoter was low in S phase, started to increase after the S/$G_2$ boundary, and reached its maximum at the $G_2$/M transition and during M phase. Transcript levels decreased after completion of mitosis, were low during the second S phase, and started to increase again afterwards. In contrast, *cyc3aAt* transcription was low in the $G_1$ phase, slowly increased during S phase, and peaked at the $G_2$ phase and $G_2$/M transition [84].

The expression patterns of *cyc3aAt* and *cyc1At* correspond to those of the mammalian A- and B-type cyclins, respectively [85,86]. As for *cyc3aAt*, human cyclin A transcription is initiated at the onset of the S phase and reaches its maximum at the $G_2$/M transition. In contrast, transcription of B-type cyclin starts at the end of the S phase and peaks at M phase, as for *cyc1At*. These results show that, although there is no clear promoter sequence homology, the transcriptional mechanisms controlling cyclin expression seem to be conserved among unrelated species. They also suggest that *Arabidopsis* cyclin promoters contain all information needed to confer a cell cycle phase-specific expression pattern. It is of great interest to identify the particular sequences which direct the cell-cycle-dependent appearance of the cyclin transcripts and, eventually, to isolate the *trans*-acting factors that they bind. To obtain a more complete picture of cyclin expression, it will also be necessary to study other putative levels of regulation, such as control of mRNA stability, mRNA translation, and protein stability.

Day and Reddy [87] identified three additional *Arabidopsis* cyclins using a similar PCR strategy. The resulting clones have been partially sequenced and their classification awaits further data.

## The isolation of D-type cyclins of *Arabidopsis*

Human D-type cyclins were isolated by their ability to rescue yeast mutants deficient in $G_1$ cyclins (CLN mutants) [75]. Their expression is strongly dependent on mitogenic signals, such as growth factors and, therefore, D-type cyclins have been suggested to mediate mitogenic stimuli with the release from quiescence ($G_0$) [88,89]. This is believed to be regulated through the hyperphosphorylation of the retinoblastoma protein (Rb) by CDK/cyclin D complexes [90,91]. Rb is a tumour-suppressor protein which plays an important role in controlling the onset of cell division. In its hypophosphorylated form, Rb is complexed with E2F-type transcription factors which are known to promote expression of S-phase-specific genes [92]. Binding of Rb to E2Fs thereby prevents S-phase induction. Phosphorylated Rb is unable to form complexes with E2F transcription factors and allows DNA synthesis. All D-type cyclins show a specific amino acid motif (LXCXE) permitting them to bind Rb.

By complementing a yeast mutant strain deficient in its CLN cyclins, Soni et al. [93] isolated three different *Arabidopsis* cyclin genes showing strongest homology to the human D-type cyclins, named *cycδ1*, *cycδ2* and *cycδ3*. These δ-type cyclins share approximately 30% sequence identity to each other. In each sequence, several potential PEST sequences were identified; moreover, they contain a LXCXE Rb interaction motif, suggesting that a plant Rb homologue protein might exist (see J.A.H. Murray et al., Chapter 5).

RNA gel blot analysis revealed the presence of δ-type cyclins transcripts in all *Arabidopsis* tissues examined. To determine their window of expression during the cell cycle, *Arabidopsis* cell suspensions were treated with either hydroxyurea or colchicine, blocking cells in early S phase or at metaphase, respectively. With both treatments a great reduction of *cycδ3* transcript was observed. Upon release from the hydroxyurea block, induction of *cycδ3* mRNA slightly preceded the induction of histone H4 gene expression in the S phase, indicating that this cyclin has a function at the onset or continuation of the S phase. *cycδ1* and *cycδ2* transcripts were low in either hydroxyurea- or colchicine-treated cells, and remained low after the release from the hydroxyurea block, suggesting that these cyclins are expressed during the S phase before the point at which hydroxyurea blocks cell cycle progression.

The possibility that, similar to mammalian D-type cyclins, the expression of the δ cyclins is governed by growth signals was studied in *Arabidopsis* cell suspension cultures. Cultures were first depleted of the required plant hormones [2,4-dichlorophenoxyacetic acid (2,4-D) and kinetin] and the sucrose carbon source. Transcript levels of the various δ cyclins were determined upon readdition of various combinations of these compounds. *cycδ3* expression was induced by kinetin alone, and more strongly by kinetin plus sucrose, although the absence of 2,4-D did not allow these cells to enter S phase. The presence of 2,4-D reduced the level of this induction. In contrast, induction of *cycδ2* transcription was solely dependent on sucrose supply. These results show that *Arabidopsis* cyclins *cycδ2* and *cycδ3* are involved in linking the presence of growth regulators and the nutritional status, with cell cycle progression. Moreover, the differential responsiveness of the two δ cyclins towards distinct mitogenic signals suggests that both cyclins are involved in different signal transduction pathways (see J.A.H. Murray et al., Chapter 5).

Plants may have evolved a cell cycle reactivation mechanism similar to that of yeasts. *S. cerevisiae* contains three different $G_1$ cyclins, termed CLN1, CLN2 and CLN3. CLN1 and CLN2 transcript levels fluctuate throughout the cell cycle, but that of CLN3 is related to the nutritional status and size of the yeast cells [94]. When cells are ready to enter a new cell cycle, the CDC28/CLN3 complex triggers transcription of the CLN1 and CLN2 genes, which in turn participate in the activation of genes necessary for DNA replication [95]. Similarly, the δ *Arabidopsis* cyclins could function as monitors of the nutritional state of plant cells. Their activation triggered by appropriate mitogenic signals could induce transcription of other cyclin genes, leading to the phosphorylation of the putative *Arabidopsis* Rb homologue. The release of E2F-type transcription factors would initiate the activation of S-phase-specific promoters of genes necessary for cell cycle progression. The isolation of *Arabidopsis* Rb and E2F homologues could help to test this hypothesis.

## Developmental expression of the mitotic cyclins

*Cyc1At* expression during development was studied in detail using *in situ* hybridization techniques and transgenic plants harbouring the *cyc1At* promoter fused to the *gus* reporter gene [96]. In root and shoot apical meristems, and during embryogenesis, *cyc1At* expression is exclusively restricted to dividing cells. In roots, expression is high in the apical meristem, but is absent in the root cap and the quiescent centre, which show no mitotic activity. Expression in the pericycle is restricted to those cells where lateral root formation is initiated. These data show that *cyc1At* expression is restricted to dividing cells, in contrast to *cdc2aAt* which is expressed in dividing tissues and tissues with an increased competence to divide. Accordingly, the root apical meristems of plants transformed with the *cyc1At* promoter fused to *gus* show a patchy pattern of GUS staining. Some cells have a more intense staining than their neighbours, presumably because at the time of histochemical staining these cells are in the $G_2$ phase, in which *cyc1At* is expressed. Further support for the correlation between cyclin expression and cell division was obtained with mesophyll protoplasts of transgenic tobacco plants containing the *cyc1At* promoter *gus* fusion. Only at the appropriate concentrations of auxin and cytokinin that allow cell division was GUS activity observed.

The expression in different tissues of *cyc2aAt*, *cyc2bAt*, *cyc3aAt*, and *cyc3bAt* was studied by RT-PCR [82]. *Cyc2aAt* mRNA is observed in similar amounts in all tissues examined (roots, stems, leaves and flowers), whereas *cyc2bAt* mRNA is only detected in roots. *Cyc3bAt* is present in all tissues, but most prominently in roots. *Cyc3aAt* is not expressed in leaves and is hardly detectable in flowers. These results show that plant cyclin genes are not only transcribed in a cell cycle phase-dependent manner, but can also show a tissue-specific expression pattern.

The tight correlation between cyclin expression and cell proliferation was further demonstrated in studies of actively dividing *Arabidopsis* cell suspensions, in which cyclin expression is much higher than in mature tissues. Cultures depleted of auxin (a treatment which did not cause a reduction in cell division during the course of the experiment), or grown in low sucrose concentration (which arrested cell division) showed a drastic reduction of the transcription level of the class 2 and 3 cyclins, whereas that of *cyc1At* decreased only moderately. However, the complete arrest of such suspension cultures in early S or metaphase, by the addition of cell cycle inhibitors, caused a dramatic reduction in the *cyc1At* transcript level.

Treatment of roots with naphthalene-2-acetic acid, causing the induction of lateral root initiation, triggers *cyc1At* expression in all sites where new root primordia are formed. The induction of *cyc1At* by auxin in the founder cells was not blocked by pre-treatment of the roots with oryzalin, demonstrating that *cyc1At* transcription precedes the first round of division. This indicates that *cyc1At* expression could be one of the rate-limiting factors for the initiation of the cell cycle [96]. In mammalian cell cultures overexpression of specific cyclins shortens the length of the cell cycle [97,98]. It will be exciting to see whether overexpression of plant cyclins has the same effect on the overall duration of the cell cycle.

## Perspectives

In mammals and yeasts it has been shown that specific cyclins associate with certain CDK catalytic subunits at precise time points of the cell cycle, suggesting that each CDK complex might have its own specific task during cell division. In *Arabidopsis*, up to now, two CDKs and eleven cyclin genes were isolated. It is of great interest to know which CDK associates with which cyclin during the cell cycle, and what the precise functions of the different CDK/cyclin complexes might be. One way to approach this problem is the biochemical purification of CDK/cyclin complexes from actively dividing tissues, and the subsequent determination of their subunit composition. Also the specific roles and the putative substrates of each complex would benefit from the knowledge of the cell cycle dependent cellular localization pattern of the different CDKs and cyclins.

The high conservation of the cell cycle machinery in eukaryotes suggests that other components involved in the regulation of cell division in yeasts and mammals might also be present in plants. A most interesting discovery was families of low molecular mass proteins in mammals and yeasts which, depending upon external signals, can inhibit CDK kinase activity by direct binding to the CDK complexes (reviewed in [99]). The association of these molecules with CDK/cyclin complexes has been shown to be associated with the onset of cell differentiation

[100,101]. The availability of many yeast mutants could help to identify the plant genes encoding such regulatory proteins. One way to do so could be the functional complementation of yeast cell cycle mutants. These mutants can be transformed with an *Arabidopsis* cDNA library cloned in a yeast expression vector, and clones recovering the mutant phenotype can be selected. The identification of novel cell cycle genes can also be achieved by the use of the two-hybrid system [102]. This yeast genetic strategy has been developed to identify proteins able to interact *in vivo* with a known protein. Interactions between proteins are detected through the reconstitution of the activity of a transcription activator and the subsequent expression of a reporter gene.

The identification of *Arabidopsis* plants with conditional mutations affecting cell division could also help to identify new genes involved in cell cycle regulation. Plants able to grow at a permissive temperature (e.g. 16°C), but not at an elevated restrictive temperature (e.g. 28°C), could bear a temperature-sensitive mutation in a cell cycle regulatory gene.

The availability of *Arabidopsis* lines containing the *cdc2aAt* and *cyc1At* promoters fused to the *gus* reporter gene can benefit the isolation of *Arabidopsis* mutants showing altered *cdc2aAt* or *cyc1At* expression patterns. With the recent availability of the Green Fluorescent Protein system in plants, it will even be possible to score for such mutants in a non-destructive way [103,104]. Alternatively, the *cdc2aAt* and *cyc1At* expression patterns can be studied in developmental mutants, as markers for cell division.

A precise study of the expression pattern of the different cell cycle regulatory proteins may reveal how cell division in plants is integrated into specific developmental processes. It is of considerable interest to know how plants adapt to environmental changes by altering their regulation of cell division. A successful approach to this problem could be the overproduction of CDK and cyclin-negative and cyclin-positive mutants in plants. In yeasts, a great number of such mutations in *cdc2* and cyclin genes have been described. Some of these mutants cause a block or speed up the cell cycle, or even uncouple the S phase from M phase [58]. We are currently testing whether these mutations could also affect cell division in plants. Since some of these mutations might interfere with normal plant development, it will be necessary to place the mutant *cdc2* and cyclin alleles under the control of chemically inducible promoters. Alternatively, these mutant genes could be driven by tissue-specific promoters, allowing us to study the role of cell division during particular developmental processes.

*The authors thank Sylvia Burssens, Arnould Savouré and Antje Rohde for critical reading, and Martine De Cock for help preparing the manuscript. This work was supported by grants from the Belgian Programme on Interuniversity Poles of Attraction (Prime Minister's Office, Science Policy Programming, #38), the Vlaams Actieprogramma Biotechnologie (ETC 002), the International Human Frontier Science Program (IHFSP no. RG-*

*434/94M), and in part by the European Communities' BIOTECH Programme, as part of the Project of Technological Priority 1993-1996. L.D.V. is indebted to the Vlaams Instituut voor de Bevordering van het Wetenschappelijk-Technologisch Onderzoek in de Industrie for a predoctoral fellowship. DI is a Research Director of the Institut National de la Recherche Agronomique (France).*

## References

1.  Scheres, B., Wolkenfelt, H., Willemsen, V., Terlouw, M., Lawson, E., Dean, C. and Weisbeek, P. (1994) Development **120**, 2475–2487
2.  Irish, V.F. and Sussex, I.M. (1992) Development **115**, 745–753
3.  Lörincz, A.T. and Reed, S.I. (1984) Nature (London) **307**, 183–185
4.  Jacobs, T.W. (1995) Annu. Rev. Plant Physiol. Plant Mol. Biol. **46**, 317–339
5.  Pines, J. (1994) Semin. Cell Biol. **5**, 399–408
6.  Ferreira, P.C.G., Hemerly, A.S., Villarroel, R., Van Montagu, M. and Inzé, D. (1991) Plant Cell **3**, 531–540
7.  Hirayama, T., Imajuku, Y., Anai, T., Matsui, M. and Oka, A. (1991) Gene **105**, 159–165
8.  Imajuku, Y., Hirayama, T., Endoh, H. and Oka, A. (1992) FEBS Lett. **304**, 73–77
9.  Ducommon, B., Brambilla, P., Félix, M.-A., Franza, B.R., Jr, Karsenti, E. and Draetta, G. (1991) EMBO J. **10**, 3311–3319
10. Gould, K.L. and Nurse, P. (1989) Nature (London) **342**, 39–45
11. Gould, K.L., Moreno, S., Owen, D.J., Sazer, S. and Nurse, P. (1991) EMBO J. **10**, 3297–3309
12. Krek, W. and Nigg, E.A. (1991) EMBO J. **10**, 305–316
13. Krek, W. and Nigg, E.A. (1991) EMBO J. **10**, 3331–3341
14. Norbury, C., Blow, J. and Nurse, P. (1991) EMBO J. **10**, 3321–3329
15. Solomon, M., Lee, T. and Kirschner, M. (1992) Mol. Biol. Cell **3**, 13–27
16. Poon, R.Y.C., Yamashita, K., Adamczewski, J.P., Hunt, T. and Shuttleworth, J. (1993) Cell **58**, 833–846
17. Solomon, M.J., Harper J.W. and Shuttleworth J. (1993) EMBO J. **12**, 3133–3142
18. Fisher, R.P. and Morgan, D.O. (1994) Cell **78**, 713–724
19. Mäkalä, T.P., Tassan, J.-P., Nigg, E.A., Frutiger, S., Hughes, G.J. and Weinberg, R.A. (1994) Nature (London) **371**, 254–256
20. Krek, W. and Nigg, E.A. (1989) EMBO J. **8**, 3071–3078
21. Lehner, C.F. and O'Farrell, P.H. (1990) EMBO J. **9**, 3573–3581
22. Martinez, M.C., Jørgensen, J.E., Lawton, M.A., Lamb, C.J. and Doerner, P.W. (1992) Proc. Natl. Acad. Sci. USA **89**, 7360–7364
23. Hemerly, A.S., Ferreira, P.C.G., de Almeida Engler, J., Van Montagu, M., Engler, G. and Inzé, D. (1993) Plant Cell **5**, 1711–1723
24. Colasanti, J., Tyers, M. and Sundaresan, V. (1991) Proc. Natl. Acad. Sci. USA **88**, 3377–3381
25. Hirt, H., Páy, A., Györgyey, J., Bakó, L., Németh, K., Bögre, L., Schweyen, R.J., Heberle-Bors, E. and Dudits, D. (1991) Proc. Natl. Acad. Sci. USA **88**, 1636–1640
26. Fobert, P.R., Coen, E.S., Murphy, G.J.P. and Doonan, J.H. (1994) EMBO J. **13**, 616–624
27. Theologis, A. and Ray, P.M. (1982) Proc. Natl. Acad. Sci. USA **79**, 418–421
28. Alliotte, T., Tiré, C., Engler, G., Peleman, J., Caplan, A., Van Montagu, M. and Inzé, D. (1989) Plant Physiol. **89**, 743–752
29. John, P.C.L., Zhang, K., Dong, C., Diederich, L. and Wightman, F. (1993) Aust. J. Plant Physiol. **20**, 503–526
30. Chung, S.K. and Parish, R.W. (1995) FEBS Lett. **362**, 215–219
31. Pines, J. and Hunter, T. (1991) J. Cell Biol. **115**, 1–17
32. Jackman, M., Firth, M. and Pines, J. (1995) EMBO J. **14**, 1646–1654
33. Bailly, B., Dorée, M., Nurse, P. and Bornens, M. (1989) EMBO J. **8**, 3985–3995
34. Nigg, E.A. (1993) Curr. Opin. Cell Biol. **5**, 187–193
35. Jerzmanowski, A. and Cole, R.D. (1992) J. Biol. Chem. **267**, 8514–8520

36.  Colasanti, J., Cho, S.-O., Wick, S. and Sundaresan, V. (1993) Plant Cell **5**, 1101–1111
37.  Gunning, B.E.S and Wick, S.M. (1985) J. Cell Sci. 2, (suppl. 2), 157–179
38.  Wick, S.M. (1991) Curr. Opin. Cell Biol. **3**, 253–260
39.  Hemerly, A., de Almeida Engler, J., Bergounioux, C., Van Montagu, M., Engler, G., Inzé, D. and Ferreira, P. (1995) EMBO J. **14**, 3925–3936
40.  Amon, A., Surana, U., Muroff, I. and Nasmyth, K. (1992) Nature (London) **355**, 368–371
41.  Sorger, P.K. and Murray, A.W. (1992) Nature (London) **355**, 365–368
42.  Bell, M.H., Halford, N.G., Ormrod, J.C. and Francis, D. (1993) Plant Mol. Biol. **23**, 445–451
43.  De Bondt, H.L., Rosenblatt, J., Jancarik, J., Jones, H.D., Morgan, D.O. and Kim, S.-H. (1993) Nature (London) **363**, 595–602
44.  Mendenhall, M.D., Richardson, H.E. and Reed, S.I. (1988) Proc. Natl. Acad. Sci. USA **85**, 4426–4430
45.  van den Heuvel, S. and Harlow, E. (1993) Science **262**, 2050–2054
46.  Traas, J., Bellini, C., Nacry, P., Kronenberger, J., Bouchez, D. and Caboche, M. (1995) Nature (London) **375**, 676–677
47.  van den Berg, C., Willemsen, V., Hage, W., Weisbeek, P. and Scheres, B. (1995) Nature (London) **378**, 62–65
48.  Barton, M.K. and Poethig, R.S. (1993) Development **119**, 823–831
49.  Foard, D.E. and Haber, A.H. (1961) Am. J. Bot. **48**, 438–446
50.  Haber, A.H. (1962) Am. J. Bot. **49**, 583–589
51.  Guerche, P., Tiré, C., Grossi De Sa, F., De Clercq, A., Van Montagu, M. and Krebbers, E. (1990) Plant Cell **2**, 469–478
52.  Mayer, U., Büttner, G. and Jürgens, G. (1993) Development **117**, 149–162
53.  Berleth, T. and Jürgens, G. (1993) Development **118**, 575–587
54.  Carr, A.M., MacNeill, S.A., Hayles, J. and Nurse, P. (1989) Mol. Gen. Genet. **218**, 41–49
55.  Fleig, U.N. and Nurse, P. (1991) Mol. Gen. Genet. **226**, 432–440
56.  MacNeill, S.A., Creanor, J. and Nurse, P. (1991) Mol. Gen. Genet. **229**, 109–119
57.  MacNeill, S.A. and Nurse, P. (1993) Mol. Gen. Genet. **236**, 415–426
58.  Labib, K., Craven, R.A., Crawford, K. and Nurse, P. (1995) EMBO J. **14**, 2155–2165
59.  Connell-Crowley, L., Solomon, M.J., Wei, N. and Harper, J.W. (1993) Mol. Biol. Cell **4**, 79–92
60.  Jeffrey, P.D., Russo, A.A., Polyak, K., Gibbs, E., Hurwitz, J., Massagué, J. and Pavletich, N.P. (1995) Nature (London) **376**, 313–320
61.  Brown, N.R., Noble, M.E.M., Endicott, J.A., Garman, E.F., Wakatsuki, S., Mitchell, E., Rasmussen, B., Hunt, T. and Johnson, L.N. (1995) Structure **3**, 1235–1247
62.  Richardson, H.E., Wittenberg, C., Cross, F. and Reed, S.I. (1989) Cell **59**, 1127–1133
63.  Wittenberg, C., Sugimoto, K. and Reed, S.I. (1990) Cell **62**, 225–237
64.  Bueno, A., Richardson, H., Reed, S.I. and Russell, P. (1991) Cell **66**, 149–159
65.  Surana, U., Robitsch, H., Price, C., Schuster, T., Fitch, I., Futcher, A.B. and Nasmyth, K. (1991) Cell **65**, 145–161
66.  Richardson, H., Lew, D.J., Henze, M., Sugimoto, K. and Reed, S.I. (1992) Genes Dev. **6**, 2021–2034
67.  Schwob, E. and Nasmyth, K. (1993) Genes Dev. **7**, 1160–1175
68.  Connolly, T. and Beach, D. (1994) Mol. Cell. Biol. **14**, 768–776
69.  Swenson, K.I., Farrell, K.M. and Ruderman, J.V. (1986) Cell **47**, 861–870
70.  Minshull, J., Blow, J.J. and Hunt T. (1989) Cell **56**, 947–956
71.  Murray, A.W. and Kirschner, M.W. (1989) Nature (London) **339**, 275–280
72.  Lehner, C.F. and O'Farrell, P.H. (1990) Cell **61**, 535–547
73.  Bai, C., Richman, R. and Elledge, S.J. (1994) EMBO J. **13**, 6087–6098
74.  Girard, F., Strausfeld, U., Fernandez, A. and Lamb, N.J.C. (1991) Cell **67**, 1169–1179
75.  Pagano, M., Peppepkok, R., Verde, F., Ansorge, W. and Draetta, G. (1992) EMBO J. **11**, 961–971
76.  Lew, D.J., Dulić, V. and Reed, S.I. (1991) Cell **66**, 1197–1206
77.  Koff, A., Giordano, A., Desai, D., Yamashita, K., Harper, J.W., Elledge, S., Nishimoto, T., Morgan, D.O., Franza, B.R. and Roberts, J.M. (1992) Science **257**, 1689–1694
78.  Murray, A.W. and Kirschner, M.W. (1989) Nature (London) **339**, 275–280
79.  Glotzer, M., Murray, A.W. and Kirschner, M.W. (1991) Nature (London) **349**, 132–138
80.  Reed, S. (1991) Trends Genetics **7**, 95–99

81. Hemerly, A., Bergounioux, C., Van Montagu, M., Inzé, D. and Ferreira, P. (1992) Proc. Natl. Acad. Sci. USA **89**, 3295–3299
82. Ferreira, P., Hemerly, A., de Almeida Engler, J., Bergounioux, C., Burssens, S., Van Montagu, M., Engler, G. and Inzé, D. (1994) Proc. Natl. Acad. Sci. USA **91**, 11313–11317
83. Nagata, T., Nemoto, Y. and Hasezawa, S. (1992) Int. Rev. Cytol. **132**, 1–30
84. Shaul, O., Mironov, V., Burssens, S., Van Montagu, M. and Inzé, D. (1996) Proc. Natl. Acad. Sci. USA **93**, 4868–4872
85. Pines, J. and Hunter, T. (1989) Cell **58**, 833–846
86. Pines, J. and Hunter, T. (1990) Nature (London) **346**, 760–763
87. Day, I.S. and Reddy, A.S.N. (1994) Biochim. Biophys. Acta **1218**, 115–118
88. Matsushime, H., Roussel, M.F., Ashmun, R.A. and Sherr, C.J. (1991) Cell **65**, 701–713
89. Baldin, V., Lukas, J., Marcote, M.J., Pagano, M. and Draetta, G. (1993) Genes Dev. **7**, 812–821
90. Matsushime, H., Quelle, D.E., Shurtleff, S.A., Shibuya, M., Sherr, C.J. and Kato, J. (1994) Mol. Cell. Biol. **14**, 2066–2076
91. Meyerson, M. and Harlow, E. (1994) Mol. Cell. Biol. **14**, 2077–2086
92. Nevins, J.R. (1992) Science **258**, 424–429
93. Soni, R., Carmichael, J.P., Shah, Z.H. and Murray, J.A.H. (1995) Plant Cell **7**, 85–103
94. Tyers, M., Tokiwa, G. and Futcher, B. (1993) EMBO J. **12**, 1955–1968
95. Nasmyth, K. (1993) Curr. Opin. Cell Biol. **2**, 166–170
96. Ferreira, P.C.G., Hemerly, A.S., de Almeida Engler, J., Van Montagu, M., Engler, G. and Inzé, D. (1994) Plant Cell **6**, 1763–1774
97. Quelle, D.E., Ashmun, R.A., Shurtleff, S.A., Kato, J.-Y., Bar-Sagi, D., Roussel, M.F. and Sherr, C.J. (1993) Genes Dev. **7**, 1559–1571
98. Ohtsubo, M. and Roberts, J.M. (1993) Science **259**, 1908–1912
99. Sherr, C.J. and Roberts, M.J. (1995) Genes Dev. **9**, 1149–1163
100. Kranenburg, O., Scharnhorst, V., Van der Eb, A.J. and Zentema, A. (1995) J. Cell Biol. **131**, 227–234
101. Nakanishi, M., Adami, G.R., Robetorye, R.S., Noda, A., Venable, S.F., Dimitrov, D., Pereira-Smith, O.M. and Smith, J.R. (1995) Proc. Natl. Acad. Sci. USA **92**, 4352–4356
102. Allen, J.B., Walberg, M.W., Edwards, M.C. and Elledge, S.J. (1995) Trends Biochem. Sci. **20**, 511–516
103. Cubitt, A.B., Heim, R., Adams, S.R., Boyd, A.E., Gross, L.A. and Tsien, R.Y. (1995) Trends Biochem. Sci. **20**, 448–455
104. Sheen, J., Hwang, S., Niwa, Y., Kobayashi, H. and Galbraith, D.W. (1995) Plant J. **8**, 777–784
105. Segers, G., Gadisseur, I., Bergounioux, C., de Almeida Engler, J., Jacqmard, A., Van Montagu, M. and Inzé, D. (1996) Plant J. **10**, 601–612
106. Fobert, P.R., Gaudin, V., Lunness, P., Coen, E.S. and Doonan, J.H. (1996) Plant Cell **8**, 1465–1476

# 2

# Cyclin-dependent and calcium-dependent kinase families: response of cell division cycle to hormone and stress signals

**D. Dudits\*, Z. Magyar\*, M. Deák\*, T. Mészáros\*, P. Miskolczi\*, A. Fehér\*†, S. Brown†, É. Kondorosi†, A. Athanasiadis‡, S. Pongor‡, L. Bakó\*§, Cs. Koncz\*§ and J. Györgyey\***

\*Institute of Plant Biology, Biological Research Centre, Szeged, Hungary, †Institut des Sciences Végétales, CNRS, Gif-sur-Yvette, France, ‡International Centre for Genetic Engineering and Biotechnology, Trieste, Italy, and §Max-Planck Institut für Züchtungsforschung, Köln, Germany

## Introduction

The development of plants depends largely on the cell division activity that is simultaneously controlled by the inherited programme of ontogenesis and by environmental conditions. In the life cycle of plants, the major transitions, such as development of embryo from zygote, formation of shoot and root meristems, switch from vegetative to inflorescence and flower meristem, differentiation of male and female reproductive cells, are all linked to the activation of cell cycle machinery. The unparalleled potential of plants for continuous organogenesis and plastic growth also relies on the competent or active state of the cell division apparatus. Consequently, the up- and down-regulation of cell division is a crucial control process ensuring the completion of the ontogenic programme.

Earlier cytological and physiological studies have provided a wealth of experimental data about the influence of plant growth regulators or environmental factors on the division activity of plant cells [1]. Nowadays, the molecular and biochemical mechanisms underlying these regulatory processes can be approached from new perspectives. Evidence continues to accumulate for complex signalling networks that mediate the action of different plant hormones such as abscisic acid (ABA) [2], auxins [3] and ethylene [4]. Furthermore, calcium-centred signalling pathways are responsible for transduction of various external signals [5,6]. Special attention can be devoted to the unique class of calcium-dependent protein kinases (CaDPKs) that contain calmodulin-like domains (for review see [7]).

Parallel with discoveries of proteins and their genes that contribute to the transmission of intracellular signals, considerable progress has been achieved in identifying major components of cell cycle regulation in plants. Isolation of plant genes encoding homologues of cyclin-dependent kinases (CDKs) and cyclins have shown the existence in plants of a regulatory system similar to that of other eukaryotes (for review see [8,9]). This was further strengthened by the results of phosphorylation experiments *in vitro* demonstrating the alteration of kinase activities in proteins separated by p13[suc1]–Sepharose affinity binding and immuno-precipitation with peptide antibody from synchronized cells [10,11].

In this review, we try to give a comprehensive analysis of the known plant CDKs represented as a family of related kinases. Summary of the recent results on the studies of kinase activities in hormone-treated or stressed cells may help to link the cell cycle regulatory system with other signalling pathways. We want to lay emphasis on the potential significance of CaDPKs in mediating responses to external factors during alteration of cell division activity. Finally, we contribute a hypothetical scheme towards a model integrating the plant cell cycle control into a network of different signalling pathways regulated by other developmental and environmental stimuli.

## Multiplicity of CDKs in plants

The CDKs are key regulators of the proper timing and co-ordination of eukaryotic cell cycle events [12]. In yeast, genetic analysis pinpointed a single gene (*cdc2* in *Schizosaccharomyces pombe* and *CDC28* in *Saccharomyces cerevisiae*) which encodes a CDK with crucial function at the principal transition points of the division cycle: 'START' and entry into mitosis [13,14]. Discovery of new relatives of CDC28, such as the PHO85 gene with a regulatory role in sensing external signals, has brought new insights on the diverse roles of these kinases [15]. In *Drosophila*, two CDK genes were identified and they are equally essential for $G_1/S$ and $G_2/M$ transitions [16]. The number of *Xenopus* CDK genes is limited [17]. In contrast, in human cells, several CDKs control cell cycle progression. Both of the two major human CDKs, *CDC2* (or *CDK1*) and *CDK2*, show significant homology to yeast *cdc2*/CDC28, but they are functionally distinct. Human *CDC2* has been implicated mainly in regulation of $G_2/M$ transition, while *CDK2* is primarily required for the control of $G_1$ and S phase events [18]. Extensive research on CDKs from organisms with different phylogenetic origins highlights not only the conservation of these kinases throughout evolution but also the differences of the complexity of CDK families in various species. We can now analyse many CDK-related proteins from flowering plants to evaluate the relationship of the plant CDK family to other eukaryotic kinases and to reveal the common and the plant-specific features of these regulatory proteins.

**Fig. I**     **Similarity tree of the _cdc2_ protein kinase family**

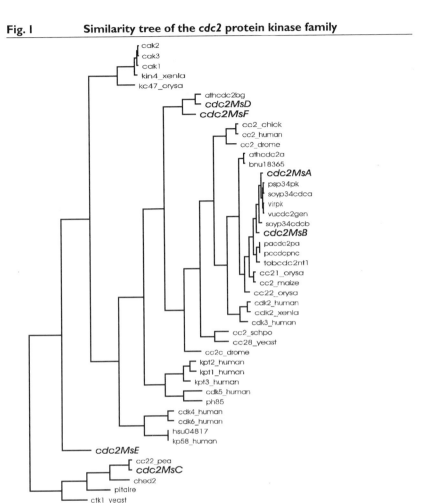

All the plant cdc2-related kinases and representative members of the other taxonomical groups available from the database were included. The UPGMA method of the Growtree program in the GCG program package was used for the comparison.

Early reports on cloning of the pea _cdc2_ homologue polymerase chain reaction (PCR) product, and the detection of plant proteins cross-reacting with p34[cdc2]-specific antibodies, have suggested the existence of a p34[cdc2]-kinase-centred regulatory system in plants [19,20]. Subsequently, the complementation of yeast _cdc2/cdc28_ mutants was achieved by expression of plant cDNAs clones isolated from alfalfa, maize and _Arabidopsis_ [21–24]. These results clearly demonstrated the functional similarity. Although none of the plant enzymes has been shown so far to possess cyclin-dependent kinase activity, this group of kinases encoded by the plant genes was defined as CDKs, based on their high amino acid sequence homology to the already known CDKs and due to their ability to complement yeast cell cycle mutants. The diversity of putative CDKs identified in plants is in agreement with a concept that emphasizes the significance of a multicomponent CDK family also in higher plants. Considering the substantial differences in characteristics of the known plant CDKs, we can expect distinct roles for the individual CDKs in relation to specific functions during cell cycle control or other cellular processes. This suggestion is supported by experimental data indicating differences between plant *cdc2* genes in their capacity to complement *cdc2*ts/*cdc28*ts mutants [25,26]. Many protein kinases related to the CDKs share conserved functional domains in combination with variable regions in the primary protein sequences deduced from cDNA sequences. Construction of an evolutionary tree of CDKs from yeasts, plants, animals and humans may reveal relationships between various members of this kinase family with diverse phylogenetic origins. Fig. 1 presents a similarity tree based on amino acid sequence data of the known CDKs. The horizontal distances from junction points reflect divergence between the various kinases.

The majority of plant CDKs belong to a separated branch of the phylogram showing a higher degree of homology between its members than to the other related CDKs. The phylogenic or functional significance of this separation is not yet known; however, it may reflect unique specificities linked to the plant enzymes. Interestingly, this group of the plant kinases is accompanied by CDKs from higher animals, and shows a considerable distance from the yeast homologues. It should be mentioned that these plant CDKs possess the well-conserved motifs in the PSTAIRE region ($\alpha 1$ helix), in the T-loop and in the catalytic domain (see Figs. 2, 3 and 4). Certain plant CDK variants can also be found in remote branches of the tree far from the major group. These representatives have an increased variability in amino acid sequences of the functional domains (see Figs. 2, 3 and 4). The separate group of two alfalfa CDK variants (*cdc2* MsD, F) and the *Arabidopsis* kinase (ath *cdc2* bg) can be considered as a unique class, since the corresponding alfalfa genes show cell-cycle-dependent activity with elevated mRNA accumulation in $G_2$/M phase [11,11a]. Other cdc2-related kinases have not been described so far as the plant enzymes exhibiting cell cycle phase-dependent transcriptional control. This can be a plant-specific regulatory component. The rice (Kc47) kinase with the *Xenopus* kin4 kinase are related to the CDK-activating kinase (CAK) enzymes. This analysis highlighted a distant position on the phylogenic tree for related kinases such as the alfalfa (*cdc2MsC*, E) and pea (cc22) CDKs. Surprisingly, the alfalfa kinase (*cdc2MsC*) shares considerable homology with the human cholinesterase cell division controller (CHED 52.8%) and with the human PITALRE protein (44%) [11a,27]. The present phylogenetic tree clearly demonstrates heterogeneity between CDKs including the plant enzymes. More extensive structural and functional studies should reveal the role of conserved and specific features of these kinases. One of the central questions is whether the increasing complexity of cell cycle control systems in yeast, plants, animals and humans correlates with a possible amplification of CDK functions and divergence in signal pathways controlling cell division. The multiplicity of the plant CDKs contradicts a close relation and suggests the involvement of these kinases in several regulatory pathways.

## Variation in the functional domains: predicted three-dimensional structure

Recent mutant analysis and biochemical studies have led to the determination of the crystal structure of the human CDK2 apoenzyme and of the human cyclin A–CDK2–ATP complex [28,29]. The proposed general model for cyclin binding and CDK activation describes key sequence determinants that can also be recognized in the plant CDKs. In human CDK2, the inhibitory Thr-14 and Tyr-15 phosphory-

**Fig. 2**      **Variability in the PSTAIRE region ($\alpha$I-helix) of CDKs**

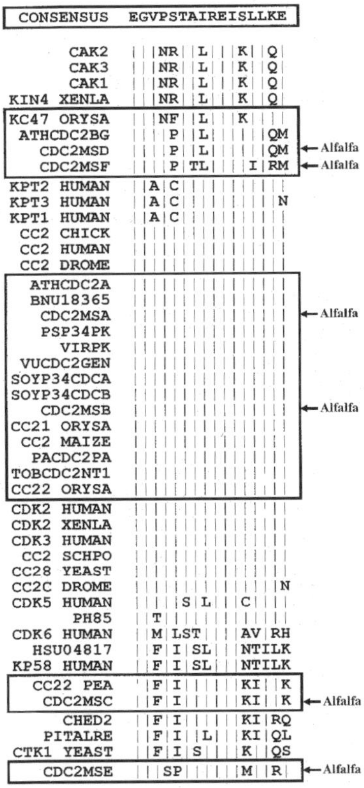

**Fig. 3**          **Variability in the T-loop region of CDKs**

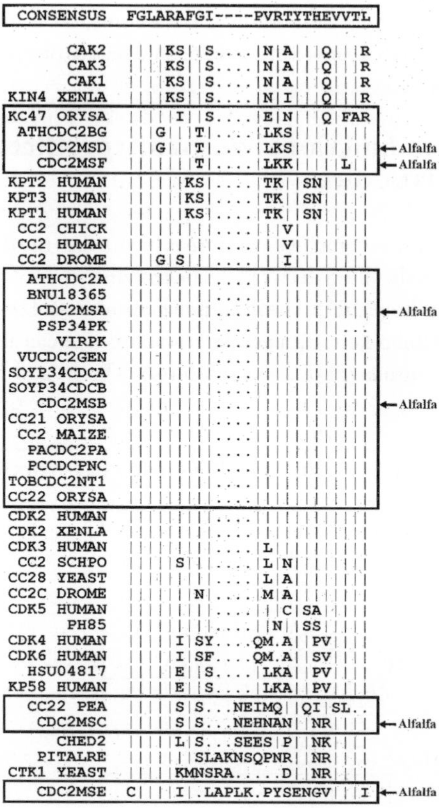

```
CONSENSUS    FGLARAFGI----PVRTYTHEVVTL

      CAK2   | | |KS| |S....|N|A| |Q| |R
      CAK3   | | |KS| |S....|N|A| |Q| |R
      CAK1   | | |KS| |S....|N|A| |Q| |R
KIN4 XENLA   | | |KS| |S....|N|I| |Q| |R

KC47 ORYSA   | | | |I| |S....|E|N| |Q|FAR
 ATHCDC2BG   | |G| | |T|....|LKS| | | | |
   CDC2MSD   | |G| | |T|....|LKS| | | | | |  ←Alfalfa
   CDC2MSF   | | | | |T|....|LKK| | | |L| |  ←Alfalfa

KPT2 HUMAN   | | | | |KS|....|TK| |SN| | |
KPT3 HUMAN   | | | | |KS|....|TK| |SN| | |
KPT1 HUMAN   | | | | |KS|....|TK| |SN| | |
 CC2 CHICK   | | | | | | |....| |V| | | | |
 CC2 HUMAN   | | | | | | |....| |V| | | | |
 CC2 DROME   | |G|S| |....| | |I| | | | | |

 ATHCDC2A    | | | | | | | |....| | | | | | |
 BNU18365    | | | | | | | |....| | | | | | |
 CDC2MSA     | | | | | | | |....| | | | | | |  ←Alfalfa
 PSP34PK     | | | | | | | |....| | | | | |...
 VIRPK       | | | | | | | |....| | | | | | |
 VUCDC2GEN   | | | | | | | |....| | | | | | |
 SOYP34CDCA  | | | | | | | |....| | | | | | |
 SOYP34CDCB  | | | | | | | |....| | | | | | |
 CDC2MSB     | | | | | | | |....| | | | | | |  ←Alfalfa
CC21 ORYSA   | | | | | | | |....| | | | | | |
 CC2 MAIZE   | | | | | | | |....| | | | | | |
 PACDC2PA    | | | | | | | |....| | | | | | |
 PCCDCPNC    | | | | | | | |....| | | | | | |
 TOBCDC2NT1  | | | | | | | |....| | | | | | |
CC22 ORYSA   | | | | | | | |....| | | | | | |

CDK2 HUMAN   | | | | | | | |....| | | | | | |
CDK2 XENLA   | | | | | | | |....| | | | | | |
CDK3 HUMAN   | | | | | | |....|L| | | | | | |
 CC2 SCHPO   | | | |S| |....|L|N| | | | | |
 CC28 YEAST  | | | | | | |....|L|A| | | | |
 CC2C DROME  | | | | |N|....|M|A| | | | | |
CDK5 HUMAN   | | | | | | |....| |C|SA| | | |
 PH85        | | | | | | |....| |N| |SS| | |
CDK4 HUMAN   | | | |I|SY....QM.A| |PV| | |
CDK6 HUMAN   | | | |I|SF....QM.A| |SV| | |
 HSU04817    | | | |E| |S....|LKA| |PV| | |
KP58 HUMAN   | | | |E| |S....|LKA| |PV| | |

 CC22 PEA    | | | |S|S...NEIMQ| |QI|SL..
 CDC2MSC     | | | |S|S...NEHNAN| |NR| | |  ←Alfalfa

 CHED2       | | | |L|S...SEES|P| |NK| | |
 PITALRE     | | | | |SLAKNSQPNR| |NR| | |
 CTK1 YEAST  | | | |KMNSRA.....D| |NR| | |

 CDC2MSE     C| | | |I|.LAPLK.PYSENGV| | |I  ←Alfalfa
```

lation sites in the middle of the glycine-rich loop may indirectly affect ATP binding. Although these critical amino acid residues are present in the plant CDKs as well, transgenic plants expressing the *Arabidopsis cdc2a* subunit mutated at these positions did not show any growth or developmental alterations ([30]; see L. De Veylder et al., Chapter 1). As shown by the crystal structure, the cyclin A–CDK2 interface comprises elements including the α1 helix with the PSTAIRE sequence motif, the T-loop with the Thr-160 regulatory phosphorylation site, part of the N-terminal β-sheet and C-terminal lobe from CDK2 and first cyclin fold repeat with the cyclin box (α3, α4 and the α5 helices), as well as the N-terminal α-helix from cyclin A. The amino acid sequence in the PSTAIRE region has been correlated with the preference in cyclin binding [31] and mutations here abolished cyclin binding [32]. Most of the plant CDKs including alfalfa kinases frequently contain the 16-amino acid domain, termed the PSTAIRE motif that is perfectly conserved in a variety of organisms (Fig.

2). The current sequence data from various CDKs also show a considerable variability in this functionally important element. A set of plant kinases contains the sequence PPTALRE that has been found only in plants so far [11,11a,33,34]. Fig. 2 presents other plant CDKs (such as the alfalfa cdc2 MsE, pea cc22, alfalfa cdc2 MsC and rice Kc47) with divergent amino acid sequence in this cyclin-binding region. Considering the numerous different cyclins encoded by plant genes (see [35] and J.P. Renaudin et al., Chapter 4) and the various classes of CDKs with or without the characteristic PSTAIRE motives, we can presume specific interactions between these partners leading to differential roles for the complexes. Identification of the collaborating CDK and cyclin partners will require extensive research using biochemical methods or the two-hybrid systems [36].

The crystallographic studies, cited above, also revealed significant interaction between helices $\alpha1$, $\alpha2$ and $\alpha3$ from the first repeat and N-terminal helix of cyclin A and N-terminal portion of the T-loop that contains a conserved threonine residue (Thr-160 or Thr-161) as a phosphorylated site in the fully active kinase [37]. In yeast, the homologous site is Thr-167. Fig. 3 shows the amino acid alignment from the T-loop of several CDKs. In most of the plant CDKs, we can identify a conserved threonine residue at the position 161 that can be a potential candidate as a phosphorylation-dependent regulatory site. In vertebrate cell cycle control, the CAK is responsible for the phosphorylation of the T-loop threonine (for review see [38]). Fig. 3 also lists several related kinases in plants and animals without this residue. The lack of this phosphorylation site may result in different post-translational control and/or divergent functions of these kinases.

The catalytic loop containing two $\beta$-strands ($\beta6$, $\beta7$) and a linker region (L10) shows high conservation with only limited variation in amino acid sequences (Fig. 4). In the centre of the linker, the proline and the glutamic acid residues are replaced by glycine–serine in both pea (cc22) and alfalfa (*cdc2*) kinases. The functional significance of these differences is unknown. The first experiments to test the physiological consequence of the ectopic expression of a mutated plant *cdc2a* (147-Asp→Asn) kinase resulted in new information about the involvement of these kinases in cell size determination [30].

The alignment of deduced amino acid sequences of alfalfa cdc2MsA and human CDK2 indicated significant homology between the two kinases (Fig. 5). Identical amino acids compose the PSTAIRE motif and the catalytic domain. In residues involved in cyclin binding, three conservative changes can be detected in the 154-Val→Ile, 159-Tyr→Phe and 278-Lys→Arg. Considering this high degree of sequence similarity, attempts were made to generate a hypothetical space-filling model of alfalfa kinase on the basis of the crystal structure of the human CDK2. In Fig. 6, the variable regions (deep brown) represent differences between the two CDK sequences. Residues in their neighbourhood (light brown) are also affected by the alterations. The large homologous regions are coloured pink. The two regulatory sites are within this conserved region and highlighted in blue (with Thr-14, Tyr-15) and turquoise (Thr-161). These regulatory sites are situated within a

**Fig. 4          Variability in the catalytic region of CDKs**

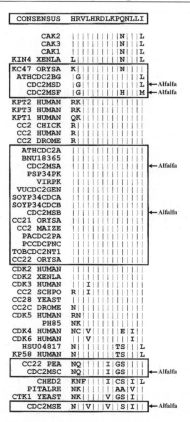

groove which harbours the active sites. The two plant-specific replacements (Ile-154 and Phe-159) are located at the entrance of this groove. They may influence the regulation and activity of the alfalfa enzyme. It is difficult to foresee the date when any plant CDK will be successfully crystallized and the structure determined to test the model predicted here. This hypothetical model may guide further characterization of functional elements in plant CDKs through generating and testing various mutant enzymes.

## Differential expression of plant CDK genes

The cell-cycle-dependent oscillation in CDKs is directed by complex mechanisms involving transcriptional as well as post-translational events. In yeast, the level of

**Fig. 5**      **Alignment of deduced amino acid sequences of alfalfa cdc2Ms and human cdk2**

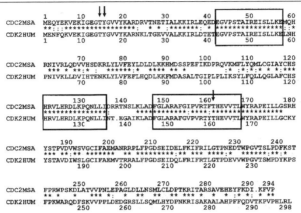

*Boxes mark the PSTAIRE, catalytic and T-loop regions respectively; (\*) identical residues; (:) similar ones.*

*cdc2* mRNA and the encoded protein kinase are fairly constant throughout the cell cycle and the Cdc2 kinase activity appears to be controlled solely by binding to regulatory cyclin subunits, and by phosphorylation at positive and negative regulatory sites [13]. On the contrary, the abundance of the CDC2 mRNA in mammalian cells was shown to vary during the cell cycle and as cells exit and re-enter the cycle (for review see [39]). The human *cdc2* gene is transcribed in S and $G_2$ phases, but not in $G_1$, until just before the $G_1$/S transition. Furthermore, the retinoblastoma susceptibility gene product (Rb) is proposed as a negative regulator of the *cdc2* gene promoter in non-cycling and cycling $G_1$ cells [40]. Shimizu et al. [41] have identified the enhancer element [-276]AAGTTACAAA[-267] in the 5'-flanking region of the rat *cdc2* gene which conferred strong inducibility at the $G_1$/S boundary on the basal *cdc2* promoter. Considering the substantial differences between yeast and mammalian cells in the regulatory role of cell-cycle-dependent transcription of *cdc2* genes, analyses of activity of plant CDK genes in various cell cycle phases and different organs and tissues may have special significance in revealing plant-specific mechanisms in comparison to other eukaryotic cells. The results of Northern hybridization studies so far clearly indicate various alternations in the level of *cdc2* transcripts in plant cells that suggest similarities to the animal cell systems in this respect.

     Only a limited number of experiments were carried out with synchronized plant cells to test the cell cycle dependence of *cdc2* mRNA accumulation. The initial Northern hybridization data did not show variation in abundance of alfalfa *cdc2* transcript after release from a hydroxyurea block [10]. In this case, full-length *cdc2* cDNA was used as hybridization probe. Later, a set of alfalfa CDK

**Fig. 6**　　　　　Postulated structure of the alfalfa CDK (*cdc2MsA*)

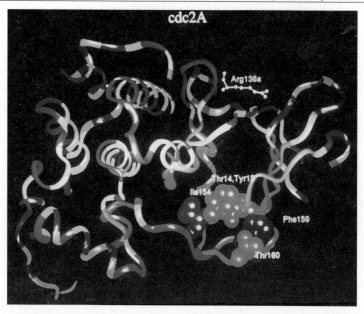

*The ribbon model was created by using the structural coordinates for human CDK2 and CDK2 complexed with ATP (kindly provided by Dr Sung-Hou Kim; see De Bondt et al. [28]). Residue positions with altered side-chains between the cdc2MsA and CDK2 are shown in deep brown, while light brown denotes conserved residues in the vicinity of the mutations. Conserved residues in conserved environments are shown in pink. Residues Ile-154 and Phe-159 are the only mutated residues at/within a distance L5 Å from the phosphorylation sites. Arg-136 is a unique insertion in cdc2MsA.*

genes was identified and the hybridization with variant specific probes revealed a complex regulatory system ensuring characteristic expression patterns for the individual alfalfa CDK genes [11,11a]. Based on a series of synchronization experiments, the authors defined two CDK genes (*cdc2* Ms B and E) with low and constant mRNA level in all cell cycle phases studied. The *cdc2* Ms D and F genes were highly activated in $G_2/M$ cells. Both of these gene variants contain an atypical PSTAIRE motif (see Fig. 2). The two additional CDK variants (*cdc2* Ms A, C) were active throughout the cycle, and the corresponding mRNA accumulated in variable amounts in different phases. The significance of these differences remains to be determined by further experiments. Expression analysis of a *Petunia cdc2* gene from sorted nuclei revealed a higher transcript level in 4C compared with 2C nuclei [42].

　　　　Growth-stimulated mammalian cells show increased transcription of CDK genes [43,44]. In plant tissue cultures, auxins and cytokinins can activate cell division. Indeed, the start of the DNA synthetic activity could be correlated with an elevated level of *cdc2* mRNA in mesophyll protoplast-derived alfalfa cells [10].

Auxins such as indole-3-acetic acid (IAA), 2,4-dichlorophenoxyacetic acid (2,4-D) and naphthylacetic acid (NAA) were also shown to stimulate the accumulation of *cdc2* transcripts in various plant species [10,22,26,45]. Hormonal control of the promoter from *Arabidopsis cdc2*a was also demonstrated by transformation experiments using the GUS reporter gene constructs [30]. Both Northern and *in situ* hybridization studies indicated a correlation between the expression levels of various CDK genes and the proliferative state of different organs and tissue components. In general, the activity of plant homologues of *cdc2* is higher in mitotically active tissues such as root tips, flower buds and embryos, and reduced in mature non-proliferating organs such as leaves [22,23,25,26,42,46,47]. *In situ* hybridization of longitudinal sections of different plant organs, such as the *Antirrhinum* inflorescences or the *Arabidopsis* vegetative and reproductive meristems, emphasized the non-uniform expression of *cdc2* transcripts in relation to mitotic activity ([30,34,45] and see J.H. Doonan, Chapter 10). Transcriptional activation of the *cdc2* gene accompanied by an active histone, mitogen-activated protein kinase (MAPK) or cyclin B genes was reported during the growth stimulation of dormant buds of pea by decapitating the terminal bud [48]. In the future, we can expect a more refined picture of spatial and temporal expression of *cdc2* genes, if variant-specific probes will be used for hybridization experiments.

## CDK complexes in various phases of the cell cycle

Before the molecular cloning of the first plant homologues of *cdc2*/CDC28 genes, John et al. [19] detected 34 kDa proteins in different plant species that were recognized by antibody against the internal PSTAIRE peptide. Subsequently, the sequence information from the cloned *cdc2* cDNAs of pea, alfalfa, maize and *Arabidopsis* confirmed the presence of this motif in the plant kinases and further experiments supported the existence of a p34[cdc2] homologue in higher plants [20–22,42,49]. The amounts of p34 protein cross-reacting with the PSTAIRE peptide antibody were found to be higher in basal segments of wheat leaves which comprise dividing cells than in fully differentiated zones [49]. With p13[suc1]–Sepharose chromatography it is possible to demonstrate the histone H1 kinase activity of maize p34[cdc2] that also correlated with the frequency of mitotic cells in these tissues [21]. A 13 kDa protein encoded by the *suc1* gene in *S. pombe* binds to p34[cdc2] kinase as shown by genetic and biochemical data [50,51]. Therefore, p13[suc1]–Sepharose affinity columns have been used extensively to isolate p34[cdc2]-kinase activity from various species [50,52]. The p13[suc1]–Sepharose-bound complexes may comprise several kinase components as indicated by the recent phosphorylation experiments comparing proteins extracted by p13[suc1]–Sepharose binding or immunoprecipitation with CDK-specific antibodies [11a].

The method of p13[suc1]–Sepharose affinity binding was also used in the first attempts to follow the changes of p34[cdc2] kinase activities during cell cycle progression in plant cells [10,11]. Irrespective of the applied synchronization methods based on the use of either hydroxyurea or aphidicolin, two characteristic p34$^{cdc2}$-kinase activities could be detected in alfalfa cells by assay *in vitro* (see also Fig. 7). Cells at the G$_1$/S boundary or in early S phase exhibited elevated p34$^{cdc2}$-kinase activity as measured by histone H1 kinase assays or by the level of phosphorylation of endogenous proteins bound to the p13$^{suc1}$–Sepharose matrix. A dominant phosphoprotein close to 40 kDa molecular mass was repeatedly present in S phase cells. The nature of this protein is currently unknown. The G$_1$/S phase kinase complexes may play a role during the reactivation of the cell cycle in mesophyll protoplast-derived cells [10]. A second, frequently higher, kinase activity peak is characteristic for the G$_2$/M cells. These biochemical data are in agreement with immunofluorescence microscopy localizing the p34$^{cdc2}$ kinase to the plane of the preprophase band [53,54]. The presence of different and cell cycle phase-related p34$^{cdc2}$ kinase complexes in plants has recently also been demonstrated by immunoprecipitation experiments [11,11a]. The G$_1$/S or early S and G$_2$/M kinase components can be recognized in protein fractions immunoprecipitated by antibodies against C-terminal peptide of *cdc2Ms* A/B proteins. A characteristic kinase complex is immunoprecipitated by anti-*cdc*2Ms F antibodies in G$_2$/M cells. These kinase complexes differ also in their substrate preference as was demonstrated in peptide phosphorylation experiments (Z. Magyar and T. Mészáros, unpublished work). Characterization of various alfalfa kinase complexes, separated by size fractionation chromatography, underlines the differences between these regulatory complexes (L. Bakó, unpublished work).

It is widely accepted that, in eukaryotic cells, cyclins are the activating partners of the various kinase complexes co-ordinating the cell cycle events. Since we can detect kinase activities in protein fractions separated by p13$^{suc1}$–Sepharose affinity binding or immunoprecipitation, the plant cyclins are expected to be present in the analysed samples. Little data are currently available about the cyclin components. Initially, heterologous antibodies were used to detect cyclins in plant cell extracts. Bakó et al. [55] reported a 62 kDa alfalfa protein recognized by antibodies raised against the fission yeast cyclin B encoded by the *cdc13* gene from fission yeast. Later, Magyar et al. [10] tested anti-human cyclin A antibodies against alfalfa extracts. These antibodies stained two proteins (42 and 65 kDa) on Western blots and could immunoprecipitate protein fractions with histone H1 kinase activities. As the sequence of several alfalfa cyclin genes became available [56], plant-specific peptide antibodies could be generated and tested in synchronized cell systems. The antibodies against the C-terminal peptide of a mitotic alfalfa cyclin (cycMs2) immunoprecipitated protein complexes with a dominant kinase activity from G$_2$/M phase cells. A lower histone H1 phosphorylation activity was also detected in cell extracts from the S phase cells [11a]. This is the first proof of the existence of CDK–cyclin complexes in plants. This anti-cyclin antibody cross-reacted with two different alfalfa proteins of 50 and 58 kDa molecular mass on Western blots (L. Bakó, unpublished work). In accordance with the sequence and

Northern hybridization data indicating the differential function of multigene families with several plant CDK and cyclin gene variants, the present biochemical studies also show the multicomponent nature and phase-specificity of the regulatory complexes. Identification and functional characterization of CDK–cyclin partners in these protein complexes of various cell cycle phases are central issues in current cell cycle research with plants. Further efforts are expected to be focused on the role of phosphatases in the regulation of the plant cell cycle. Recently, Magyar et al. [57] reported the detection of phosphatase activity in protein fractions bound to p13[sucl]–Sepharose columns. Application of various phosphatase inhibitors revealed that the PP1-type activity was constant in cells at different cell cycle phases. Preliminary data indicate an elevated PP2 activity in $G_2$/M cells (see Fig. 7). Parallel with the identification of plant phosphatase genes, the biochemical characterization of these enzymes would help to clarify the significance of protein dephosphorylation in the control of plant cell division.

## Cell division and CDK activity under hormonal and stress control

Developmental and environmental control of cell division activity, of transition from quiescence to active proliferation, or from dividing, meristematic phase to senescence, are mediated by structurally different chemicals—the so-called plant growth regulators. Both activator and inhibitor compounds can directly influence the division cycle of plant cells as demonstrated by extensive experimentation. Cytokinins are considered as one of the major controlling factors of cell division in cultured cells as well as in meristems of intact plants [58]. Cytokinin treatment can shorten the S phase of DNA replication and activate latent DNA-replication origins [59,60]. Inhibition of tobacco BY-2 cell growth by lovastatin, a blocker of the 3-hydroxy-3-methylglutaryl coenzyme A reductase, could be overcome by zeatin [61]. Little is known about the direct action of cytokinins on the cell cycle control elements. Hemerly et al. [30] showed activation of the promoter from *Arabidopsis cdc2a* both in transgenic *Arabidopsis* roots and in transgenic tobacco protoplasts by benzyl-6-aminopurine or kinetin. These cytokinins induced β-glucuronidase (GUS) activity under this promoter without stimulation of [³H]thymidine incorporation. Soni et al. [61a] reported an increase in the mRNA level of the cyclin δ3 gene in kinetin-treated *Arabidopsis* suspension cultures. John et al. [62] analysed the level of p34[cdc2]-like protein with PSTAIRE antibody in different segments of pea roots. The amount of cross-reacting protein was significantly induced by IAA within 24 h, but the subsequent treatment with zeatin riboside reduced the elevated level of this protein. In alfalfa suspension cultures grown in the presence of 10 μM NAA, the increase in kinetin concentration from 1 μM to 10 μM did not result in a significant change of the histone H1 kinase activity of the p13[sucl]–Sepharose bound complex (T. Mészáros, unpublished work). Current experimental data are insufficient to propose biochemical mediators that might link the action of cytokinins to the response of the cell cycle control system.

In contrast, the experimental findings convincingly show the activation of CDK complexes by auxin treatment [10, 63]. In alfalfa suspension cultures exposed to 10 μM or 30 μM 2,4-D for 24 h, the histone H1 kinase activity was increased by 6–10 times over the control level in the p13$^{suc1}$–Sepharose-bound fraction. Bimodal fluctuation of p34$^{cdc2}$ kinase activity can be recognized with an early peak (within minutes) and a late peak (after a few hours). In this respect, the p34$^{cdc2}$ kinase response shares similarities with other auxin-activated processes [64]. Induction of cell division by auxins belongs to the late responses that require significant reprogramming of the transcriptional pattern and alteration of the cellular structure (for review, see [65,66]). Up-regulation of *cdc2* genes and p34$^{cdc2}$ kinase function in gibberellin-treated nodes of rice plants also reflects the complexity of interaction between plant hormones and cell division control system [66a].

Other molecules can also activate the cell cycle machinery. Recently, the nodulation factors, the lipo-chito-oligosaccharides (LCOs), were shown to stimulate CDK activity and cell cycle progression in alfalfa microcallus suspension [67]. Synthetic LCOs could replace auxin and cytokinin in the initiation of cell division and could sustain growth of cultured tobacco protoplasts [68]. Currently, it is not known whether LCOs act through the same or different signal transduction pathways as plant hormones. Different signals with similar effects upon cell cycle progression may be mediated by the same cellular messengers. Among several mechanisms, the significance of variation in cytoplasmic pH should be considered as a regulatory factor. Pichon and Desbiez [69] reported recently that mechanical stimulation of *Bidens* hypocotyl under blue light induced an alkalinization of cytoplasm with simultaneous stimulation of DNA duplication and mitotic activity. Under white light, the cytoplasmic pH was reduced in combination with inhibition of division. The range of metabolites influencing the cell cycle in plants can be extended to ascorbic acid [70,71]. Ascorbic acid can stimulate the onset of cell proliferation, as seen as a $G_0$–$G_1$ transition in cells of germinating pea embryos [72]. The authors suggest that ascorbic acid is required for cell cycle progression, although CDK activity was not monitored in this case. Other stress factors, including electroporation or heat shock, can stimulate the division of different plant protoplasts [73].

In quiescent embryos or dormant axillary buds, cell cycle progression is arrested and the cells are blocked in various cell cycle phases. For full characterization of differentiated or quiescent cells, we require biochemical and molecular markers defining the state of the cell cycle machinery. The cytometric assay of nuclear DNA is only a modest classification, especially in plants where tissue tends to be mixoploid and the cell commitment is relatively plastic. In maize embryos, the majority of cells accumulate in the $G_1$ phase during seed maturation [75,76]. As shown by the cytometric data, dormant pea buds contain $G_1$ and $G_2$ cells. Devitt and Stafstrom [48] proposed that these cells were arrested in mid $G_1$ at the $G_1$/S boundary and near the $S/G_2$ boundary. Fully differentiated mesophyll cells in *Petunia* leaves are not cycling: nuclear DNA amounts may be 2C or 4C, corresponding to $G_1$ and $G_2$ or possibly to the $G_1$ phase at the start of an endoreduplicating cycle [77].

Despite the expected complexity of molecular events resulting in differential inhibition of cell cycle progression, there is a general consensus that ABA can be one of the key players in generating a dormant stage and mediating stress-induced inhibition. Reduction of cell division and/or DNA synthesis by ABA has been described in root tips [78], in apical and axillary shoot meristems [79–81] and in embryos [82]. If pea embryo axes were cultured in the presence of ABA (100 $\mu$M), the frequency of $G_0/G_1$ cells remained high reflecting a delay in the onset of DNA synthesis [83]. Similarly, ABA treatment reduced the rate of cell division in the developing endosperm of cultured maize kernels [84]. Furthermore, post-anthesis drought decreased the maximum number of endosperm cells by 30–40% in developing wheat grains [85]. Accumulation of ABA under water-deficient conditions is a well-documented stress response (for review, see [86]). In the light of these observations, a major question is how ABA can interact with various components of the cell cycle regulatory system. Mészáros et al. [63] analysed the p34[cdc2] kinase level in alfalfa suspension culture treated with ABA or exposed to different stresses as such osmotic or heat shock. Monitoring the phosphorylation of histone H1 in the p13[suc1]–Sepharose-bound fraction revealed an ABA concentration-dependent inhibition above 50 $\mu$M. These results were confirmed by measurement of kinase activities after immunoprecipitation with antibody against the C-terminal peptide of Ms *cdc2* A/B. Alfalfa cells grown in the presence of 0.4 $\mu$M mannitol or exposed to 42°C for 30 min exhibited a significantly lower kinase activity. These are the first attempts to identify the molecular basis of ABA-induced alteration of cell cycle parameters.

## Calcium signalling and cell division: role for CaDPKs

The previous discussion has emphasized the pivotal role of plant hormones such as auxins or ABA in regulating expression of cell cycle control genes or in alteration of kinase activities related to CDK complexes. The present models of auxin- or ABA-dependent signal transduction pathways point out $Ca^{2+}$ as a second messenger with a central function in transmitting the hormone-activated intracellular signals (for reviews, see [87–89]). Based on these considerations, we may also postulate that $Ca^{2+}$ and $Ca^{2+}$-modulated proteins are integral components in cell cycle control in plants, especially in sensing developmental or environmental signals. Several supportive results and conclusions can be cited from studies on animal and human cells as reviewed by Takuwa et al. [90]. Some of these may help the progress of research with plants:

— entry into the S phase, but not the maintenance of DNA-synthesis, is dependent on $Ca^{2+}$ influx
— calmodulin is the major mediator of $Ca^{2+}$-dependent mitogenesis notably at the transition from $G_1$ to S and $G_2$ to M phase
— serum-induced activation of Cdc2 kinase is potentially inhibited by active CAM antagonists [91].

$Ca^{2+}$-mediated events influence the role of Cdc2 kinase activation and are required to initiate cyclin degradation [92]. Many review articles summarize the basic features of $Ca^{2+}$-based signal transduction and the influence of hormones on cytosolic $Ca^{2+}$ in plants [5,93,94]. However, only sporadic data deal with the role of $Ca^{2+}$ during the plant cell cycle [95,96].

The regulatory function of intracellular $Ca^{2+}$ is frequently mediated by effector proteins, so-called $Ca^{2+}$-modulated proteins such as CAM or the $Ca^{2+}$-dependent (CAM-independent) protein kinases (CaDPKs), reviewed by Roberts and Harmon [97]. Here we prefer to use the abbreviation CaDPK instead of CDPK as this scarcely differs from CDK. Calmodulins possessing four functional EF-hand $Ca^{2+}$-binding domains are coded by gene families in higher plants. Expression studies showed elevated CAM mRNA accumulation in meristematic tissues comprising proliferating cell populations [98–100]. Analysis of CAM protein levels indicated a similar profile with higher CAM concentrations in shoot and root meristems than in mature tissues [101,102]. It may be important to mention the results from Hernández-Nistal et al. [103], who showed that ABA treatment reduced, and the higher temperature (30°C) increased, the concentration of CAM in embryonic axes of the chickpea. The interaction of CAM with microtubules in the mitotic apparatus has been demonstrated by immunochemistry ([104,105]; see M. Vantard et al., Chapter 15).

The dominant role of CaDPKs in transmitting $Ca^{2+}$ signals in plants represents a characteristic regulatory system distinct from those found in most eukaryotes. These kinases with unique structure have been extensively studied by both molecular and biochemical methods in a variety of plant species (reviewed in [7]). In CaDPKs, the catalytic kinase domain with significant homology to the mammalian $Ca^{2+}$/CAM-dependent protein kinase type II (CAMKII) is linked to a CAM-like domain (four EF-hand motives) through a conservative junction region. CaDPKs vary in size, subcellular localization, substrate specificity and response to $Ca^{2+}$ and lipids [7]. The first indication of a potential role of CaDPKs in cell division was provided by immunocytological localization experiments [106]. Soybean CaDPK–monoclonal antibody recognized the association of CaDPK with the F-actin network in three different plant species. As reflected by sequence variation, CaDPKs are encoded by gene families and expression of these gene variants is differentially controlled in various plants. As an example, the activity of a rice gene for CaDPK has been shown to be spatially and temporally regulated during seed development [107]. Pollen-specific CaDPK gene has been cloned from maize [108].

Involvement of CaDPK in stress responses of plants has been demonstrated by cloning two *Arabidopsis* CaDPK genes activated by drought and high-salt stress, but not by ABA or low-temperature or heat stress [109].

Out of several possible physiological functions of the plant CaDPKs, biochemical and molecular cloning experiments with an alfalfa CaDPK have provided further support for a link between CaDPK-mediated processes and cell cycle regulatory systems. A $Ca^{2+}$-dependent protein kinase (52–54 kDa) with autophosphorylation capability was partially purified from cultured alfalfa cells [110,111]. This CaDPK preferentially phosphorylated alfalfa histone H3 protein and showed a significantly increased level of autophosphorylation in protoplast-derived cells which are activated by hormones to divide (see [87]). Subsequently, an alfalfa cDNA encoding a homologue of CaDPKs has been cloned by using a carrot CDPK cDNA as hybridization probe [112] and characterized by Northern hybridization [112a]. Cyclic changes in the mRNA level of CaDPK were detected in alfalfa cells released from either aphidicolin or hydroxyurea blocks. Samples with increased frequency of S phase cells showed higher transcript accumulation. A second expression peak was recognized in mid $G_2$ cells after hydroxyurea treatment. These expression data are in accordance with the results of the previous protoplast culture experiments. This alfalfa CaDPK gene was also activated by auxin (2,4-D) and heat shock, but not by ABA, or high-salt or cold stress. The cDNA corresponding to the full-size alfalfa CaDPK protein was cloned into a bacterial expression vector and the recombinant kinase was purified from *Escherichia coli*. Importantly, its autophosphorylation and histone H1 phosphorylation activities were inhibited by two known CAM antagonists (W7 or trifluoropherazine dimaleate, TFP) [112a]. Another plant CaDPK was also blocked in cells treated with these inhibitors [113]. These data indicate that both $Ca^{2+}$-modulated proteins, CAM and CaDPK, can be inhibited by these compounds, observed in analogous comparisons [97]. If we analyse the 2,4-D activation of the CDK complex detected by p13[suc1]–Sepharose affinity binding in cells exposed simultaneously to antagonist, W7, a complex response can be recognized through histone H1 phosphorylation. The original CDK peaks were abolished from W7-treated cells. However, a strong peak of activity was observed after 30 min (T. Mészáros, unpublished work). These preliminary observations suggest that CAM or CAM-mediated processes, as alternatives to CaDPKs, can modulate p34[cdc2]-related protein complexes. We think that these first results will stimulate further studies to clarify the potential mechanisms of interaction between the $Ca^{2+}$ signalling and the cell cycle control pathways.

# Cell cycle control integrated to development and environmental adaptation in higher plants

As a consequence of the continuous extension of molecular tools including DNA probes and antibodies, significant progress has been achieved in the identification and characterization of cell cycle control elements in plants. However, the list is far from being complete. The accumulated information may provide the initial steps to identify molecular pathways that ensure the control of developmental and environmental signals over the cell division machinery. During a life cycle of a plant, the first activation of cell division is triggered by fertilization of the egg cell. The sperm and egg cell interaction generates a signal transduction cascade resulting in the dividing zygote. The contributing components in this signalling are well characterized in relation to cell cycle control in animal experimental systems [114,115]. Analogous information about the biochemical events in fertilized plant cells is missing and notably the possible role of CDKs and cyclins has not been studied so far. Auxins and ethylene are considered as activators during pollination and fertilization [116,117]. These experimental data support a central role for auxin-activated signals during the start of the zygotic embryogenesis. The recent concept of somatic embryogenesis considers the auxin-induced asymmetric cell division and transcriptional reprogramming as primary requirements for the transition from somatic to embryogenic cell type (for review, see [65,66]). In somatic cell cultures, the embryogenic induction through 2,4-D shock is linked to the activation of *cdc2* and cyclin genes ([22] and D. Dedeoglu, unpublished work). Cell cycle length is reduced through induction of somatic embryogenesis in different experimental systems [118–120]. Asymmetry of the first cell division is also characteristic of zygotic embryogenesis in plants [121]. If this pattern of cell division is disturbed by mutations, the subsequent embryo development is defective and abnormal seedlings are formed [122,123].

Pattern formation in the plant embryos and during organogenesis of root, shoot and flower meristems is determined by the orientation of the division plane, the rate and number of cell divisions, and the direction of cell growth [124–127]. The cell cycle control genes and their products are expected to determine directly the frequency of dividing cells, as well as the duration and synchrony of cell division. In vegetative meristems, the growth activation of axillary buds by decapitation was reflected by the elevated transcript level of cell cycle control genes [48]. During vegetative growth, leaf initiation is also linked to changes in cell division characteristics that are suggested as permissive components interacting with other regulatory systems [124]. The interplay between the developmental programme and cell cycle control has been most clearly outlined by several studies on cell cycle parameters during transition from a vegetative to a floral meristem (reviewed in [127]). Shortening the cell cycle and an increase in synchrony are the characteristic changes during this crucial developmental switch (see D. Francis, Chapter 9). But few and only indirect observations address molecular mediators linking the developmental signals and cell cycle regulation. From the expression patterns of the *cdc2* gene in *Arabidopsis* meristems, the p34[cdc2]-related system is clearly involved. Martinez et al. [45] noted that the distribution of *cdc2* transcripts followed a pattern reciprocal in comparison to expression of several floral homeotic genes. In the attempt to define possible ways of communication between developmental factors and cell cycle control, one of the central questions is how homeotic gene products can influence cell division and its regulatory system. Homeodomain proteins encoded by the knotted (KN1) gene of maize and other homologues such as KNAT1 and KNAT2 in *Arabidopsis*, or SBH1 in soybean, may play a role in the determination of cell fate through interaction with cell cycle control elements [128–130]. Dominant mutation of the maize KN1 gene or overexpression of the corresponding cDNA in tobacco or *Arabidopsis* resulted in an extended cell division activity and generated ectopic meristems [129,131,132]. In maize, the highest levels of KN1 protein occurred in the shoot apical meristem in all cell layers. The peripheral zone cells that are specified for initiation of organ primordia lacked the KN1 protein. It was present in the cells of floral meristems until development of ovule primordium terminating the meristem activity [133]. The *Arabidopsis* KNAT1 was active until the transition from vegetative to reproductive development and showed the reciprocal pattern of expression in comparison to the floral meristem identity genes [129]. In our experiments, ectopic expression of the soybean homoeobox-containing gene (Sbh1, 130) in tobacco caused similar modifications of leaf morphology to those shown in transformants with the maize gene [132]. The expression of cell cycle control genes and the p34[cdc2]-kinase activities are currently being analysed in the transformed and control tobacco plants (P. Miskolczi, unpublished work). The preliminary results indicate an increased histone H1 kinase activity in p13[suc1]–Sepharose-bound fractions from tobacco leaves expressing the soybean Sbh1 cDNA that also causes curled and wrinkled leaves. These findings encourage a more comprehensive study of the interaction between the homeotic regulatory systems and various components of the plant cell cycle control.

Environmental factors can directly influence the cell division cycle. The nutritional supply is also expected to regulate cell division just as phosphate starvation is known to block cultured cells in $G_1$ phase [10,135]. The duration of different cell cycle phases may vary significantly at low or high temperature [136]. Extreme conditions can also block the cell cycle progression. Recently, an interesting example has been reported by Logemann et al. [134], who showed transcriptional repression of several cell cycle genes encoding histone, p34[cdc2] and cyclin in UV-irradiated or fungal elicitor-treated parsley cells. The two stress responses followed different time kinetics. In the previous sections, we cited results on repression of kinase activity of p34[cdc2] protein complexes in osmotically shocked alfalfa cells. Here again a central question concerns how the biochemical pathways mediate stress signals towards the cell division machinery.

**Fig. 7**          **Presumptive scheme of interplay between the cell cycle control system and intracellular signalling pathways mediating activation or inhibition of cell division by hormonal, stress and developmental factors**

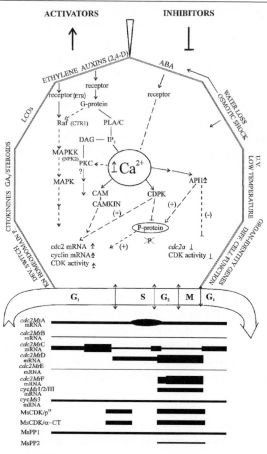

The details and references are given in the text.

Abbreviations: 2,4-D, dichlorophenoxyacetic acid; ETRI, dimeric plasma membrane-localized ethylene receptor; CTRI, homologue of Raf family of serine/threonine kinases; MAPKK, mitogen-activated protein kinase-kinase (NPK2); MAPK, mitogen-activated protein kinase; G-protein, GTP-binding protein; PL, phospholipase; $IP_3$, inositol 1,4,5-trisphosphate; DAG, 1,2-diacylglycerol; PKC, protein kinase C; CAM, calmodulin; CAMKIN, calmodulin-dependent protein kinase; ABA, abscisic acid; CDPK, $Ca^{2+}$-dependent, CAM-independent protein kinase; API1, phosphatase type 2C; Ms CDK/p13, p13[sucl]–Sepharose-bound histone H1 kinase activity in alfalfa cell extract; Ms CDKI-CT, histone H1 kinase activity of alfalfa protein complexes immunoprecipitated by antibody against C-terminal peptide of cdc2Ms A/B; Ms PP1, alfalfa phosphatase type 1; Ms PP2, alfalfa phosphatase type 2; LCOs, lipo-chito-oligosaccharides.

Given the advances achieved in identification and functional characterization of cell cycle control genes, and protein complexes and their hormonal and stress regulation, it should be possible to find components of interacting signalling pathways. As characteristic examples, we previously introduced auxins as activators and ABA as an inhibitor of the cell cycle progression. As shown by Fig. 7, in both pathways $Ca^{2+}$ is present as a central regulator that modifies the activity of different kinase and phosphatase components. Since these pathways are not completely elucidated yet, we can only postulate an interaction between the hormone-regulated components and transcription of cell cycle control genes or alteration of the functional constituents of protein complexes. Fig. 7 also illustrates a fairly complicated pattern of expression of many cell cycle control genes and fluctuation of p34$^{cdc2}$-kinase and phosphatase activity during the cycle. This complexity must provide a context for differential regulation of cell division by different external and developmental factors. The auxin action may be transferred through the MAP-kinase pathway either indirectly through ethylene signalling or directly. In the $Ca^{2+}$-centred pathway, several effector proteins can differentially sense the $Ca^{2+}$ level. As shown by inhibitor experiments based on W7, the prompt activation of CaDPK kinase activity by 2,4-D is a CAM- or CaDPK-linked procedure. Yet the second peak of kinase activity can be further stimulated by W7 inhibitor. These observations point out the potential involvement of other factors with a dominating contribution over the CAM or CaDPK pathway. Although, as yet, the plant homologues of yeast or mammalian protein kinase C (PKC) have not been identified, we cannot exclude the contribution of a $Ca^{2+}$/phospholipid-activated plant kinase in sensing hormonal and stress signals. The inhibition produced by ABA may be realized directly by the known phosphatase (API1) or indirectly through additional components. Despite its incompleteness, the proposed scheme should stimulate progress in identification of regulatory molecules that transmit developmental and environmental signals towards cell cycle regulatory elements (see Fig. 7).

*The authors thank Mrs Keczán, Zsuzsa Czakó, for preparation of the manuscript. The research projects in the Laboratory for Cell Division and Differentiation in the Institute of Plant Biology, Biological Research Centre, Szeged, were supported by grants from Körber Foundation, Hamburg, International Centre for Genetic Engineering and Biotechnology, Trieste and Hungarian Grant Agency, OTKA.*

## References

1. Bryant, J.A. (1976) in Molecular Aspects of Gene Expression in Plants (Bryant, J.A., ed.), pp. 177–216, Academic Press, London, New York, San Francisco
2. Giraudat, J. (1995) Curr. Opin. Cell Biol. **7**, 232–238
3. Millner, P.A. (1995) Curr. Opin. Cell Biol. **7**, 224–231
4. Ecker, J.R. (1995) Science **268**, 667–675

5.   Poovaiah, B.W. and Reddy, A.S.N. (1993) Crit. Rev. Plant Sci. **12**, 185–211
6.   Gilroy, S. and Trewavas, A.J. (1994) BioEssays **16**, 677–682
7.   Roberts, D.M. (1993) Curr. Opin. Cell Biol. **5**, 242–246
8.   Staiger, C. and Doonan, J. (1993) Curr. Opin. Cell Biol. **5**, 226–231
9.   Doerner, P.W. (1994) Plant Physiol. **106**, 823–827
10.  Magyar, Z., Bakó, L., Bögre, L., Dedeoglu, D., Kapros, T. and Dudits, D. (1993) Plant J. **4**, 151–161
11.  Magyar, Z., Mészáros, T., Fehér, A., Brown, S., Deák, M., Bottka, S., Györgyey, J., Pongor, S. and Dudits, D. (1995) in Phosphorylation in Plants (Shewry, P.R., Halford, N.G. and Hooley, R., eds.), pp. 197–209, Oxford University Press, Oxford
11a. Magyar, Z., Mészáros, T., Miscolczi, P., Deák, M., Fehér, A., Brown, S., Kondorosi, É., Athanasiadis, A., Pongor, S., Bilgin, M., Bako, L., Koncz, C., and Dudits, D. (1997) Plant Cell, **9**, 223–225
12.  Morgan, D.O. (1995) Nature (London) **374**, 131–134
13.  Norbury, C. and Nurse, P.A. (1992) Rev. Biochem. **61**, 441–470
14.  Nasmyth, K. (1993) Curr. Opin. Cell Biol. **5**, 166–179
15.  Kaffman, A., Herskowitz, I., Tjian, R. and O'Shea, E.K. (1994) Science **263**, 1153–1154
16.  Stern, B., Ried, G., Clegg, N.J., Grigliatti, T.A. and Lehner, C.F. (1993) Development **117**, 219–232
17.  Paris, J., Le Guellec, R., Couturier, A., Le Guellec, K., Omilli, F., Camonis, J., MacNeil, S. and Philippe, M. (1991) Proc. Natl. Acad. Sci. U.S.A. **88**, 1039–1043
18.  Van den Heuvel, S. and Harlow, E. (1993) Science **262**, 2050–2054
19.  John, P.C.L., Sek, F.J. and Lee, M.G. (1989) Plant Cell **1**, 1185–1193
20.  Feiler, H.S. and Jacobs, T. W. (1990) Proc. Natl. Acad. Sci. U.S.A. **87**, 5397–5401
21.  Colosanti, J., Tyers, M. and Sundaresan, V. (1991) Proc. Natl. Acad. Sci. U.S.A. **88**, 3377–3381
22.  Hirt, H., Páy, A., Györgyey, J., Bakó, L., Németh, K., Bögre, L., Schweyen, R.J., Heberle-Bors, E. and Dudits, D. (1991) Proc. Natl. Acad. Sci. U.S.A. **88**,1636–1640
23.  Ferreira, P.C.G., Hemerly, A.S., Villarroel, R., Van Montagu, M. and Inzé, D. (1991) Plant Cell **3**, 531–540
24.  Hirayama, T., Imajuku, Y., Anai, T., Matsui, M. and Oka, A. (1991) Gene **105**, 159–165
25.  Hirt, H., Páy, A., Bögre, L., Meskiene, I. and Heberle-Bors, E. (1993) Plant J. **4**, 61–69
26.  Miao, G.H., Hong, Z. and Verma, D.P.S. (1993) Proc. Natl. Acad. Sci. U.S.A. **90**, 943–947
27.  Grana, X., De Luca, A., Sang, N., Fu, Y., Claudio, P.P., Rosenblatt, J. and Morgan, D.O. (1994) Proc. Natl. Acad. Sci. U.S.A. **91**, 3834–3838
28.  De Bondt, H.L., Rosenblatt, J., Jancarik, J., Jones, H.D., Morgan, D.O. and Kim, S.H. (1993) Nature (London) **363**, 595–602
29.  Jeffrey, P.D., Russo, A.A., Polyak. K., Gibbs, E., Hurwitz, J., Massagué, J. and Pavletich, N.P. (1995) Nature (London) **376**, 313–320
30.  Hemerly, A.S., Ferreira, P., de Almeida Engler, J., Van Montagu, M., Engler, G. and Inzé, D. (1993) Plant Cell **5**, 1711–1723
31.  Pines, J. (1994) Semin. Cancer Biol. **5**, 305–313
32.  Ducommun, B., Brambilla, P., Felix, M.A., Franza, Jr, B.R., Karsenti, E. and Draetta, G. (1991) EMBO J. **10**, 3311–3319
33.  Imajuku, Y., Hirayama, T., Endoh, H. and Oka, A. (1992) FEBS Lett. **304**, 73–77
34.  Fobert, P.R., Coen, E.S., Murphy, G.J.P. and Doonan, J.H. (1994) EMBO J. **13**, 616–624
35.  Szarka, S., Fitch, M., Schaerer, S. and Moloney, M. (1995) Plant Mol. Biol. **27**, 263–275
36.  Fields, S. and Song, O.-K. (1989) Nature (London) **340**, 245–246
37.  Desai, D., Gu, Y. and Morgan, D.O. (1992) Mol. Biol. Cell **3**, 571–582
38.  Pines, J. (1995) Biochem. J. **308**, 697–711
39.  Müller, R., Mumberg, D. and Lucibello, F.C. (1993) Biochim. Biophys. Acta **1155**, 151–179
40.  Dalton, S. (1992) EMBO J. **11**, 1797–1804
41.  Shimizu, M., Ichikawa, E., Inoue, U., Nakamura, T., Nakajima, T., Nojima, H., Okayama, H. and Oda, K. (1995) Mol. Cell. Biol. **15**, 2882–2892
42.  Bergounioux, C., Perennes, C., Hemerly, A., Qin, L.X., Sarda, C., Inzé, D. and Gadal, P. (1992) Plant Mol. Biol. **20**, 1121–1130
43.  Tsai, L., Lees, E., Faha, B., Harlow, E. and Riabowol,K. (1993) Oncogene **8**, 1593–1602

44. Pagano, M., Pepperkok, R., Lukas, J., Baldin, V., Ansorge, W., Bartek, J. and Draetta, G. (1993) J. Cell Biol. **121**, 101–111
45. Martinez, M.C., Jørgensen, J.E., Lawton, M.A., Lamb, C.J. and Doerner, P.W. (1992) Proc. Natl. Acad. Sci. U.S.A. **89**, 7360–7364
46. Hashimoto, J., Hirabayashi, T., Hayano, Y., Hata, S., Ohashi, Y., Suzuka, I., Utsugi, T., Toh-E, A. and Kikuchi, Y. (1992) Mol. Gen. Genet. **233**, 10–16
47. Kvarnheden, A., Tandre, K. and Engström, P. (1995) Plant. Mol. Biol. **27**, 391–403
48. Devitt, M.L. and Stafstrom, J.P. (1995) Plant. Mol. Biol. **29**, 255–265
49. John, P.C.L., Sek, F.J., Carmichael, J.P. and McCurdy, D.W. (1990) J. Cell Sci. **97**, 627–630
50. Hindley, J., Phear, G., Stein, M. and Beach, D. (1987) Mol. Cell. Biol. **7**, 504–511
51. Moreno, S., Hayles, J. and Nurse, P. (1989) Cell **58**, 361–372
52. Brizuela, L., Draetta, G. and Beach, D. (1989) Proc. Natl. Acad. Sci. U.S.A. **86**, 4362–4366
53. Mineyuki, Y., Yamashita, M. and Nagahama, Y. (1991) Protoplasma **162**, 182–186
54. Colosanti, J., Cho, S.O., Wick, S. and Sundaresan, V. (1993) Plant Cell **5**, 1101–1111
55. Bakó, L., Bögre, L. and Dudits, D. (1991) in NATO Advanced Studies on Cellular Regulation by Protein Phosphorylation (Heilmayer, L., ed.), pp. 435–439, Springer Verlag, Berlin
56. Hirt, H., Mink, M., Pfosser, M., Bögre, L., Györgyey, J., Jonak, C., Gartner, A., Dudits, D. and Heberle-Bors, E. (1992) Plant Cell **4**, 1531–1538
57. Magyar, Z., Mészáros, T., Fehér, A., Brown, S., Athanasiadis, A., Pongor, S., Deák, M., Vissy, E., Dombrádi, P.V., Gergely, P., Miskolczi, P., Kondorosi, É., Bakó, L., Koncz, Cs. and Dudits, D. (1995) in Abstract Book of EMBO Workshop on Control of Cell Division Cycle in Higher Plants (Dudits, D., ed.), pp. 16, 5–7 Oct. 1995, Szeged, Hungary
58. Jacqmard, A., Houssa, C. and Bernier, G. (1994) in Cytokinines: Chemistry, Activity, and Function (Mok, D.W.S. and Mok, M.C., eds.), pp. 197–215, CRC Press, Boca Raton
59. Houssa, C., Jacqmard, A. and Bernier, G. (1990) Planta **181**, 324–326
60. Houssa, C., Bernier, G., Pieltain, A., Kinet, J.M. and Jacqmard, A. (1994) Planta **193**, 247–250
61. Crovell, D.N. and Salaz, M.S. (1992) Plant Physiol. **100**, 2090–2095
61a. Soni, R., Carmichael, J.P., Shah, Z.H. and Murray, A.H. (1995) Plant Cell **7**, 85–103
62. John, P.C.L., Zhang, K. and Dong, D. (1993) in Molecular and Cell Biology of the Plant Cell Cycle (Ormrod, J.C. and Francis, D., eds.), pp. 9–34, Kluwer Academic Publishers, Dordrecht
63. Mészáros, T., Miskolczi, P., Dedeoglu, D., Setenci, F., Bakó, L., Koncz, Cs., Deák, M. and Dudits, D. (1995) in Abstract Book of EMBO Workshop on Control of Cell Division Cycle in Higher Plants (Dudits, D., ed.), pp 59, 5–7 Oct. 1995, Szeged, Hungary
64. Brummell, D.A. and Hall, J.L. (1987) Plant Cell Environ. **10**, 523–543
65. Dudits, D., Bögre, L. and Györgyey, J. (1991) J. Cell Sci. **99**, 475–484
66. Dudits, D., Györgyey, J., Bögre, L. and Bakó, L. (1995) in *In Vitro* Embryogenesis in Plants (Thorpe, T.A., ed.), pp. 267–308, Kluwer Press, Dordrecht
66a. Sauter, M., Makhedov, S.L and Kende, H. (1995) Plant J. **7**, 623–632
67. Savouré, A., Magyar, Z., Pierre, M., Brown, S., Schultze, M., Dudits, D., Kondorosi, Á. and Kondorosi, É. (1994) EMBO J. **13**, 1093–1102
68. Röhrig. H., Schmidt, J., Walden, R., Czaja, I., Miklasevics, E., Wieneke, U., Schell, J. and John, M. (1995) Science **269**, 841–843
69. Pichon, O. and Desbiez, M.O. (1994) Physiol. Plant. **92**, 261–265
70. Liso, R., Calabrese, G., Bitonti, M.B. and Arrigoni, O. (1984) Exp. Cell Res. **150**, 314–320
71. Liso, R., Innocenti, A.M., Bitonti, M.B. and Arrigoni, O. (1988) New Phytol. **110**, 469–471
72. Citterio, S., Sgorbati, S., Scippa, S. and Sparvoli, E. (1994) Physiol. Plant. **92**, 601–607
73. Gupta, H.S., Rech, E.L., Cocking, E.C. and Davey, M.R. (1988) J. Plant Physiol. **133**, 457–459
74. Reference deleted
75. Deltour, R. and Jacqmard, A. (1974) Ann. Bot. **38**, 329–334
76. Georgieva, E.J., López-Rodas, G., Hittmair, A., Feichtinger, H., Brosch, G. and Loidl, P. (1994) Planta **192**, 118–124
77. Bergounioux, C., Brown, S.C. and Petit, P. (1992) Physiol. Plant. **85**, 374–386
78. Barlow, P.W. and Pilet, P.E. (1994) Physiol. Plant. **62**, 125–132
79. Fraser, L. G. and Matthews, R.E.F. (1983) Plant Sci. Lett. **29**, 67–72
80. Nougaréde. A., Rondet, P., Landré, P. and Rembur, J. (1987) Can. J. Bot. **65**, 907–915
81. Jacqmard, A., Houssa, C. and Bernier, G. (1995) J. Exp. Bot. **46**, 663–666
82. Bouvier-Durand, M., Real, M. and Come, D. (1989) Plant Physiol. Biochem. **27**, 511–518

83.  Levi, M., Brusa, P., Chiantante, D. and Sparvoli, E. (1993) In Vitro Cell. Dev. Biol. **29**, 47–50
84.  Myers, P.N., Setter, T.L., Madison, J.T. and Thompson, J.F. (1990) Plant Physiol. **94**, 1330–1336
85.  Nicolas, M.E., Gleadow, R.M. and Dalling, M.J. (1985) Ann. Bot. **55**, 433–444
86.  Bray, E.A. (1993) Plant Physiol. **103**, 1035–1040
87.  Dudits, D., Bögre, L., Bakó, L., Dedeoglu, D., Magyar, Z., Kapros, T., Felföldi, F. and Györgyey, J. (1993) in Molecular and Cell Biology of the Plant Cell Cycle (Omrod, J.C. and Francis, D., eds.), pp. 111–131, Kluwer Academic Publishers, Dordrecht
88.  Bowler, C. and Chua, N.H. (1994) Plant Cell **6**, 1529–1541
89.  Giraudat, J., Parey, F., Bertauche, N., Gosti, F. and Leungi, J. (1994) Plant Mol. Biol. **26**, 1557–1577
90.  Takuwa, N., Zhou, W. and Takuwa, Y. (1995) Cell. Signal. **7**, 93–104
91.  Hiaka, H., Sasaki, Y., Tanaka, T., Endo, T., Ohno, S., Fujii, Y. and Nagata, T. (1981) Proc. Natl. Acad. Sci. U.S.A. **78**, 4354–4357
92.  Lindsay, H.D., Whitaker, M.J. and Ford, C.C. (1995) J. Cell Sci. **108**, 3557–3568
93.  Trewavas, A. and Gilroy, S. (1991) Trends Genet. **7**, 356–361
94.  Bush, D.S. (1995) Annu. Rev. Plant Physiol. Plant Mol. Biol. **46**, 95–122
95.  Hepler, P.K. and Callaham, D.A. (1987) Cell Biol. **105**, 2137–2143
96.  Lino, M., Endo, M. and Wada, M. (1989) Plant Physiol. **91**, 610–616
97.  Roberts, D.M. and Harmon, A.C. (1992) Annu. Rev. Plant Physiol. Plant Mol. Biol. **43**, 375–414
98.  Zielinski, R.E. (1987) Plant Physiol. **84**, 937–943
99.  Ling, V. and Zielinski, R.E. (1989) Plant Physiol. **90**, 714–719
100. Breton, C., Chabound, A., Matthys-Rochon, E., Bates, E.E.M., Cock, J.M., Fromm, H. and Dumas, C. (1995) Plant Mol. Biol. **27**, 105–113
101. Muto, S. and Miyachi, S. (1984) Z. Pflanzenphysiol. **114**, 421–431
102. Allan, E. and Trewavas, A. (1985) Planta **165**, 493–501
103. Hernández-Nistal, J., Rodriguez, D., Nicolás, G. and Aldasoro, J.J. (1989) Physiol. Plant. **75**, 255–260
104. Vantard, M., Lambert, A.M., De Mey, J., Picquot, P. and Van Eldik, L. (1985) J. Cell Biol. **101**, 488–499
105. Wick, S.M., Muto, S. and Duniec, J. (1985) Protoplasma **126**, 198–206
106. Putman-Evans, C., Harmon, A.C., Palevitz, B.A., Fechheimer, M. and Cormier, M.J. (1989) Cell Motil. Cytoskel. **12**, 12–22
107. Kawasaki, T., Hayashida, N., Baba, T., Shinozaki, K. and Shimada, H. (1993) Gene **129**, 183–189
108. Estruch, J.J., Kadwell, S., Merlin, E. and Crossland, L. (1994) Proc. Natl. Acad. Sci. U.S.A. **91**, 8837–8841
109. Urao, T., Katagiri, T., Mizoguchi, T., Yamaguchi-Shinozaki, K., Hayashida, N. and Shinozaki, K. (1994) Mol. Gen. Genet. **244**, 331–340
110. Bögre, L., Oláh, Z. and Dudits, D. (1988) Plant Sci. **58**, 135–144
111. Oláh, Z., Bögre, L., Lehel, Cs., Faragó, A., Seprôdi, J. and Dudits, D. (1989) Plant Mol. Biol. **12**, 453–461
112. Suen, K.L. and Choi, J.H. (1991) Plant. Mol. Biol. **17**, 581–590
112a. Deák, M., Mészáros, T., Davletova, S.H., Dedeoglu, D., Oberschall, A., Miskolczi, P., Török, K. and Dudits, D. (1995) in Abstract Book of EMBO Workshop on Control of Cell Division Cycle in Higher Plants (Dudits, D., ed.), pp. 16, 5–7 Oct. 1995, Szeged, Hungary
113. Abo-El-Saad, M. and Wu, R. (1995) Plant Physiol. **108**, 787–793
114. Epel, D. (1990) Cell Differ. Dev. **29**, 1–12
115. Lorca, T., Cruzalegni, F.H., Fesquet, D., Cavadore, J.C., Means, A. and Doree, M. (1993) Nature (London) **366**, 270–273
116. Zhang, X.S. and O'Neill, S.D. (1993) Plant Cell **5**, 403–418
117. O'Neill, S.D., Nadeau, J.A., Zhang, X.S., Bui, A.Q. and Halevy A.H. (1993) Plant Cell **5**, 419–432
118. Warren, G.S. and Fowler, M.W. (1978) Experientia **34**, 356–357
119. Fujimura, T. and Komamine, A. (1980) New Phytol. **86**, 213–218

120. Bögre, L., Stefanov, I., Ábrahám, M., Somogyi, I. and Dudits, D. (1990) in Progress in Plant Cellular and Molecular Biology (Nijkamp, H.J.J., Vander Plas, L.H.W. and Van Aartrijk, J., eds.), pp. 427–436, Kluwer Academic Publishers, Dordrecht, Boston, London
121. Mansfield, S.G. and Briarty, L.G. (1991) Can. J. Bot. **69**, 461–476
122. Mayer, U., Büttner, G. and Jürgens, G. (1993) Development **117**, 149–162
123. Schevell, D.E., Leu, W. M., Gillmor, C.S., Xia, G., Feldmann, K.A. and Chua, N.H. (1994) Cell, **77**, 1051–1062
124. Lyndon, R.F. (1990) Plant Development: The Cellular Basis, Unwin Hyman, London
125. Jacobs, T. (1992) Dev. Biol. **153**, 1–15
126. Barlow, P.W. (1993) in Molecular and Cell Biology of the Plant Cell Cycle (Ormrod, J.C. and Francis, D., eds.), pp.179–199, Kluwer Academic Publishers, Dordrecht
127. Francis, D. and Herbert, R.J. (1993) in Molecular and Cell Biology of the Plant Cell Cycle (Ormrod, J.C. and Francis, D., eds.), pp. 201–210, Kluwer Academic Publishers, Dordrecht
128. Vollbrecht, E., Veit, B., Sinha, N. and Hake, S. (1991) Nature (London) **350**, 241–243
129. Lincoln, C., Long, J., Yamaguchi, J., Serikawa, K. and Hake, S. (1994) Plant Cell **6**, 1859–1876
130. Ma, H., McMullen, M.D. and Finer, J.J. (1994) Plant Mol. Biol. **24**, 465–473
131. Freeling, M. and Hake, S. (1985) Genetics **111**, 617–634
132. Sinha, N.R., Williams, R.E. and Hake, S. (1993) Genes Dev. **7**, 787–795
133. Smith. L.G., Greene, B., Veit, B. and Hake, S. (1992) Development **116**, 21–30
134. Logemann, E., Wu, S.C., Schröder, J., Schmelzer, E., Somssich, J.E. and Hahlbrock, K. (1995) Plant J. **8**, 865–876
135. Amino, S.I., Fujimura, T. and Komamine, A. (1983) Physiol. Plant. **59**, 393–396
136. Francis, D. and Barlow, P.W. (1988) in Plants and Temperature (Long, S.P. and Woodward, F.I., eds.), pp. 181–201, Company of Biologists, Cambridge

# 3

# Cyclin-dependent kinases in higher plants: spatial and temporal control of cell division

**Venkatesan Sundaresan\* and Joseph Colasanti**

Cold Spring Harbor Laboratory, 1 Bungtown Road,
Cold Spring Harbor, NY 11724, U.S.A.

## Introduction

The discovery of a common mechanism underlying the regulation of the cell cycle in yeasts and animals has led to efforts to extend these findings to the plant kingdom, particularly towards the study of problems specific to the division of plant cells. It is worthwhile, therefore, to discuss the findings from the yeast and animal systems first, and then consider the current knowledge of cell division kinases and their activities in higher plants. Since there have been several recent reviews on the yeast and animal cell cycles, only a brief summary will be given, with the focus on details pertinent to understanding some of the recent research on plants discussed in this chapter.

## Cell cycle regulation by cyclin-dependent kinases in yeasts and animals: similarities and differences

The cell cycle in yeasts and animals is driven by protein kinases called cyclin-dependent kinases (Cdk proteins), because their kinase activity depends upon their association with cyclins [1]. Cyclins are proteins that were first identified by their dramatic fluctuations in levels during the cell cycle [2]. A large number of cyclins have been identified in many organisms [3]. The different cyclins within an organism can be characterized by their peaks of expression in the cell cycle. Progression through each phase of the cell cycle is the result of phosphorylation of specific substrates by the cdks, with the specificity determined through association with different cyclins for each phase.

Currently the best understood transition of the cell cycle is the transition from $G_2$ into mitosis. In both yeasts and animals, the $G_2/M$ transition is driven by a Cdk called p34$^{cdc2}$, a 34 kDa protein first identified as the product of the *Schizo-*

*\*To whom correspondence should be addressed at the present address: Institute of Molecular Agrobiology, Singapore Science Park, 1 Science Park Drive, Singapore 118240.*

*saccharomyces pombe cdc2* gene. In association with mitotic cyclins, p34[cdc2] phosphorylates a set of substrates at $G_2$/M, driving the cells into mitosis (reviewed [4–6]). The changes that occur at the onset of M phase, i.e. chromatin condensation, nuclear envelope breakdown, disassembly of the nucleolus, re-organization of the cytoskeleton, and spindle pole assembly, are the result of p34[cdc2] kinase activation. It has been possible to identify some of the substrates and associated functions at $G_2$/M of cdc2–cyclin B [7]. These include the nuclear lamins, the phosphorylation of which by p34[cdc2]–cyclin B results in nuclear envelope breakdown; the actin-binding protein, caldesmon, the phosphorylation of which could result in microfilament disassembly; the myosin II regulatory light chain, the phosphorylation of which may delay contractile ring formation before anaphase; a microtubule-associated protein (MAP) which dissociates from the microtubules upon phosphorylation, possibly resulting in the depolymerization of the interphase microtubules at M phase; and many other substrates [7].

A general picture has emerged on the sequence of events leading to activation of p34[cdc2] at $G_2$/M (reviewed in [8]). During S and $G_2$, p34[cdc2] becomes associated with cyclin B, the levels of which are rising at this time of the cell cycle. The p34[cdc2] in this complex is phosphorylated at Thr-161 by the pMO15 kinase [9], which is now known as Cdk7 [10]. This phosphorylation is necessary for activation of p34[cdc2], and also increases its affinity for the cyclin subunit. However, the p34[cdc2] kinase remains inactive due to tyrosine phosphorylation by the Wee1 kinase. At late $G_2$, dephosphorylation of Tyr-15 by the Cdc25 phosphatase occurs, resulting in the activation of p34[cdc2]–cyclin B kinase. Interestingly, Cdc25 phosphatase is itself activated by p34[cdc2]–cyclin B phosphorylation, which probably accounts for the rapid activation of p34[cdc2]–cyclin B at $G_2$/M [11]. This general picture (with some complexities that have been omitted, e.g. Thr-14 phosphorylation, mik1 kinase, etc.) appears to be valid in both animals and yeasts [12].

However, there are a number of significant differences between yeasts and animals in the regulation of the $G_1$ and S phases. In the yeasts *Saccharomyces cerevisiae* and *S. pombe*, the engine that drives the cell cycle is a single protein kinase encoded by the *CDC28* gene in *S. cerevisiae* and by the *cdc2* gene in *S. pombe*. The activity of the Cdc2/CDC28 kinase is regulated by its association with different sets of cyclins at different points of the cell cycle. The level of Cdc2/CDC28 remains constant, but the levels of the cyclins fluctuate during the cell cycle, in a predictable and characteristic manner for each cyclin. Since cyclins regulate the substrate specificity, as well as the overall kinase activity of Cdc2/CDC28, the result is that specific substrates are phosphorylated at different times, driving the various transitions of the cell cycle. For example, in *S. cerevisiae*, two $G_1$ cyclins (called CLN1 and CLN2) are expressed at their highest levels in $G_1$, and CLN-CDC28 complexes drive the cells from $G_0$ into $G_1$ (reviewed by Nasmyth [13]). The CLNs 1 and 2 are degraded during S phase, and progression from S through M is regulated by B-type cyclins (called CLBs 1–6). The expression of CLB5 and CLB6 peaks in early S, that of CLB3 and CLB4 peaks in $G_2$, and that of CLB1 and CLB2 peaks in

$G_2/M$. Mutational analysis shows that different CLBs vary in their time of action, with CLBs 5 and 6 acting during the beginning of S, and CLBs 1 and 2 acting late in $G_2$ [14,15] (and see P. Sabelli et al., Chapter 12). It is concluded that CDC28 drives progression through S phase by forming an active kinase complex with the CLB5 and CLB6 cyclins, through $G_2$ phase by the CLB3 and CLB4 cyclins, and through M phase by the CLB1 and CLB2 cyclins.

In animal cells, a number of Cdk proteins have been identified. These include Cdc2 homologues, as well as Cdc2-related proteins that have been demonstrated to associate with cyclins. Next to p34$^{cdc2}$, the best-characterized Cdk in animals is the Cdk2 kinase (reviewed by Pines and Hunter [1]). Remarkably human CDK2 is as distant from human CDC2 as it is from yeast Cdc2 (65% identity in both cases), but closely related homologues of human CDK2 are found in other vertebrates [16]. Subsequently several new CDKs have been identified, called CDK3, CDK4, CDK5, CDK6 and CDK7 [10,17,18]. Of these, only CDK3 is a Cdc2 homologue, with a conserved PSTAIRE motif and the *CDK3* gene has the ability to complement yeast *cdc2/cdc28* mutations [19]. The *cdks* 4–7 are Cdc2-related genes that cannot complement yeast *cdc2/cdc28* mutations, and encode kinases that have weak conservation of the archetypical PSTAIRE motif of the Cdc2 family [17–20].

Several groups of animal cyclins have also been identified, exhibiting different levels of expression through the cell cycle. No homologues of the CLNs have been found, but a number of cyclins called cyclins C, D and E are expressed during $G_1$ [17,18]. In addition, the 'mitotic cyclins', cyclin A and cyclin B, are expressed during S and $G_2$, with the expression of cyclin A preceding that of cyclin B [21]. The formation of active CDK–cyclin complexes involves specific interactions. The D-type cyclins can be found in association with CDK2, CDK4 and CDK6, but not CDC2 [18]. Cyclin A associates with CDK2 and CDC2, but not CDK4 or CDK6, while cyclin B is only found in association with CDC2. Progression through the cell cycle in early $G_1$ is regulated by the association of the D-type cyclins with CDK2, CDK4, and CDK6 [18,22]. The CDK2–cyclin E complex appears to be required for the initiation of DNA replication at $G_1/S$ [23]. The CDK2–cyclin A complex regulates progression through S phase, and, finally, the CDC2–cyclin B complex drives the cell through $G_2/M$ and mitosis. CDC2 also associates with cyclin A, and there is some evidence that CDC2–cyclin A plays a synergistic role with CDC2–cyclin B in the onset of M phase [24]. In accordance with their different functions and times of activation during the cell cycle, there are specific substrates phosphorylated by each type of CDK–cyclin complex [25]. For example, CDK2–cyclin A phosphorylates the retinoblastoma-related protein p107, which is involved in the initiation of S phase, but CDC2–cyclin A and CDC2–cyclin B do not phosphorylate this substrate [26]. The M phase substrates are phosphorylated only by CDC2–cyclin B. Therefore, the vertebrate cell cycle is driven by several CDK–cyclin complexes, in which both the cyclin and CDK partners are different for each of the transitions of the cell cycle.

# Cdk genes in plants

## Are there functionally distinct cdks regulating the plant cell cycle?

In higher plants, homologues of Cdc2 kinase have now been cloned from several species (see below). In addition, several cyclins which are related to the animal A and B cyclins have been cloned from plants, and, more recently, cyclins related to the animal D cyclins have also been cloned ([27,28]; J. P. Renaudin et al., Chapter 4). It is now clear that the central features of cell cycle regulation are conserved in plants. However, a fundamental question that still needs to be answered is whether the plant cell cycle operates as in yeast, with one Cdk driving the cell cycle in co-operation with multiple cyclins, or whether the plant cell cycle is more similar to the animal cell cycle, with multiple Cdk proteins and their cyclin partners regulating different steps of the cell cycle. In addition to a general discussion of plant *cdk* genes, the evidence relating to this question is summarized in this section.

Homologues of the *cdc2* gene have been cloned from several plant species including alfalfa, *Arabidopsis*, maize, pea, rice, snapdragon and soybean [29–36]. In many of these cases, functionality of the plant Cdc2 was demonstrated by complementation of a yeast mutation (a *cdc2* mutation of *S. pombe* or a *cdc28* mutation of *S. cerevisiae*) (see H. Hirt et al., Chapter 6). All of these functional complementing plant Cdc2 proteins show an equally high degree of homology (*ca.* 65% amino acid identity) to the yeast Cdc2 and Cdc28 proteins, and to human CDC2 and CDK2, and have complete conservation of the hallmark PSTAIRE motif. Genes encoding other Cdc2-related proteins from plants, carrying altered versions of the PSTAIRE motif, have also been cloned (reviewed in Jacobs [37]). Since these altered motifs are different from the motifs found in animal CDK4, CDK5 or CDK6, these plant proteins are not obvious structural homologues of any of the known animal CDKs (Table 1). In the absence of functional information, the possible role of these Cdc2-related proteins in regulation of the plant cell cycle remains speculative. A possible exception to the above generalization should be noted. It has been pointed out that the Cdc2-related protein R2 from rice shows some similarity to animal pMO15 kinase, which is an activator of cdc2, suggesting that R2 might be the plant homologue of pMO15 ([37]; see Table 1). Although pMO15 has now been designated as CDK7 (owing to its association with cyclin H), it is likely to be a secondary regulator of the cell cycle, exerting its action through the regulation of CDC2 activity [10]. In this respect, CDK7 resembles other secondary regulators like Wee1 kinase or Cdc25 phosphatase, rather than the central cell cycle engines such as Cdk2 or Cdk4.

Multiple *cdc2* genes have been identified in many plants, but it is not clear whether these genes represent proteins with different functions in the plant cell cycle. We have previously found two *cdc2* genes in maize, which we have called *cdc2–ZmA* and *cdc2–ZmB*. However, their close relationship (98% similar

**Table I**       **Sequence motifs of animal cdks and some plant cdc2-related proteins**

| Sequence motif | Proteins |
| --- | --- |
| PSTAIRE | cdc2, cdk2, cdk3 (animals); all plant cdc2 homologues shown to complement yeast mutants |
| PISTVRE | cdk4 (human) |
| PSSALRE | cdk5 (human) |
| PLSTIRE | cdk6 (human) |
| NRTALRE | cdk7 (human pMO15 kinase) |
| PPTALRE | cdc2b (*Arabidopsis*); cdc2c (*Antirrhinum*) |
| PPTTLRE | cdc2d (*Antirrhinum*) |
| PITAIRE | cdkPs3 (*Pisum sativum*) |
| NFTALRE | R2 (*Oryza sativa*) |

For references see text, and review by Jacobs [37].

in protein sequence, and 96% similar in DNA sequence) suggests that they are simply duplicated genes, and not different functionally [29]. Consistent with this idea, they behave identically in complementation tests using yeast mutants, and have similar patterns of expression in development, although *cdc2–ZmA* appears to be expressed at higher levels than *Cdc2–ZmB* (our unpublished work). So far, we have not identified any more Cdc2 homologues from maize. Hirayama et al. [34] cloned two genes from *Arabidopsis*, *cdc2a* and *cdc2b*, but later found that only *cdc2a* is functional in complementing yeast mutants, and that the protein encoded by *cdc2b* is altered in the PSTAIRE motif ([38]; see Table 1). Hashimoto et al. [35] have cloned two rice *cdc2* genes, *cdc2Os-1* and *cdc2Os-2*, which are 83% similar. *cdc2Os-1* can functionally complement a $G_1/S$ mutation in yeast, while *cdc2Os-2* cannot complement the same mutation. Miao et al. [36] found two *cdc2* genes in soybean that were 90% similar, and behaved similarly in yeast, but showed different responses to hormones and to *Rhizobium* infection. Finally, Hirt et al. [39] cloned two alfalfa *cdc2s*, *cdc2MsA* and *cdc2MsB*, which are 89% similar, but function differently in yeast (see H. Hirt et al., Chapter 6). Interestingly, *cdc2MsA* complemented only the $G_2/M$ budding yeast *cdc28* mutant, while *cdc2MsB* complemented only the $G_1/S$ yeast *cdc28* mutant. Based on these results, Hirt et al. [39] have made the plausible suggestion that *cdc2MsA* and *cdc2MsB* may function at different points in the plant cell cycle, analogous to animal *CDC2* and *CDK2* respectively.

The present evidence for functionally different Cdc2 proteins in higher plants is not conclusive. In animals, CDKs of the same type from different species are closely related, while different CDKs from the same species are very distantly related (e.g. human CDK2 and *Xenopus* CDK2 are 89% similar, while human CDK2 and human CDC2 are only 65% similar). In contrast, the cloned plant Cdc2 proteins within a species are quite closely related (83% or higher; Table 2).

**Table 2    Comparison of cdc2 homologues from plants with multiple cdc2 genes**

|       | Ata | GmS5 | GmS6 | MsA | MsB | Os1 | Os2 | ZmA |
|-------|-----|------|------|-----|-----|-----|-----|-----|
| Ata   | 100 |      |      |     |     |     |     |     |
| GmS5  | 84  | 100  |      |     |     |     |     |     |
| GmS6  | 84  | 94   | 100  |     |     |     |     |     |
| MsA   | 82  | 94   | 90   | 100 |     |     |     |     |
| MsB   | 83  | 90   | 91   | 89  | 100 |     |     |     |
| Os1   | 82  | 83   | 84   | 80  | 84  | 100 |     |     |
| Os2   | 83  | 86   | 85   | 82  | 83  | 83  | 100 |     |
| ZmA   | 83  | 84   | 85   | 82  | 85  | 94  | 83  | 100 |

Numbers indicate percentage identity at the amino acid level. At, Arabidopsis thaliana; Os, Oryza sativa; Zm, Zea mays; Ms, Medicago sativa; Gm; Glycine max. The maize and Arabidopsis cdc2 proteins were selected for reference. All of the above cdc2 homologues have been demonstrated to functionally complement a yeast cdc2 or cdc28 mutation, with the exception of cdc2Os-2. See text for references.

Furthermore, there are no particularly close relationships between individual plant Cdc2 proteins from different species. Comparison of the published Cdc2 protein sequences from plants shows that all of them are about equally related to each other (identities of 82–94%; Table 2). The type of complementary functions in the yeast assay found by Hirt et al. [39] for the alfalfa *cdc2* genes does not appear to extend to the other cases of multiple *cdc2* genes in plants (e.g. both of the soybean *cdc2* genes could complement both types of yeast mutants; [36]), suggesting that it is not a general phenomenon. Therefore, the observations of Hirt et al. [39] with the alfalfa *cdc2* genes in yeast, while interesting, may not completely describe their functions in the plant. An alternative possibility is that the multiple *cdc2* genes found in some plants may be involved in responses to different types of developmental signals, as suggested by Miao et al. [36].

## Study of cdk function by complementation analysis in yeast

In this connection, it is worthwhile examining more closely the interpretation of the results from complementation of yeast mutants. Based upon early studies that demonstrated the universality of the cell-cycle control mechanism, it had been assumed for some time that the results from the complementation of a yeast *cdc2* or *cdc28* mutation by a gene from a heterologous species provide definitive evidence of its function in that species. The cloning of several other *cdk* genes has made it evident that the picture is more complex. For example, in human cells it is now clear that CDC2 is exclusively a $G_2/M$ kinase, whereas CDK2 is primarily a $G_1/S$ and S-phase kinase. Nevertheless, human *CDK2* can complement both $G_1/S$ and $G_2/M$ *cdc28* mutants of *S. cerevisiae*, in the study by Meyerson et al. [19]. To complicate matters further, in another study, the same human *cdk2* could not complement a $G_1/S$ *cdc28* mutation unless a second uncharacterized yeast mutation was also present [16]. These different outcomes are very likely owing to differences in the strain backgrounds, and point to the difficulty of drawing unambiguous conclusions from these types of experiments (see H. Hirt et al., Chapter 6).

Experiments using the maize *cdc2–ZmA* gene to complement different yeast mutations illustrate some of these difficulties. The results from these experiments are summarized in Table 3. We find that *cdc2–ZmA* complements the *S. cerevisiae* *cdc28–1N* mutation, which is defective in $G_2/M$ [29], but not the *cdc28–4* and the *cdc28–13* mutations, which are defective in $G_1/S$, nor the null mutation *cdc28::TRP*, which is defective in both $G_1/S$ and $G_2/M$ (Table 3). The complementation results from these three mutants can be interpreted as evidence of an exclusively $G_2/M$ function for *cdc2–ZmA*. Confusingly however, the maize *cdc2–ZmA* gene does show weak complementation of the *cdc28–17* mutation, which is defective at $G_1/S$ (Table 3). From this result, it remains possible that the function of *cdc2–ZmA* in maize is to drive both the $G_1/S$ and the $G_2/M$ transitions. The likely explanation for these apparently contradictory results is that the yeast mutations are in different backgrounds, and that the ability of a particular heterologous *cdc2* gene to complement a *cdc28* mutation is at least partially dependent upon the background. For this reason, failure of a *cdc2* homologue to complement a yeast mutation (e.g. as in the case of the rice *cdc2Os-2* gene; [35]), should not be taken as evidence for non-functionality in the endogenous species.

| Table 3 | Complementation of *S. cerevisiae* cdc28 mutants by cdc2–ZmA | | |
|---|---|---|---|
| | **Defective in:** | | **Complementation** |
| **mutant allele** | **G1/S** | **G2/M** | **by cdc2–ZmA** |
| *cdc28–4* | Yes | No | − |
| *cdc28–13* | Yes | No | − |
| *cdc28–17* | Yes | No | + |
| *cdc28–1N* | No | Yes | +++ |
| *cdc28::TRP* | Yes | Yes | − |

An additional complication that can arise in the interpretation of yeast complementation experiments is that a heterologous *cdc2* gene can behave very differently in *S. cerevisiae* and *S. pombe*. An illustration of this difference is provided by the maize *cdc2–ZmA* gene, which does not complement the *S. pombe cdc2* mutation. In fact, overexpression of *cdc2–ZmA* is lethal to *S. pombe*, as shown in Fig. 1. In this experiment, *cdc2–ZmA* was introduced into *S. pombe* under the control of a thiamine-repressible promoter. When uninduced, the cells are normal, but upon induction of *cdc2–Zma* expression the cells die. Microscopic examination reveals that cell death is due to the uncoupling of mitosis from cytokinesis, resulting in cells that either have no nuclei, or have multiple nuclei (Fig. 1). Dominant–negative lethal mutants of *S. pombe cdc2* have been used recently to dissect *cdc2* function [40], but the phenotype obtained by expressing maize *cdc2* in *S. pombe* does not resemble any of these mutants. Furthermore, the lethality that we observe is not a non-specific effect of overexpressing the maize Cdc2 protein, but is related to Cdc2 function, since it can be rescued by overexpression of the *S. pombe cdc2* gene in the same cells (data not shown). These results suggest that *cdc2–ZmA* functions improperly in *S. pombe*, possibly because it interacts poorly with proteins that link cytokinesis and mitosis (e.g. *cdc16*; see review by Simanis [41]). Therefore, expression of the maize *cdc2–ZmA* gene has very different effects in *S. cerevisiae* and *S. pombe*, with the ability to complement mutations in the former, but causing lethality in the latter. The conclusion is that considerable caution should be exercised when attempting to infer the function of a plant *cdk* gene by studying the effects of its expression in yeast (see H. Hirt et al., Chapter 5).

**Fig. 1**        Cultures of *S. pombe* cells carrying the maize *cdc2–ZmA* gene
                 under the control of the thiamine repressible *nmt1* promoter

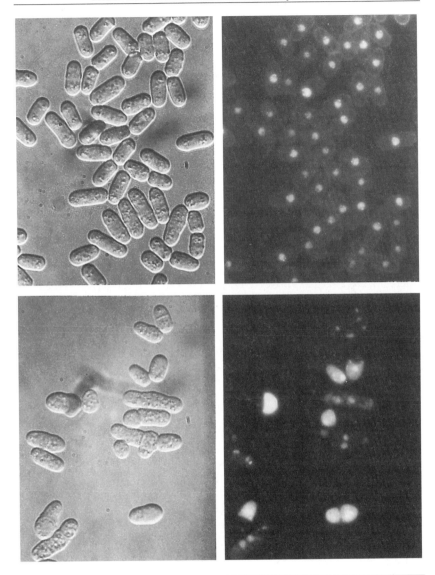

*Cells in the top two panels are grown in the presence of 10 mM thiamine and have a normal*
*appearance. The bottom two panels show the same strain grown in the absence of thiamine, which*
*results in over-expression of the maize Cdc2–ZmA protein. Panels on the left show cells visualized under*
*Nomarski optics; panels on the right show the same cells stained with 4-6 diamido-2-phenylindole (DAPI).*

## Universality of cell cycle regulation: $G_1/S$ versus $G_2/M$ in plants

The differences between the yeast and vertebrate cell cycle regulators are primarily in $G_1$ control, and there appears to be relatively little difference in the control of $G_2/M$. It is plausible that the evolution of multicellular organisms with complex developmental programmes has resulted in the need for cells in $G_1$ to respond to many diverse developmental and extracellular cues. By contrast, in the yeasts, control of cell proliferation is primarily a response to nutritional deficiencies/starvation or to mating factors. Therefore, the regulation of $G_1$ progression in animals and yeasts have evolved differently from each other. On the other hand, the regulation of $G_2/M$, which is controlled primarily by intracellular checkpoints (e.g. completion of DNA synthesis), may be expected to be much less divergent in all species.

By this reasoning, we may expect that the $G_2/M$ transition in higher plants is likewise driven by a single Cdk–cyclin complex, related to Cdc2–cyclin B. Although plants do not have 'typical' A or B cyclins, they have both A-like and B-like cyclins (see chapter 4 by J.P. Renaudin et al.), and it is likely that the B-like plant cyclins are regulators of $G_2/M$. This argument leaves unanswered the situation with respect to the regulators of $G_1$ and S phase in plant cells. The discovery of D-like $G_1$ cyclins in plants [28] points to the existence of a kinase with which they must associate during $G_1$. In vertebrates, D-type cyclins associate with three quite different CDC2-related kinases: CDK2, CDK4 and CDK6. CDK2 is the most closely related to CDC2, with conservation of the PSTAIRE motif, and the ability to functionally complement yeast $cdc2/cdc28$ mutations. On the other hand, CDK4 and CDK6 will not complement yeast $cdc2/cdc28$ mutations, and they also show weak conservation of the PSTAIRE motif. Therefore, the discovery of D-like plant cyclins by itself does not give any clue as to the identity of the $G_1$ kinase. It is currently an open question as to whether the currently characterized plant Cdks will interact with both the D-like cyclins and the mitotic cyclins, or whether there are additional plant Cdks that are specific for interaction with the D-like cyclins.

Results from a recent study by Grafi and Larkins [42] has raised the possibility that two distinct Cdks exist in maize. They have studied kinase activities in developing maize endosperm at a stage when endosperm cells are undergoing endoreduplication. During endoreduplication, these cells can be considered to have a truncated cell cycle with an S phase, but no M phase, resulting in polyploidy. The study found no detectable decrease in the levels of Cdc2–ZmA protein after the endosperm switches to growth primarily by endoreduplication. Interestingly, a significant decrease in p13[suc1]-associated histone H1 activity (a standard assay for Cdc2 kinase activity), concurrent with a significant increase in a kinase activity that could phosphorylate the S-phase-specific animal transcription factors E2F and E1A, was found. The decrease in histone H1 activity appears to be due to an inhibitor of p34[cdc2] activity present in the endoreduplicating cells. Western blotting with antibodies indicates that the H1 kinase fraction bound to p13[suc1] contains Cdc2–ZmA. Upon elution from the p13[suc1] column, it was found to have no affinity for E1A, suggesting that the S-phase kinase is a distinct Cdk from Cdc2–ZmA.

In vertebrate cells, phosphorylation of S-phase transcription factors is due to CDK2 [18]. Since CDK2 binds to p13$^{suc1}$ with an affinity comparable to that of Cdc2 [19], we might have expected that the plant equivalent of Cdk2, if it exists, will also be recovered in the p13$^{suc1}$ fractions. Therefore, the absence of the presumptive S-phase kinase from the p13$^{suc1}$-bound fractions deserves further investigation. One possibility is that the S-phase kinase in plants is only distantly related to Cdc2/Cdk2, and does not bind p13$^{suc1}$ (as is the case with the animal G$_1$ kinases CDK4 and CDK6; [19]). Consistent with this possibility is that, so far, only a single Cdc2/Cdk2 homologue has been detected in maize (see above). An alternative possibility is that the assay for S-phase kinase activity (phosphorylation of animal S-phase transcription factors) detects a kinase which is not actually a different cdk, but is a downstream S-phase target of cdc2–ZmA. If so, the decrease in p13$^{suc1}$-associated H1 kinase activity observed in endoreduplicating endosperm might be due to the suppression of activity from the M-phase kinase complex, without significant change in the S-phase kinase activity. Notably, in animal and yeast cells, the total p13$^{suc1}$-associated histone H1 kinase activity is several-fold higher in G$_2$/M than in S or G$_1$ [43,44]. In any case, the results of Grafi and Larkins are intriguing, and make it clear that further progress in this field will require much more detailed biochemical characterization of the Cdk–cyclin complexes in plant cells. This task is now being attempted in several laboratories [42,45].

## Cdc2 and the spatial control of cell division

### Localization of Cdc2 in plant cells

Since there is no cell migration in developing plants, the plane of cell division fixes the spatial relationship between the daughter cells. Consequently, division planes follow a predictable programme during plant morphogenesis (reviewed by Lyndon and Cunninghame [46]). The establishment of the plane of cell division in plant cells is accomplished through structural reorganization of the cytoskeleton during the cell cycle (reviewed by Staiger and Lloyd [47], Wick [48] and Palevitz [49]).

The reorganization of the plant cytoskeleton that occurs during mitosis and cytokinesis is very different from that in either animals or fungi (reviewed by Wick [48,50]). Plant cells lack discernible centrosomes or spindle pole bodies; and, at mitosis, no centrioles or asters are visible. There are, however, two microtubule-containing structures found only in cells of higher plants that are important to the process of cell division. Plant cells form a preprophase band (PPB), a dense array of microtubules that aggregate in the cell cortex at the position where the future cell wall will join parent cell walls at cytokinesis [50,51]. The PPB normally forms before or as chromosome condensation begins and before nuclear envelope breakdown is apparent, and disappears before appearance of the metaphase spindle. In many plant species, the PPB forms when cells are still in interphase, but persists and reaches its maturity close to the onset of mitosis. The position of the PPB is somehow imprinted in cellular memory, as its location marks where the new cell wall will form at cytokinesis, when another microtubule-containing structure, the phragmoplast, forms between the separated daughter nuclei in the central body of the cell [52]. This structure, which is aligned in the plane or curved surface of the cell previously outlined by the PPB, is believed to coordinate the movement of vesicles containing cell wall components to be deposited at the phragmoplast equator for the formation of the cross wall that separates the daughter nuclei. The mechanism that determines the positioning of the PPB and the phragmoplast is not known at present, but it is known that microtubules and microfilaments play important roles in this process [47,49,50].

Owing to the fundamental role played by the Cdc2 kinase in reorganizing intracellular structures for mitosis, it is of great interest to examine the localization of the p34cdc2 kinase within dividing plant cells. Mineyuki et al. [53] used antibodies directed against the PSTAIRE epitope present in Cdc2 and cdc2-related proteins to examine dividing root cells from onion. They found predominantly cytoplasmic staining in interphase cells, and staining of the PPB in cells in late $G_2$. This association between cdc2-like proteins and the PPB points to a role for cell division kinases in the establishment of the division plane.

In a separate study, we raised antibodies that are specific to the C-terminal peptide of maize *cdc2–ZmA*, and used immunofluorescence microscopy to examine its intracellular location within dividing cells of the root apex and the progenitor cells of the stomatal complexes of maize [54]. We found that Cdc2 protein is localized mainly to the nucleus in plant cells at interphase and early prophase. In a sub-population of root cells at early prophase, p34cdc2 protein is also distributed in a band bisecting the nucleus. Double labelling with anti-p34cdc2 and anti-tubulin antibodies revealed that this band co-localizes with the PPB of microtubules, which predicts the future division site. In general, cells that displayed co-localization of p34cdc2 with the PPB were those cells that had mature PPBs, and were likely to be further along in the $G_2$/M transition. Root cells, in which microtubules had been disrupted with oryzalin, did not contain a band of p34cdc2 protein, suggesting that formation of the microtubule PPB is necessary for localization of p34cdc2 kinase to the site occupied by the PPB. In contrast to most root apical cells and leaf epidermal cells, both of which commonly divide symmetrically and in a plane transverse to the long axis of the organ, cells dividing to give rise to the stomatal complex in maize undergo highly asymmetric divisions and divisions longitudinal to the organ axis ([55,56]; see Fig. 2). In these stomatal complex progenitor cells, the p34cdc2 protein is again localized to the nucleus and the PPB (Fig. 3; [54]). As the cells progress through mitosis, from late prophase through anaphase, the anti-cdc2 staining becomes diffuse and uniform. In late telophase, the anti-cdc2 staining is concentrated within the newly forming daughter nuclei, and no significant staining of the phragmoplast can be observed [54].

The difference between our observation that p34cdc2 is primarily nuclear in interphase cells, and the predominantly cytoplasmic staining observed by Mineyuki

**Fig. 2**   **Schematic representation of PPB formation in a sudsidiary mother cell (SMC) with an accompanying guard mother cell (GMC)**

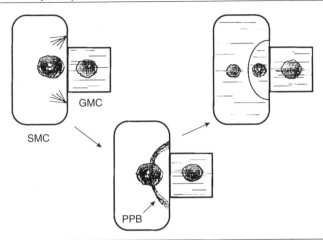

GMC

SMC

PPB

*The mature PPB (centre) marks the future division site, where the new cell plate will join to generate a lens-shaped subsidiary cell (right). Figure based on data in Cho and Wick [55].*

et al. [53], might be owing to differences between the epitopes detected by an anti-PSTAIRE antibody and a specific antibody to a cdc2 homologue. In yeasts, animals and plants there are many Cdc2-like protein kinases that have motifs related to the PSTAIRE motif [19,37,57], and which can cross-react with anti-PSTAIRE antibodies [19]. We have also found extensive cytoplasmic staining and relatively weak nuclear staining in cells stained with a monoclonal antibody to the PSTAIRE peptide; the same cells, when stained with the maize p34[cdc2]-specific antibody, displayed predominantly nuclear staining [54].

    More recently, Hepler et al. [58] examined the localization of labelled p13[suc1] protein microinjected into stamen hair cells. In this study, Suc1 is found to be predominantly nuclear until the onset of mitosis, when it appears to become distributed throughout the cell. No co-localization with the PPB could be detected in these experiments. While the observed nuclear localization of p13[suc1] is consistent with the previously observed localization of p34[cdc2] [54], the reason for the absence of suc1 in the PPB is not clear, since Suc1 associates tightly with p34[cdc2]. However, it should be pointed out that the function of Suc1 is still controversial [59], and *in vivo* only a fraction of p34[cdc2] might be associated with it [60]. Since Suc1 appears to inhibit Cdc2 kinase activity [59], the Cdc2 associated with the PPB may be in a conformation that is inaccessible to Suc1. Therefore, the interpretation of the Suc1 localization experiments must await more information on the intracellular functions of p13[suc1].

**Fig. 3**     **Immunofluorescence staining of leaf epidermis showing a developing stomatal complex**

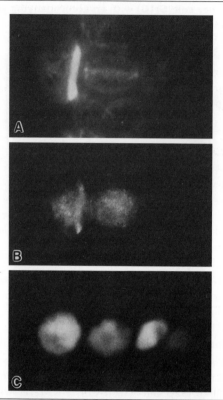

*Cells are stained with tubulin antibody, which reveals the tranverse interphase band in the guard mother cell (GMC) and the more brightly stained preprophase band in the subsidiary mother cell (SMC) on the left (A), maize p34<sup>cdc2</sup> antibody showing nuclear localization and co-localization with the PPB of the SMC (B), and with DAPI to show the DNA (C). Magnification 1600 ×. Figure reproduced from Colasanti et al. [54].*

## Model for the role of Cdc2 in the establishment of the division site

The distribution of p34$^{cdc2}$ in plant cells suggests a role in the imprinting of the plane of cell division. The plane of cell division is accurately determined by the PPB, as the new cell plate formed by the expanding phragmoplast contacts the cortex at the same site where the microtubules of the PPB made contact before mitosis. The mechanism by which the site of PPB formation is 'imprinted' before mitosis, and remains in cellular memory after the PPB disappears, is currently unknown. A band of F-actin has been shown to remain associated with the division site leading to models in which division plane memory may be maintained by F-actin [61–63]. However, the findings of other workers that F-actin in some cells assumes other conformations and is not always associated with the division site suggests that previous observations on localization of F-actin may not apply to all cell types (reviewed in Wick [48]; Goddard et al. [64]). Although no F-actin was found to be associated with the PPB in the divisions of the stomatal complex progenitor cells [56], we observed that association of p34cdc2 with the PPB occurs in these cells [54].

The observations on Cdc2 localization suggest an alternative model for the mechanism of imprinting the plane of cell division. A number of studies have found that the site previously occupied by the PPB appears to exert an attractive force on the edges of the growing phragmoplast [65–68]. The formation of this 'zone of attraction' may be the result of changes occurring at the division site generated by the PPB [69]. A major function of the PPB might be to direct p34cdc2 kinase to specific sites on the cell cortex, where it can modify the PPB contact point by phosphorylation of cortical substrates. Actin may be required for the narrowing of the PPB as suggested by drug studies [70,71]. A second function of p34cdc2 can now come into play. It has been shown that kinase inhibitors can block the disappearance of the PPB which normally occurs by the end of prophase, suggesting that a kinase activated at $G_2/M$ is responsible for this disappearance [72]. Clearly, p34cdc2 would be a good candidate for this kinase, and its association with the PPB is consistent with this role. A likely mechanism for dissociation of the PPB is through phosphorylation of MAPs that might stabilize the microtubules of the PPB. In animal cells, phosphorylation of MAPs by p34cdc2 at $G_2/M$ has been demonstrated [73,74]. While there is little detailed information on plant MAPs, animal MAP2 has been shown to bind to plant tubulin [75], and recently some MAPs have been isolated from maize [76].

The above discussion can be summarized in the model shown in Fig. 4. Through much of interphase, Cdc2 is primarily nuclear, although a cytoplasmic component cannot be ruled out. In late $G_2$, the PPB predicting the future division site appears. The positioning of the PPB by the cell is governed by the position of the nucleus, as well as by developmental cues received by the cell, and interpreted by the cytoskeleton [77–79]. As the cells approach $G_2/M$, a fraction of the Cdc2 leaves the nucleus, and associates with the PPB at the cortex. Since the transport of Cdc2 from the nucleus to the division site is sensitive to the microtubule-depoly-merizing reagent oryzalin [54], it is likely to be mediated by microtubules that connect the nucleus with the cortex [47,80]. At the stage of the cell cycle when p34cdc2 is transported to the cell cortex (the $G_2/M$ transition), Cdc2 kinase activity is approaching its peak, resulting in phosphorylation of cortical substrates at the future division site. At the same time, phosphorylation of MAPs associated with the PPB also occurs, resulting in dissociation of the PPB. The PPB, having served its purpose, is now dispensable; its disappearance is acheived through the same p34cdc2 kinase molecules that it recruited to the division site. In this model, phosphory-lation of the division site determined by the PPB and the cortex would be responsible for guiding the expanding phragmoplast in cytokinesis (see M. Vantard et al., Chapter 15).

This model raises several questions which require further investigation. The mechanism by which Cdc2 kinase is recruited to the division site from the nucleus is one such question to be resolved. Since microtubules are required for this

**Fig. 4**      **Model for the role of Cdc2 kinase in the spatial control of cell division**

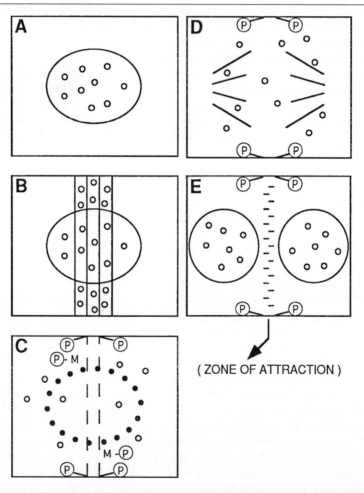

Panels represent different stages of mitosis. A, interphase cell; B, cell in late $G_2$ with PPB; C, early prophase with breakdown of nuclear envelope and disappearance of PPB; D, metaphase cell with spindle and no PPB; and E, late telophase with the phragmoplast between the daughter nuclei. The small open circles represent Cdc2 kinase molecules, the straight lines represent microtubules in different structures (the PPB, the mitotic spindle, and the phragmoplast), M represents MAPs, and P represents phosphorylation.

transport [54], the mechanism probably involves '+ end' directed motors; recent studies have found evidence for such motors in plants [81]. A second question concerns the determinants for targeting Cdc2 to the PPB. Studies in animal systems show that the specificity of localization for a Cdk probably resides within the associated cyclin. Now that several mitotic cyclins have been cloned from plants (see Chapter 4 by J.P. Renaudin et al.), it should be possible to establish if one or more of these cyclins co-localizes with the PPB. Finally, the identity of the hypothetical cortical substrates for the Cdc2 kinase–cyclin complex at the division site should be investigated. Knowledge of these cortical substrates should help elucidate the mechanism by which the 'zone of attraction' attracts the expanding phragmoplast. As with many of the other questions concerning the plant cell cycle, the answers to these questions are likely to emerge from future studies that will emphasize biochemical approaches.

*We are grateful to Susan Wick and Bruce Futcher for discussions, and to Sadie Arana for help with preparation of the manuscript. This research was supported by grant no. 94-37304-1004 from the United States Department of Agriculture.*

## References

1.  Pines, J. and Hunter, T. (1991) Trends Cell Biol. **1**, 117–121
2.  Kirschner, M. (1992). Trends Biochem. Sci. **17**, 281–285
3.  Xiong, Y. and Beach, D. (1991) Curr. Biol. **1**, 362–364
4.  Lewin, B. (1990) Cell **61**, 743–752
5.  Norbury, C. and Nurse, P. (1992) Annu. Rev. Biochem. **61**, 441–470
6.  King, R.W., Jackson, P.K. and Kirschner, M.W. (1994) Cell **79**, 563–571.
7.  Nigg, E.A. (1993) Trends Cell Biol. **3**, 296–301
8.  Solomon, M.J. (1993) Curr. Opin. Cell Biol. **5**, 180–186
9.  Fesquet, D., Labbe, J.-C., Derancourt, J., Capony, J.-P., Galas, S., Girard, F., Lorca, T., Shuttleworth, J., Doree, M. and Cavadore, J.-C. (1993) EMBO J. **12**, 3111–3121
10. Fisher, R.P. and Morgan, D.O. (1994) Cell **78**, 713–724
11. Hoffmann, I., Clarke, P.R., Marcote, M.J., Karsenti, E. and Draetta, G. (1993) EMBO J. **12**, 53–63
12. Draetta, G. (1993) Trends Cell Biol. **3**, 287–289
13. Nasmyth, K. (1993) Curr. Opin. Cell Biol. **5**, 166–179
14. Fitch, I., Dahmann, C., Surana, U., Amon, A., Nasmyth, K., Goetsch, L., Byers, B. and Futcher, B. (1992) Mol. Biol. Cell **3**, 805–815
15. Kuhne, C. and Linder, P. (1993) EMBO J. **12**, 3437–3447
16. Elledge, S.J. and Spottswood, M.R. (1991) EMBO J. **10**, 2653–2659
17. Sherr, C.J. (1993) Cell **73**, 1059–1065
18. Sherr, C.J. (1994) Cell **79**, 551–555
19. Meyerson, M., Enders, G.H., Wu, C.-L., Su, L.-K., Gorka, C., Nelson, C., Harlow, E. and Tsai, L.-H. (1992) EMBO J. **11**, 2909–2917
20. Shuttleworth, J., Godfrey, R. and Colman, A. (1990) EMBO J. **9**, 3233–3240
21. Pines, J. and Hunter, T. (1991) Cold Spring Harbor Symp. Quant. Biol. **56**, 449–464
22. Sherr, C.J. (1995) Trends Biochem. Sci. **20**, 187–190
23. Heichman, K.A. and Roberts, J.M. (1994) Cell **79**, 557–562
24. Knoblich, J.A. and Lehner, C.F. (1993) EMBO J. **12**, 65–74
25. Nigg, E.A. (1993) Curr. Opin. Cell Biol. **5**, 187–193

26. Peeper, D.S., Parker, L.L., Ewen, M.E., Toebes, M., Hall, F.L., Xu, M., Zantema, A., van der Eb, A.J. and Piwnica-Worms, H. (1993) EMBO J. **12**, 1947–1954
27. Renaudin, J.-P., Colasanti, J., Rime, H., Yuan, Z. and Sundaresan, V. (1994) Proc. Natl. Acad. Sci. U.S.A. **91**, 7375–7379
28. Soni, R., Carmichael, J.P., Shah, Z.H. and Murray, J.A.H. (1995) Plant Cell **7**, 85–103
29. Colasanti, J., Tyers, M. and Sundaresan, V. (1991) Proc. Natl. Acad. Sci. U.S.A. **88**, 3377–3381
30. Feiler H.S. and Jacobs, T.W. (1990) Proc. Natl. Acad. Sci. U.S.A. **87**, 5397–5401
31. Ferreira, P., Hemerly, A.S., Villarroel, R., Van Montagu, M. and Inze, D. (1991) Plant Cell **3**, 531–540
32. Fobert, P.R., Coen, E.S., Murphy, G.J.P. and Doonan, J.H. (1994) EMBO J. **13**, 616–624
33. Hirt, H., Pay, A., Gyorgyey, J., Bako, L., Nemeth, K., Bogre, L., Schweyen R.J., Heberle-Bors, E. and Dudits, D. (1991) Proc. Natl. Acad. Sci. U.S.A. **88**, 1636–1640
34. Hirayama, T., Imajuka, Y., Anai, T., Matsui, M. and Oka, A. (1991) Gene **105**, 159–165
35. Hashimoto, J., Hirabayashi, T., Hayano, Y., Hata, S., Ohashi, Y., Suzuka, I., Utsugi, T., Toh-E, A. and Kikuchi, Y. (1992) Mol. Gen. Genet. **233**, 10–16
36. Miao, G.H., Hong, Z. and Verma, D.P.S. (1993) Proc. Natl. Acad. Sci. U.S.A. **90**, 943–947
37. Jacobs, T.W. (1995) Annu. Rev. Plant Physiol. Plant Mol. Biol. **46**, 317–339
38. Imajuku, Y., Hirayama, T., Endoh, H. and Oka, A. (1992) FEBS Lett. **304**, 73–77
39. Hirt, H., Pay, A., Bogre, L., Meskiene, I. and Heberle-Bors, E. (1993) Plant J. **4**, 61–69
40. Labib, K., Craven, R.A., Crawford, K. and Nurse, P. (1995) EMBO J. **14**, 2155–2165
41. Simanis, V. (1995) Semin. Cell Biol. **6**, 79–87
42. Grafi, G. and Larkins, B.A. (1995) Science **269**, 1262–1264
43. Pagano, M., Pepperkok, R., Verde, F., Ansorge, W. and Draetta, G. (1992) EMBO J. **11**, 961–971
44. Tyers, M., Tokiwa, G. and Futcher, B. (1993) EMBO J. **12**, 1955–1968
45. Magyar, Z., Bako, L., Bögre, L., Dedeoglu, D., Kapros, T. and Dudits, D. (1993) Plant J. **4**, 151–161
46. Lyndon, R.F. and Cunninghame, M.E. (1986) in Plasticity in Plants (Jennings, D.H. and Trewavas, A.J., eds., Symposia of the Society for Experimental Biology vol. **40**, pp. 233–255
47. Staiger, C.J. and Lloyd, C.W. (1991) Curr. Opin. Cell Biol. **3**, 33–42
48. Wick, S.M. (1991) Curr. Opin. Cell Biol. **3**, 253–261
49. Palevitz, B.A. (1993) Plant Cell **5**, 1001–1009
50. Wick, S.M. (1991) in The Cytoskeleton in Plant Growth and Development (Lloyd, C.W., ed.) Academic Press, London
51. Pickett-Heaps, J.D. (1969) J. Ultrastructure Res. **27**, 24–44
52. Gunning, B.E.S. (1982) in The Cytoskeleton in Plant Growth and Development (Lloyd, C.W., ed.), pp. 229–292, Academic Press, London,
53. Mineyuki, Y., Yamashita, M. and Nagahama, Y. (1991) Protoplasma **162**, 182–186
54. Colasanti, J., Cho, S.-O., Wick, S. and Sundaresan, V. (1993) Plant Cell **5**, 1101–1111
55. Cho, S.-O. and Wick, S.M. (1989) J. Cell Sci. **92**, 581–594
56. Cho, S.-O. and Wick, S.M. (1990) Protoplasma **157**, 154–164
57. Toh-e, A., Tanaka, K., Uesono, Y. and Wickner, R.B. (1988) Mol. Gen. Genet. **214**, 162–164
58. Hepler, P.K., Sek, F.J. and John, P.C.L. (1994) Proc. Natl. Acad. Sci. U.S.A. **91**, 2176–2180
59. Basi, G. and Draetta, G. (1995) Mol. Cell. Biol. **15**, 2028–2036
60. Brizuela, L., Draetta, G. and Beach, D. (1987) EMBO J. **6**, 3507–3514
61. Palevitz, B.A. (1987) J. Cell Biol. **104**, 1515–1519
62. Traas, J.A., Doonan, J.H., Rawlins, D.J., Shaw, P.J., Watts, J. and Lloyd, C.W. (1987) J. Cell Biol. **105**, 387–395
63. Lloyd, C.W. and Traas, J.A. (1988) Development **102**, 211–221
64. Goddard, R.H., Wick, S.M., Silflow, C.D. and Snustad, D.P. (1994) Plant Physiol. **104**, 1–6
65. Ota, T. (1961) Cytologia **26**, 428–447
66. Gunning, B.E.S. and Wick, S.M. (1985) J. Cell Sci. **2**, Suppl. 2, 157–179
67. Palevitz, B.A. (1986) Dev. Biol. **117**, 644–654
68. Apostolakos, P. and Galatis, B. (1987) Protoplasma **140**, 26–42
69. Mineyuki, Y. and Gunning, B.E.S. (1990) J. Cell Sci. **97**, 527–537
70. Mineyuki, Y. and Palevitz, B.A. (1990) J. Cell Sci. **97**, 283–295
71. Eleftheriou, E.P. and Palevitz, B.A. (1992) J. Cell Sci. **103**, 989–998

72. Katsuta, J. and Shibaoka, H. (1992) J. Cell Sci. **103**, 397–405
73. Shiina, N., Moriguchi, T., Ohta, K., Gotoh, Y. and Nishida, E. (1992) EMBO J. **11**, 3977–3984
74. Ookata, K., Hisanaga, S., Bulinski, J.C., Murofushi, H., Aizawa, H., Itoh, T.J., Hotani, H., Okumura, E., Tachibana, K. and Kishimoto, T. (1995) J. Cell Biol. **128**, 849–862
75. Hugdahl, J.D., Bokros, C.L., Hanesworth, V.R., Aalund, G.R. and Morejohn, L. (1993) Plant Cell **5**, 1063–1080
76. Schellenbaum, P., Vantard, M., Peter, C., Fellous, A. and Lambert, A.-M. (1993) Plant J. **3**, 253–260
77. Cyr, R.J. (1994) Annu. Rev. Cell Biol. **10**, 153–180
78. Cyr, R.J. and Palevitz, B.A. (1995) Curr. Opin. Cell Biol. **7**, 65–71
79. Lloyd, C. (1994) Mol. Biol. Cell **5**, 1277–1280
80. Lambert, A.M. (1993) Curr. Opin. Cell Biol. **5**, 116–122
81. Mitsui, H., Nakatani, K., Yamaguchi-Shinozaki, K., Shinozaki, K., Nishikawa, K. and Takahashi, H. (1994) Plant Mol. Biol. **25**, 865–876

# 4

# Characterization and classification of plant cyclin sequences related to A- and B-type cyclins

**J.P. Renaudin*¶, A. Savouré†, H. Philippe‡, M. Van Montagu†, D. Inzé§ and P. Rouzé§**

*Laboratoire de Biochimie et Physiologie Végétales, Institut National de la Recherche Agronomique, 2 Place Viala, F-34060 Montpellier Cedex 1, France; †Laboratorium voor Genetica, via the Department of Genetics, affiliated to the Flanders Interuniversity Institute for Biotechnology; §Laboratoire Associé de l'Institut National de la Recherche Agronomique, Universiteit Gent, K.L. Ledeganckstraat 35, B-9000 Gent, Belgium; and ‡Laboratoire de Biologie Cellulaire, CNRS URA1134, Bâtiment 444, Université de Paris-Sud, F-91405 Orsay Cedex, France

## Introduction

The proper progression of eukaryotic cells through the cell cycle requires the co-ordinated activation of a series of serine/threonine protein kinases named cyclin-dependent kinases (CDKs) [1–3]. The CDKs bind to positive regulators, which have been named cyclins, as some exhibit cyclic accumulation and destruction during embryonic development [4], and to negative regulators, which have been identified more recently in yeast and animals and named cyclin-dependent kinase inhibitors (CKIs) [5,6]. The universality of the basic machinery of cell cycle controls among eukaryotes [7] enabled the isolation of CDKs and cyclins from yeast, animal and plant cells. The CDK family is restricted to one or a few kinases only, Cdc2/CDC28, in yeast, whereas they are more numerous in plants [8–10] and in animals, e.g. there are at least seven CDKs in humans. In animals and fungi the cyclin family displays a complex organization, which relates to the timing of their appearance and therefore function during the cell cycle. Nine classes of cyclins (A–I) are expressed in animal cells. Fungi have fewer cyclin classes although they still have a large number of cyclins, e.g. six B-type cyclins [11,12] and three $G_1$/S cyclins named *CLN* in *Saccharomyces cerevisiae*, and five cyclins mostly of the B-type in *Schizosaccharomyces pombe* [13]. In plants, cyclins have initially been found

¶ *To whom correspondence should be addressed.*

by sequence homology with the best conserved A- and B-type cyclins [14–17]. Recently, plant homologues to the $G_1$/S vertebrate D-type cyclins [18,19] have been found by complementation of *CLN* mutants of *S. cerevisiae*. Most of the plant cyclins cloned to date show homology to the cyclin A/B classes, but they display structural characteristics that have prevented their proper classification and naming. This novelty has also prevented the number of classes of plant cyclins to be assessed. These drawbacks can be now circumvented by a comparative analysis of the large number of sequences which have become available in the cyclin A/B family. We report here a sequence analysis of plant cyclins with special reference to the structure–function relationship of cyclins A and B and demonstrate three distinct cyclin A groups and two cyclin B groups in plants. For each group consensus sequences have been established and a new, rational naming system is proposed.

# The function and structure of A- and B-type cyclins

## Function of cyclin A and cyclin B

A- and B-type cyclins are often quoted as mitotic cyclins because they are required for progression from $G_2$ into M phase. As a consequence, both of them can drive $G_2$-arrested *Xenopus* oocytes into meiotic metaphase [20,21]. However, the two types of cyclins display profound differences in their precise functions. This stems partly from the fact that in animals cyclin A can bind either to CDK1 (p34[cdc2] kinase) or to the related CDK2 kinase, whereas cyclin B binds exclusively to CDK1.

Cyclin A and cyclin B form stable, active complexes with CDK1 in $G_2$/M phases. The functions of these complexes are not redundant [22,23]. The cyclin A kinase is activated before the cyclin B kinase, and cyclin A is degraded in metaphase, before cyclin B [24,25]. The two cyclins differentially modulate the substrate specificity of their CDK counterpart [26]. The activation of CDK1 by cyclin B is required to initiate prophase. The cyclin B–CDK1 complex itself is subject to activation, the best-studied mechanism for this being the release of inhibitory tyrosine phosphorylation of CDK1 [7,27]. One function of the cyclin A–CDK1 complex might be to switch the equilibrium of cyclin B–CDK1 to the dephosphorylated form through inhibition of the tyrosine kinase that phosphorylates CDK1 [28].

Entry into S phase requires the presence of active cyclin A–CDK2 complexes which are formed at that time [22,29,30]. The function of the cyclin A–CDK2 complex in S phase is not yet totally understood. Cyclin A accumulates in the nucleus as soon as it is synthesized [22,30–32], and it is found at sites of DNA replication [33]. This indicates a role for cyclin A in regulating gene expression by controlling the phosphorylation of transcription factors. This model is reinforced because p107, a protein related to pRB, a negative transcriptional regulator, and E2F, a transcription factor regulating the expression of a number of growth-promoting genes, have also been demonstrated to interact with the cyclin A–CDK2 complex [34–36]. A cyclin A–p107 complex binds to and activates a 25-bp element in the human thymidine kinase promoter [37]. The cyclin A–CDK2 complex could also be involved in gene repression, because E2F can be phosphorylated by the cyclin A–CDK2–p107 complex in S phase and this phosphorylation represses E2F-induced gene expression [38,39].

In yeast, which lack cyclin A, B-type cyclins are active in more steps of the cell cycle than in animal. The cyclins cig1 and cig2 of *S. pombe* act only in $G_1$ phase [40–42] and in *S. cerevisiae*, CLB5 and CLB6 are B-type cyclins which are functionally important in S phase [12,43].

## Structural organization of cyclin A and cyclin B

A central region encompassing approximately 250 amino acids is highly conserved among A- and B-type cyclins [44]. This region is less conserved in other cyclins, the closest cyclin A and B homologue being cyclin F [45,46] then cyclins E, D, C to I, and *CLN*s. This domain has been restricted in some works to the 150 residues which are the best conserved in A-, B-, D- and E-type cyclins [47]. Hereafter we will refer to the 250-amino-acid domain as the cyclin core.

Recent results have shed light on the molecular basis for the strong conservation of the cyclin core. The analyses of the crystal structure of the bovine cyclin A core [48] and of the human cyclin A core–CDK2 complex [49] have revealed that cyclin A has a rigid tertiary structure that is not modified by complex formation. The cyclin core, which is sufficient for binding to, and activation of, CDK1 and CDK2 [50–52], is organized as two domains with a strikingly similar tertiary structure, the cyclin fold (Fig. 1) [48, 49, 53]. Each fold is composed of five helices, helix H3 being surrounded by helices H1 and H2, which are involved in the tight packing of the two repeated domains, and helices H4 and H5 which are exposed to the solvent and contact protein partners, i.e. CDK for the first fold. Since the five helices cross each

---

**Fig. I**          **The cyclin fold**

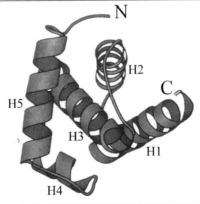

*Adapted from [49].*

other, the compactness of the domain depends on the small size of the side-chain of the residues at the crossing points, most of these residues being alanine. The actual cyclin box has been defined structurally from these studies [48] as the 100 residues of the first fold, which define the CDK-binding domain, and which are the best conserved in all cyclins. The position of helices in the two folds of the cyclin core is indicated in Figs. 2 and 8 which show the alignment and consensus sequences of the cyclin core of A- and B-type cyclins. In B-type cyclins, the beginning of the cyclin box has also been implicated in the activation of Cdc25, the threonine/tyrosine phosphatase which activates the cyclin B–CDK1 complex at the onset of mitosis [52]. Although the conservation of residues forming the second fold is less, this domain displays a remarkably similar helical structure to that of the first fold, but the site that interacts directly with CDKs is absent.

A prominent feature observed in the N-terminal region of A- and B-type animal and yeast cyclins is the occurrence of a small conserved motif of nine amino acids which is required for the degradation of cyclins during mitosis, and which is called the destruction box [54]. In animal, the proteolysis of cyclin A and of cyclin B2, but not that of cyclin B1, requires their binding to CDK1 [25]. In addition to the destruction box, the degradation of cyclin A requires another structural feature, not yet identified, but situated between the destruction motif and the cyclin box [25]. Cyclin F, which has a cyclin core most closely related to A- and B-type cyclins, lacks a destruction box. This largest cyclin, with a molecular mass of 87 kDa, has an extensive PEST-rich C-terminus which is probably involved in the degradation of this cyclin at mitosis [45,46]. PEST-rich sequences are also present in $G_1$ cyclins such as D cyclins and CLNs.

# Structural characterization of plant cyclins related to A- and B-type cyclins

## Plants possess a large number of cyclins

Initially, the search for cyclins in plants exploited the strong conservation of two motifs of the cyclin box. The first motif, MR-IL(IV)DW, forms part of helix H1 of the first cyclin fold which shows the best conserved pattern in the cyclin A/B family (Fig. 2). In this motif, the tryptophan residue is part of a surface site probably involved in cyclin–protein interaction [48]. The second motif, KYE---P, is at the end of helix H3, the two acidic residues being part of the CDK-binding site [48]. About 60 full-length or nearly complete plant cyclin cDNAs have been isolated from 13 species within the last five years. Among these, 51 are related to A- and B-type cyclins. Nine D-type cyclins have been found by complementation of G1/S cyclin-deficient yeast cells or by hybridization with probes derived from the clones obtained in this way (Table 1). As the extent of the screening procedures has not yet been fully examined, there is little doubt that this number will expand in the near future.

**Table I.    Plant cyclins**

The sequences have been named according to the classification into five structural groups of CycA and CycB plant cyclins proposed in this chapter. Gene names are thus CycA1, CycA2, CycA3, CycB1, CycB2, CycD1, CycD2 and CycD3. Gene names are followed by the initials of the species. When there are several sequences from the same species in a given group, they are considered as part of a multigene family, and member numbers are given sequentially (1, 2, 3, ...) after the species. Accession numbers in Genbank/EMBL/DDBJ are indicated.

| Plant cyclin | Old name | Species | Accession | Reference |
|---|---|---|---|---|
| **CycA1 group** | | | | |
| CycA1;bn | bncyc2 | Brassica napus | L25406 | [92] |
| CycA1;gm | cyc3gm | Glycine max | D5O870 | [76] |
| CycA1;nt;1 | ntcyc25 | Nicotiana tabacum | D50735 | [86] |
| | ntcycA19[1] | | X92966 | [97] |
| CycA1;nt;2[2] | ntcycA30 | Nicotiana tabacum | X92967 | [97] |
| CycA1;os;1 | cycOs1 | Oryza sativa | – | J.Hashimoto et al., unpublished work |
| CycA1;os;2 | cycOs3 | Oryza sativa | – | J.Hashimoto et al., unpublished work |
| CycA1;zm;1 | cycllzm | Zea mays | U10077 | [17] |
| CycA1;zm;2 | cycZm2w | Zea mays | U50064 | W.L. Hsieh and S.M.Wolniak, unpublished work |
| **CycA2 group** | | | | |
| CycA2;at;1 | cyc3aAt | Arabidopsis thaliana | Z31589 | [16] |
| CycA2;at;2 | cyc3bAt | Arabidopsis thaliana | Z31402 | [16] |
| CycA2;at;3 | cyc3c-at | Arabidopsis thaliana | U17890 | G. Lu and R.J. Ferl, unpublished work |
| CycA2;at;4[3] | ATU17889 | Arabidopsis thaliana | U17889 | G. Lu and R.J. Ferl, unpublished work |
| CycA2;bn | bncyc1 | Brassica napus | L25405 | [92] |
| CycA2;gm | cyc2gm | Glycine max | D5O869 | [76] |
| CycA2;ms[4] | cycMs3 | Medicago sativa | X85783 | [87] |
| CycA2;nt | Ntcyc27 | Nicotiana tabacum | D50736 | [86] |
| CycA2;ps;1 | cycps3 | Pisum sativum | – | T.Jacobs et al., unpublished work |
| CycA2;ps;2[5] | cycps2 | Pisum sativum | – | T.Jacobs et al., unpublished work |

Table1 (Contd.)

| Plant cyclin | Old name | Species | Accession | Reference |
|---|---|---|---|---|
| **CycA3 group** | | | | |
| CycA3;am | - | Antirrhinum majus | - | V. Gaudin et al., unpublished work |
| CycA3;dc | C13-1 | Daucus carrota | S49312 | [14] |
| CycA3;gm | cyc1gm | Glycine max | D5O868 | [76] |
| CycA3;nt;1 | ntcycA105 | Nicotiana tabacum | X92964 | [97] |
| CycA3;nt;2 | ntcycA59 | Nicotiana tabacum | X92965 | [97] |
| CycA3;nt;3 | ntcycA13 | Nicotiana tabacum | X93467 | [97] |
| **CycB1 group** | | | | |
| CycB1;am;1 | amcycl1 | Antirrhinum majus | X76122 | [61] |
| CycB1;am;2 | amcycl2 | Antirrhinum majus | X76123 | [61] |
| CycB1;at;1 | cyc1at | Arabidopsis thaliana | X62279 | [91] |
| CycB1;at;2 | cyc1bAt | Arabidopsis thaliana | L27223 | [93] |
| CycB1;at;3 | cycx3-ara | Arabidopsis thaliana | L27224 | [93] |
| CycB1;gm;1 | cyc5gm | Glycine max | X62820 | [76] |
| | S13-6 | | | [14] |
| | gmcyclin[6] | | Z26331 | J. Deckert and P.M. Gresshoff, unpublished work |
| CycB1a;gm;2[7] | S13-7 | Glycine max | X62303 | [14] |
| CycB1;gm;3 | cyc4gm | Glycine max | D50871 | [76] |
| CycB1;nt;1 | Ntcyc1 | Nicotiana tabacum | Z37978 | [94] |
| CycB1;nt;2 | Ntcyc29 | Nicotiana tabacum | D50737 | [86] |
| CycB1;pc | pummicy | Petroselinum crispum | L34207 | [95] |
| CycB1;ps | cycps1 | Pisum sativum | - | T. Jacobs et al., unpublished work |
| CycB1;zm;1 | cyclazm | Zea mays | U10079 | [17] |
| CycB1;zm;2 | cyclbzm | Zea mays | U10078 | [17] |

**Table I (Contd.)**

| Plant cyclin | Old name | Species | Accession | Reference |
|---|---|---|---|---|
| **CycB2 group** | | | | |
| CycB2;at;1 | cyc2aAt | Arabidopsis thaliana | Z31400 | [9] |
| CycB2;at;2 | cyc2bAt | Arabidopsis thaliana | Z31401 | [9] |
| CycB2;ms;1 | cycMs1 | Medicago sativa | X82039 | [87] |
|  | cycMs1[8] |  | X68740 | [15] |
| CycB2;ms;2 | cycMs2 | Medicago sativa | X82040 | [87] |
|  | cycMs2[9] |  | X68741 | [15] |
| CycB2;ms;3[10] | cycIIIms | Medicago sativa | X78504 | [90] |
| CycB2;os;1 | cycOs1 | Oryza sativa | X82035 | [60] + J. Hashimoto, unpublished work |
| CycB2;os;2 | cycOs2 | Oryza sativa | X82036 | [60] + M. Sauter, unpublished work |
| CycB2a-zm | cycIIIzm | Zea mays | U10076 | [17] |
| **CycD1 group** | | | | |
| CycD1;am | - | Antirrhinum majus | - | V. Gaudin et al., unpublished work |
| CycD1;at | cycδ1 | Arabidopsis thaliana | X83369 | [18] |
| CycD1;ht | - | Helianthus tuberosus | - | D. Freeman and J. Murray, unpublished work |
| **CycD2 group** | | | | |
| CycD2;at | cycδ2 | Arabidopsis thaliana | X83370 | [18] |
| **CycD3 group** | | | | |
| CycD3;am | - | Antirrhinum majus | - | V. Gaudin et al., unpublished work |
| CycD3;am | - | Antirrhinum majus | - | V. Gaudin et al., unpublished work |
| CycD3;at | cycδ3 | Arabidopsis thaliana | X83371 | [18] |
| CycD3;ht | - | Helianthus tuberosus | - | D. Freeman and J. Murray, unpublished work |
| CycD3;ms | cycms4 | Medicago sativa | X88864 | [19] |

[1] 99.9% similar to D50735; [2] 95% similar to CycA1;nt;1; [3] 99% similar to CycA2;at;2 except in C-terminus; [4] complements a α-pheromone-response yeast mutant; [5] closely related to CycA2;ps;1; [6] incomplete cDNA, 100% similar to CycB1;gm;1; [7] 95% similar to CycB1;gm;1; [8] incomplete cDNA; [9] incomplete cDNA; [10] 97% similar to CycB2;ms;1.

## Fig. 2    Multiple alignment of the conserved amino acid sequences of the cyclin core from representative animal, yeast, and plant cyclins A, B and F

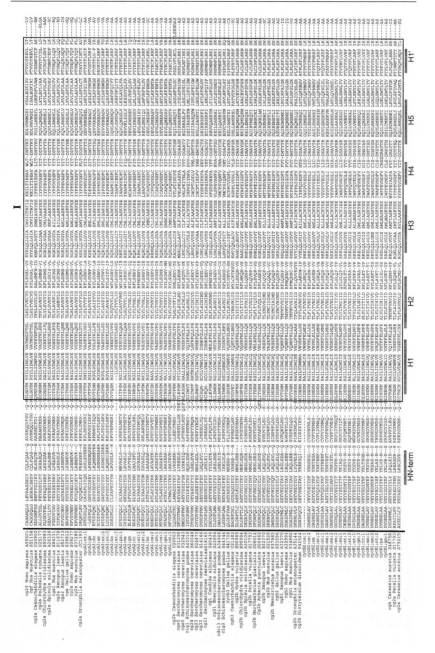

# Fig. 2 (Contd.)

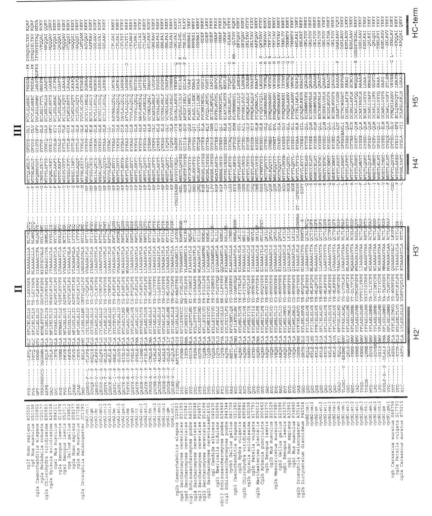

The plant cyclins are identified in Table 1 and the other eukaryotic cyclins by an accession number. The 84 sequences were initially aligned using CLUSTALW version 1.5 and then adjusted further manually. Boxes indicated by roman numerals show highly conserved regions which were used for the construction of phylogenetic trees. Helices were positioned according to their place in bovine cyclin A [48]. The dashes indicate gaps required for maximum alignment.

## Plant cyclin sequences related to animal A- and B-type cyclins belong to two classes and are clustered into five groups

We performed phylogenetic analysis to determine the relationship between plant cyclins and their animal and yeast counterparts. This study encompassed most of plant cyclin sequences, a number of animal A- and B-type cyclins and of yeast B-type cyclins, cyclin B of *Dictyostelium* and vertebrate cyclin F. The sequences which were identical by more than 90% and thus did not bring any additional information in the clustering were not considered. Fig. 2 shows a multiple alignment of the more highly conserved domains of the cyclin core of 84 cyclin sequences. For the analysis, we corrected manually the alignment according to known structural data and we discarded the most variable parts of the sequences for which a reliable alignment could not be achieved.

Phylogenetic trees have been constructed from the aligned sequences of Fig. 2 with the program PAUP [55] and with the PHYLIP package [56] for parsimony method (Fig. 3), and with the MUST package [57] for distance method (Neighbour Joining algorithm; Fig. 4) [58]. The robustness of the trees was assessed by the bootstrap method [59].

Animal, yeast and plant cyclins form distinct groups with a reasonable degree of reliability: the bootstrap values are higher than 80%, except for the monophyly of animal A- and B-type cyclins (Figs. 3 and 4). This latter result is mainly due to the *Caenorhabditis* and *Drosophila* sequences, which seem to have a higher evolutionary rate (see below). This overall comparison demonstrates that plant cyclins related to animal A- and B-type cyclins are distributed in two different classes which we propose to call CycA and CycB as explained below, due to their clustering with animal A- and B-type cyclins, respectively. It must be pointed out, nevertheless, that the positions of the nodes of plant CycA and CycB compared to animal cyclin A and B are not well resolved due to limitations of phylogenetic tree reconstruction (see below). Other criteria such as the identification of specific signatures have to be used to unravel the relationship of the different groups.

The phylogenetic analysis clearly shows that plant A-type sequences can be classified into three groups, CycA1, CycA2 and CycA3, and that plant B-type sequences form two groups, CycB1 and CycB2 (Figs. 3 and 4). The percentage of each group appearing in bootstrap replicates are 100, 88, and 84 for the CycA, and 98 and 100 for the CycB, respectively, with the parsimony method, and 100 for the five groups with the distance method, showing that this classification is robust.

Cyclins in the same group are more closely related to each other than to cyclins from the same species in different groups. The average percentage of amino acid similarity within the cyclin core of a given group is 46% for CycB and 47–52% for CycA. The average similarity between classes CycA and CycB is 31% to 36% (Table 2). The occurrence of CycA and CycB cyclins in a given species has been recorded in many plants and it is probably a general situation in both monocotyledonous and dicotyledonous plants. This suggests that the evolutionary origin of these cyclins arose before the appearance of Angiosperms. The same holds true for

## Fig. 3    A parsimony consensus bootstrap tree based on cyclin sequences

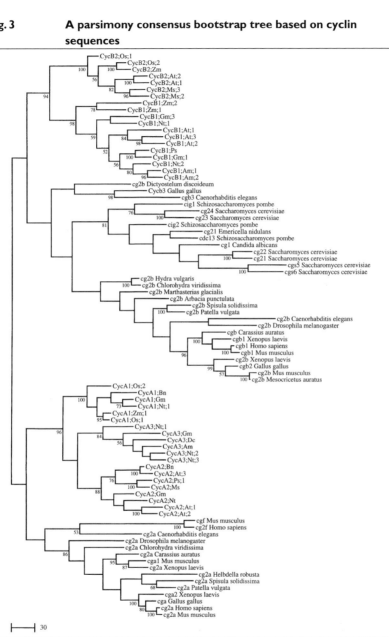

The tree was constructed using an heuristic search of the PAUP program version 3.0 [55]. The numbers below the branches indicate the percentage of times the species which are at the right grouped together in the bootstrap tree, obtained after 100 bootstrap replicates, using the PROTPARS program. The scale bar indicates the number of amino acid substitutions.

**Fig. 4**      **A distance tree with the Neighbour Joining method based on cyclin sequences**

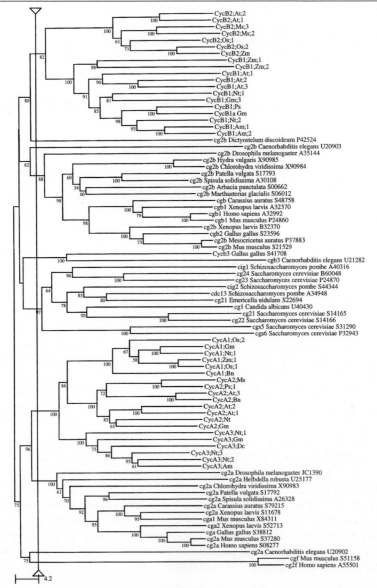

*The numbers below the branches indicate the percentage of times the divergent species grouped together in the bootstrap tree, obtained after 1000 bootstrap replicates, using the NJBOOT program. The scale bar indicates the frequency of amino acid differences. The nodes located on the left of the vertical line are not significant, due to mutational saturation.*

**Table 2        Homology of plant CycA and CycB cyclins**

The data are the rate of identity (+ conservative substitutions) (%) in the cyclin core (262–272 amino acids) among the cyclins from one given group (calculated from the number of conserved residues in each consensus sequence from Fig. 8) or among the five groups of plant cyclins, animal cyclins A, and yeast and animal cyclins B (calculated from the number of identical residues between each pair of consensus sequences of Fig. 8).

|          | CycA1 | CycA2 | CycA3 | CycB1 | CycB2 | Cyclin A | Cyclin B |
|----------|-------|-------|-------|-------|-------|----------|----------|
| CycA1    | 83    |       |       |       |       |          |          |
| CycA2    | 53    | 78    |       |       |       |          |          |
| CycA3    | 48    | 50    | 76    |       |       |          |          |
| CycB1    | 35    | 36    | 32    | 73    |       |          |          |
| CycB2    | 34    | 35    | 31    | 44    | 77    |          |          |
| Cyclin A | 37    | 36    | 37    | 31    | 31    | 66       |          |
| Cyclin B | 26    | 27    | 26    | 30    | 29    | 25       | 48       |

the occurrence of the five groups in every plant species although cyclins belonging to the five different groups have not yet been found in any species. The simultaneous occurrence of CycB1 and CycB2 cyclins has been recorded so far in *Arabidopsis* and in maize, and polymerase chain reaction cloning of partial cyclin cDNAs indicates that it is also the case in tobacco (J.P. Renaudin, unpublished work). The simultaneous occurrence of CycA1, CycA2 and CycA3 cyclins has been reported only in tobacco and in soybean (Table 1). The conservation of five structural groups of plants cyclins during evolution indicates that probably these cyclins perform different functions.

Sequence analysis has revealed that some plant species have more than one cyclin in each of the CycA and CycB groups (Table 1). Inside groups, cyclins share, in general, 70–76% identity over their complete amino acid sequences [15,16,60,61]. Nevertheless, there are cases where these cyclins from the same species, inside the same group, share more similarity with a cyclin from another species than with each other, e.g. between soybean CycB1;gm;1-2 and CycB1;gm;3, tobacco CycA3;nt;1, CycA3;nt;2 and CycA3;nt;3, rice CycB2;os;1 and CycB2;os;2, and *Arabidopsis* CycA2;at;1-2-4 and CycA2;at;3. This indicates both recent and more ancient gene duplication events in plant cyclins, a recurrent phenomenon which leads us to expect that the number of cyclins in the different subgroups defined above will continue to grow in the near future and that a further step in subclassification will probably be possible by then.

## Phylogenetic analysis of eukaryotic cyclins related to classes A, B and F

The major limitation of phylogenetic inference is the occurrence of multiple substitutions at the same position, because recent substitutions mask older ones [62]. The reversion to the ancestral state is the simplest example of this problem. When there are many multiple substitutions, the sequence similarity remains the same whatever the real number of substitutions. This phenomenon is called saturation. For this reason, we have applied a simple method [63] to estimate the level of saturation of the cyclin sequences by using the parsimony approach to determine multiple substitutions occurring at the same position. This gives an estimate of the number of substitutions between each pair of taxa, which can be compared with the number of observed differences. In the case of cyclins (Fig. 5), the number of amino acid differences remains close to 150 (except for cyclin F), whereas the number of substitutions increase to 550. In other words, when the number of amino acid differences between two sequences is close to 150, it could correspond to 150 substitutions as well as to 550 substitutions. In such a case, the similarity between two sequences is not a good indicator of the distance separating them.

The occurrence of 30 amino acid positions invariant in cyclins A/B, but changed in cyclin F, explains that the distances for cyclin F remain close to 180. Following the definition of Fitch and Markowitz [64], the covariations (i.e. the positions that can change simultaneously in a given sequence at a given time) are not

---

**Fig. 5**          **Saturation curve**

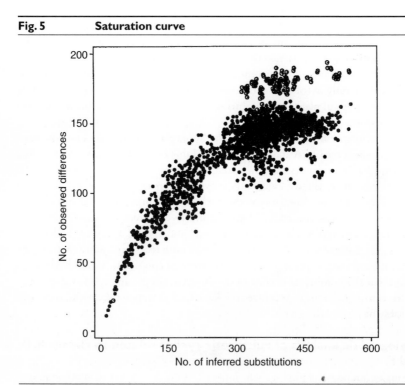

Saturation is reached when the sequences share a low level of identity. Open circles correspond to cyclin F.

the same in cyclin F and in cyclins A/B, as observed for example in the case of cytochrome c of plants and animals [65].

The major consequence of saturation is the difficulty in resolving deep branching patterns. Fig. 4 provides a good example of a tree constructed with saturated sequences by the distance method. The internal branches are very short whereas the external ones are quite long. In addition, the terminal tips, i.e. all the extant genes, are located at the same distance to the root of the tree, suggesting that all the genes accumulate substitutions at the same rate during their evolution, i.e.behave like a molecular clock. However, this was shown to be an artifact [66], because this behaviour could appear even with great differences in evolutionary rates (see below). As a consequence, all the branching points located at the left of the vertical line on Fig. 4 are not reliable because they can be due to random similarity.

Is parsimony more efficient than distance in recovering the real phylogeny? The phylogenetic pattern drawn by the parsimony method (Fig. 3) is rather similar to that reported in Fig. 4, using distances, especially for the five major groups. Nevertheless, for deep nodes, the branch lengths differ completely and the bootstrap values are very low (below 50%). Thus, the parsimony method allows us to detect multiple substitutions, especially at the base of the tree, but it is less reliable regarding their locations in the tree.

This shortcoming may be explained in several ways: (i) insufficient species sampling; the use of numerous species, regularly distributed on the tree, helps in discovering multiple substitutions. The cyclin sequences of green algae and of other major eukaryotic taxa (such as ciliates, red algae, dinoflagellates, etc.) would improve the phylogenetic reconstruction; (ii) the sequence length available is short (about 250 positions) and allows the confident resolution of only speciation or duplication events well separated in time [67]; (iii) major differences in evolutionary rates constitute an important limitation of the parsimony method, as illustrated by the long-branch attraction artifact [68].

## Evolutionary rates of A-, B- and F-related cyclins

As mentioned above, saturation generates an artificial molecular clock-like behaviour, preventing a correct estimation of the evolutionary rates. On the parsimony tree, the branches leading to the terminal taxa are of various length, suggesting difference in rates. However, estimating evolutionary rates this way is highly dependent on the quality of the parsimony inference. A better estimate is obtained using divergence times calculated from paleontology. If we assume a divergence date between diploblast and triploblast metazoans of $600 \times 10^6$ years (600 MY) [69], between nematodes and vertebrates of 530 MY [70] and between monocots and dicots of 200 MY [70], the following figures can be obtained: cg2a, 180/600=0.3 substitutions/MY; cg2b, 240/600=0.4; cgb3, 140/530=0.26; CycB1, 210/200=1.05; CycB2, 140/200=0.7; and CycA1, 90/200=0.45. A great variability of evolutionary rates is thus observed between cyclin classes, but also within classes. For example, the *Drosophila* and *Caenorhabditis* cyclins A and B evolved significantly faster than

## Fig. 6    Alignment of plant CycA cyclins

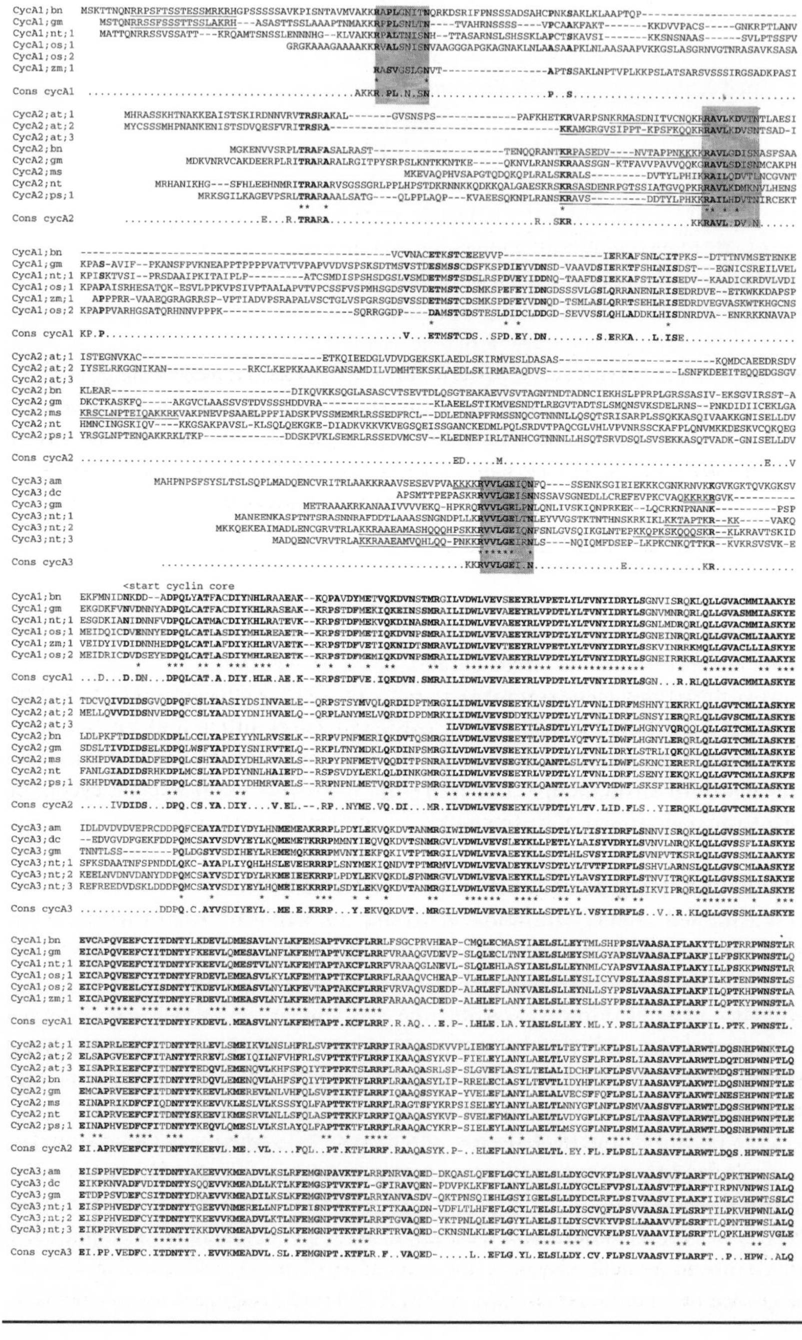

```
CycA1;bn    MSKTTNQNRRPSFTSSTESSMRKRHGPSSSSAVKPISNTAVMVAKKRAPLGNITTQRKDSRIFPNSSSADSAHCPNKSAKLKLAAPTQP----------------------
CycA1;gm    MSTQNRRSSFSSSTTSSLAKRH---ASASTTSSLAAAPTNMAKKRPPLGNLTN----TVAHRNSSSS---VPCAAKFAKT-------EGNICSREILVEL
CycA1;nt;1  MATTQNRRSVSSSATT---KRQAMTSNSSLENNNHG--KLVAKKRPALTNISNH--TTASARNSLSHSSKLAPCTSKAVSI----KKSNSNAAS-----SVLPTSSPFV
CycA1;os;1                                      GRGKAAAGAAAAKKRVALSNISNVAAGGGAPGKAGNAKLNLAASAAPVKKGSLASGRNVGTNRASAVKSASA
CycA1;os;2
CycA1;zm;1                          RASVGSLGNVT----------------APTSSAKLNPTVPLKKPSLATSARSVSSSIRGSADKPASI

Cons cycA1  ......................AKKR.PL.N.SN.............P..S.


CycA2;at;1  MHRASSKHTNAKKEAISTSKIRDNNVRVTRSRAKAL-------GVSNSPS-----------PAFKHETKRVARPSNKRMASDNITVCNQKRRAVLKDVTNTLAESI
CycA2;at;2  MYCSSSMHPNANKENISTSDVQESVVRITRSRA------------------------------------AMGRGVSIPPT-KPSFKQQKRRAVLKDVSNTSAD-I
CycA2;at;3
CycA2;bn             MGKENVVSRPLTRAFASALRAST------------------TENQQRANTKRPASEDV----NVTAPPNKKKRRAVLKDISNASFSAA
CycA2;gm    MDKVNRVCAKDEERPLRITRARARLRGITPYSRPSLKNTKKNTKE-----QKNVLRANSKRAASSGN-KTFAVVPAVVQQKGRRAVLSDISNMCAKPH
CycA2;ms             MKEVAQPHVSAPGTQDQKQPLRALSKRALS---------DVTYLPHIKRAILCDVTLNCGVNT
CycA2;nt    MRHANIKHG---SFHLEEHNMRITRARARVSGSSGRLPPLHPSTDKRNNKKQDKKQALGAESKRSASDENRPGTSSIATGVQPKRRAVLKDMKNVLHENS
CycA2;ps;1  MRKSGTILKAGEVPSRLTRARAAALSATG----QLPPLAQP---KVAEESGKRNLRANSKRAVS--------DDTYLPHKKRAILKDLTNIRCEKT

Cons cycA2  ...................E...R.TRARA........................R..SKR...........RRAVL.KD.N...N...


CycA1;bn    ------------------------------VCVNACETKSTCKEEVVP-----------------IERKAFSNLCITPKS--DTTTNVMSETENKE
CycA1;gm    KPAS-AVIF--FKANSFPVKNEAPPTPPPPPVATVTVPAFVVDVSPSKSDTMSVSTDESMSSCDSPKSPDIEYVDNSD-VAAVDSIERKTFSHLNISDST---EGNICSREILVEL
CycA1;nt;1  KPISKTVSI--PRSDAAIPKITAIPLP-------ATCSMDISPSHSDGSLVSMDETMSTSDSLRSFDVEYIDDNQ-TAAFDSIEKKAFSTLYIESDV--KAADICKRDVLVDI
CycA1;os;1  KPAPAISRHESATQK-ESVLPPKVPSIVPTAALAPVTVPCSSFVSPMHSGDSVVSVDETMSTCDSMKSPDFEYIDNQ-DSSSVLGSLQRANPNLKISEDRDVE-ETKWKKDAPSP
CycA1;zm;1  APPPPR-VAAEQGRAGRRSP-VPTIADVPSRAPALVSCTGLVSPGRSGDSVSSDETMSTCDSMKSPDFEYVDNQD-TSMLASLQRRTSEHLRISEDRDVEGVASKWTKHGCNS
CycA1;os;2  KPAPPVARHGSATQRHNNVPPPK-------------------SQRRGGDP-----DAMSTCDSTESLDIDCLDDGD-SEVVSSLQHLADDKLHISDNRDVA--ENKRKKNAVAP

Cons cycA1  KP.P.....................V...ETMSTCDS..SPD.EY.DN.......S.ERKA...L.ISE.


CycA2;at;1  ISTEGNVKAC-----------------ETKQIEEDGLVDVDGEKSKLAEDLSKIRMVESLDASAS---------------------KQMDCAAEDRSDV
CycA2;at;2  IYSELRKGGNIKAN-------------RKCLKEPKKAAKEGANSAMDILVDMHITEKSKLAEDLSKIRMAEAQDVS--------------LSNFKDERITEQQEDGSGV
CycA2;at;3
CycA2;bn    KLEAR-------------------DIKQVKKSQGLASASCVTSSEVTDLQSGTEAKAEVVVSVTAGNTNDTADNCIEKHSLPPRPLGRSSASIV-EKSGVIRSST-A
CycA2;gm    DKCTKASKFQ-----AKGVCLAASSVSTDVSSSHDDVRA----------------KLAEELSTIKMVESNDTTLREGVTADTSLSMQNSVKSDELRNS--PMKDIDICIEKLGA
CycA2;ms    KRSCLNPTEIQAKKKKVAKPNEVPSAAELPPFIADSKPVSSMEMRLRSSEDPRCL--DDLEDNAPPFRMSSNQCGTNNNLLQSQTSRISARPLSSQKKASQIVAAKKGNISRLLDV
CycA2;nt    HMNCINGSKIQV-----KKGSAKPAVSL-KLSQLQEKGKE-DIADKVKKVKVEGSQEISSGANCKEDMLPQLSRDVTPAQCGLVHLVPVNRSSCKAFPLQNVMKKDESKVCQKQEG
CycA2;ps;1  YRRSGLNPTENQAKKRKLTKP--------------DDSKPVKLSEMRLRSSEDVMCSV--KLEDNEPIRLTANHCGTNNNLLHSQTSRVDSQLSVSEKKASQTVADK-GNISELLDV

Cons cycA2  .......................ED....M.......................E...V


CycA3;am    MAHPNPSFSYSLTSLSQPLMADQENCVRITRLAAKKRAAVSESEVPVAKKKKRVVLGEICNFQ-----SSENKSGIEIEKKKCGNKRNVKKGVKGKTQVKGKSV
CycA3;dc                                          APSMTTPEPASKRRVVLGEISNNSSAVSGNEDLLCREFEVPKCVAQKKRKRGVK---------
CycA3;gm                                      METRAAAKRKANAAIVVVVEKQ-HPKRQRVVLGRLPNLQNLIVSKIQNPRKEK--LQCRKKNPNANK---------PSP
CycA3;nt;1  MANEENKASPTPSRASNNEAFDDTLAAASSNGNDPLLKKVVVLGEILNLEVVVGSTKTNTHNSKRKIKLKKTAPTKR--KK-----VAKQ
CycA3;nt;2  MKKQEKEAIMADLENCGRVTRLAKKRAAEAMASHQQQHPSNKKKVVLGEIKNFSNLGVSQIKGLNTEPKKQPKSKQQQSKR--KLKRAVTSKID
CycA3;nt;3  MADQENCVRVTRLAKKRAAEANVQHLQQ-PNKKKVVVLGEISNLS----NQIQMFDSEP-LKPKCNKQTTKR--KVKRSVSVK-E

Cons cycA3  ....................KKRVVLGEI.N............E.......KR.


                            <start cyclin core
CycA1;bn    EKFPMNIDNKDD--ADPQLYATFACDIYNHLRAAEAK--KQPAVDYMETVQKDVNSTMRGILVDWLVEVAEEYKLVPDTLYLTVNYIDRYLSGNVISRQKLQLLGVACMMIASKYE
CycA1;gm    EKGDKFVNVDNNYADPQLCATFACDIYKHLRASEAK--KRPSTDFMEKVQKEINSSMRAILIDWLVEVAEEYRLVPDTLYLTVNYIDRYLSGNVMNRQRLQLLGVASMMIASKYE
CycA1;nt;1  ESGDKIANIDNNFVDPQLCATMACDIYKHLRATEVK--KRPSTDFMEKVQKDINASMRAILIDWLVEVAEEYRLVPDTLYLTVNYIDRYLSGNLMDRQRLQLLGVACMMIASKYE
CycA1;os;1  MEIDQICDVENNYEDPQLCATLASDIYMHLREAETR--KRPSTDFMETIQKDVNPISMRAVLIDWLVEVAEEYRLVPDTLYLTVNYIDRYLSGNEINRQRLQLLGVACMIIAAKYE
CycA1;zm;1  VEIDYIVDIDNNHEDPQLCATLASDIYIHLVAETK--KRPSTDFMETVQKDINPDTSMRAVLIDWLVEVAEEYRLVPDTLYLTVNYIDRYLSKVINRRKMQLLGVACLLIASKYE
CycA1;os;2  MEIDRICDVDSEYEDPQLCATLASDIYMHLREAETK--KRPSTDFMETIQKDVNSDTSMRAILIDWLVEVAEEYRLVPDTLYLTVNYIDRYLSGNEIRRKRLQLLGVACMIIAAKYE

Cons cycA1  ....D...D.DN...DPQLCAT.A.DIY.HLR.AE.K--KRPSTDFVE.IQKDVN.SMRAILIDWLVEVAEEYRLVPDTLYLTVNYIDRYLSGN...R.


CycA2;at;1  TDCVQIVDIDSGVQDPQFCSLYAAIZYDSINVAELE--QRPSTSYMVQLQRDIDPTMRGILIDWLVEVSEEYKLVSDTLYLTVNLIDRFMSHNYIEKRKLQLLGVTCMLIASKYE
CycA2;at;2  MELLQVVDIDSNVQDPQFCSLYASDIYINIHVAELQ--QRPLANYMHLKVQKDPIDPDMRKILIDWLVEVSEEYDYKLVPDTLYLTVNLIDRFLSNSYI ERQRLQLLGVTCMLIASKYE
CycA2;at;3                                LLVDWLVEVSEEYTLASDTLYLTVYLIDWFLHGNYVQRQQLQLLGITCMLIASKYE
CycA2;bn    LDLPKFTDIDSDKDPLLCCLYAPEIYYNLRVSEEK--RRPVPNFMERIQKDVTQSMRGILIDWLVEVSEEYTLQPDTLYLTVYLIDWFLGNYLEQ RQLQLLGITCMLIASKYE
CycA2;gm    SDSLTTVDIDEKLQDPQLWSFYAPDIYSNIRVTELQ--RKPLTNYMDKLQKDINPSMRGILVDWLVEVSEEYKLVSDTLYLSTRLIQRQKLQLLGVTCMLIASKYE
CycA2;ms    SKHPDVADIDADFEDPQLCSLYAADIYLSNIRLSEK--RRPVPNFMETVQQDITPSNRAILVDWLVEVSEEYKLQANTLSLTVYLIDRFLSKNCIERERLQLLGITCMLIATKYE
CycA2;nt    FANLGIADIDSRHKDPLMCSLYAPDIYNNLHAIEFD--RSPSVDYLEKLQLDINKGMRGILIDWLVEVSEEYKLVPDTLYLTVNFLSENYIKKQLQLLGITCMLIASKYE
CycA2;ps;1  SKHPDVADIDADFEDPQLCSLYAADIYDHMRVAES--RRPVPNFMETVQKDITPSMRAILVDWLVEVSEEYKLQANTLYLAVVYMDWFLSKSFIERHKLQLLGITCMLIASKYE

Cons cycA2  .....IVDIDS....DPQ..C.YA.DIY.....V.EL.---RP..NYME.VQ.DI....MR.ILVDWLVEVSEEYKLVPDTLYLTV.LID.FLS...YIERQKLQLLGVTCMLIASKYE


CycA3;am    IDLDVDVDVEPRCDDPQFCEAYATDIYDEKLOMKMETKRRPMMNYIEQVQKDVTANMRGVLVDWLVEVAEEYKLLGSDTLYLTISYIDRFLSNNVISRQKLQLLGVSMLIASKYE
CycA3;dc    --EDVGVDPGEKFDDPQMCSAYVSDVYEYLKQMKMETKRRPMMNYIEQVQKDVTSNMRGVLVDWLVEVAEEYKLLSETLYLATSYVDRYLSVNVLNRQRLQLLGVSFLIASKYE
CycA3;gm    TNNTLSS---------PQLDGSYVSDIHEYLKEMMEQVKRRPMVNYIEKFQKIVFPTMRGILVDWLVEVAEEYTLHLHLSVSYIDRFLSVNPVTKSRLQLLGVSMLIASKYE
CycA3;nt;1  SFKSDAATNFSPNDDLQKC-AYAPLIYQELHSLEKRRRPAKRLSNYMEKLQNDVTPTMRHVLVDWLVEAAEEYKLVSDTLYLSTSVDRFLSHVIARNSDLLQLLGVSMLIASKYE
CycA3;nt;2  KEELNVNDVDANYDDPQMCSAYVSDIYDYLKKMEIEEKRRPLSDYLDKVQKDLSPQMRGILIDWLVEVAEEYKLLSDTLYLAVSYIDRFLSTNVTTMKQLQLQLGVSMLIASKYE
CycA3;nt;3  REFREEDVDSKLDDDPQMCSAYVSDIYEYLHQMEIEEKRRPLSDYLEKVQKDVTANMRGILVDWLVEVAEEYKLLSDTLYLAVAYIDRYLSIKVIPRQRLQLLGVSMLIASKYE

Cons cycA3  ...........DDPQ.C.AYVSDIYEYL..ME.E.KRRP...Y.EKVQKDVT..MRG.LVDWLVEVAEEYKLLSDTLYL.VSYIDRFLS..V..R.KLQLLGVSMLIASKYE


CycA1;bn    EVCAPQVEEFCYITDNTYFKLDKEVLDMEASAVLNYLKFEMSAPTVKCFLRRLFSGCPRVHEAP--CMQLECMASYIAELSLEYTMLSHPPSLVAASAIFLAYLTLDPTRRPWNSTLR
CycA1;gm    EICAPQVEEFCYITDNTYFKEEVLQMESAVLNFLKFEMTAPTVKCFLRRFVRAAQGVDEVP--SLQLECLLTNYIAELSLMEYSMLGYAPSLVAASAIFLAKFILFSKKPWNSTLQ
CycA1;nt;1  EICAPQVEEFCYITDNTYFKLEEVLDMEAASVLKYLKFEMTAPTVKCFLRRFVRAAQVGCHEAP--VLHLEPLANYIAELSLICVYSPSLVAASAIFLAKFF ILQPTEHPWNSTLS
CycA1;os;1  EICFPQVEEELCYISDNTYFTKDEVLKMEAASVLNYLKFEMTAPTAKCFLRRFVRAAQV CDDF--ALHLEFLANYIAELGLLEYNLSYPSLVAASAIFLAEFFLQPTKHPWNSTLA
CycA1;zm;1  EICAPQVEEFCYITDNTYFTKDEVLKMERAASVLNYLKFEMTAPTAKCFLRRFARAAQACDEDP-ALHLEFLANYIAEIGLLEYSLLSSYPPSLLAASAIFLAKFFILQ PTKYPWNSTLA

Cons cycA1  * *.......*...*** *** .** ***.. ...**** ****..****.. ..*...... ... ** **..... . .****. ..** *. ** ****.* .***..** **..** . **. **NSTL.


CycA2;at;1  EISAPRLEEFCFITTDNTYTRLEVLSMEIKVLNSLHFRLSVPTTKTFLRRFIRAAQSDKVVPLIEMKYLANYFAELTLTEYTPLKFLRLLIAASAVFLARWTLDSNHPWNKTLQ
CycA2;at;2  EISAPGVEEFCFITTANTYRRKVLSMEIQILNFVHFRLSVPTTKTFLRRYIRAAQSKYP-FIELKYLANLTELTEYSPLRFLRSLIAASAVFLAKWTLDQTDHPWNPTLQ
CycA2;at;3  EISAPRIEEFCFITTDNTYTEDQVLEMENQVLKHFSPQIYTPPTKTSLRRFLRAAQASRLSP-SLGVRFLASYLTELALDICHPLKFLPSVVAASAIFLAKWTLDQSDHPWNPTLD
CycA2;bn    EINAPHIEEFCFITTDNTYRDQVLEMENQ VLNFVHYRLSVPTTKTFLRRFLRAAQSYLIP-REELCIANYLALELYLVDYHPLKFLPSVVAASAVFLAKWTLDQSDHPWNPTLE
CycA2;gm    EINAPHKDFCFIQDNTYTREVVDKLELVLKSSSVQLFAPTTKTFLRRFIRAQJAQSYKAP-YVELEFLANYIAELSCSFFQFLPSLIAASAVFLAKWTLNESEHPWNPTLE
CycA2;ms    EINAPRIKDCFIQDNTYTKEVVDKLELVLKSSQLQ LFAPTTKTFLRRFFKRPSIELKMFLANYIAQLTLSYGFLNFLPSMVAASAVFLAKWTLDQSSHPWNTTLE
CycA2;nt    EICAPRVEEFCFITTDNTYTKEVVDKLELSVNEALLSFQLASPTTKTFLRRFARQAQASYKVP-SVRFLANYLAELTLLDTVDYGFLNFLPSMVAASAVFLARWTLDQSDHPWNPTLE
CycA2;ps;1  EINAPHVEDFCFITTDNTYKEQVLQMRSLVKSLAYQLFAPTTKTFLRRFLRAAQASYKVP-REESIEELKYLANLAELTLMSYGFLNFLPSMLAASAVFLARWTLDQSDHPWNPTLE

Cons cycA2  EI.APRVEEFCFITTDNTYT.EEVLME..VL.....PQL..PT.KTFLRRF.RAAQSYK.P...ELEFLANYLAELTL.EY.FL.FLPSLIAASAVFLARWTLDQS.HPWNPTLE


CycA3;am    EISPPHVEDFCYITDNTYAKEEVVVKMEADVLKSLRFEMGNPAVKTFLRRIYRAAQ DKQASLQFRFLGCYLAELSLLDYQCVLPSLVAASVFLAARFTLDHPWNSALQ
CycA3;dc    EIKPKNVADFVDITDNTYSQQEVVKMEADLLKTLKFEMGSPTVKTFL-GFIRAVQRN-PDVPKLKFHFLANCSDV-QKTPNSQIEHLGSYIGELSLLDYDCLEFVPSLIAASVFLARPTIRPNVNPWSIALQ
CycA3;gm    ETDPPSVEDFCSITTDNTYDKAVVMRMADILKSLKFEMGNPTVTSFFLRRYIRAVNCSDV-QKTPNSQIEHLGSYIGELSLLDYDCLEFVPSLVAASVFLAKFTINHPWTESALC
CycA3;nt;1  EISPPHVEDFCYITTDNTYIGEEVVNMRELLNPLDFEISNPTTRTFLRIFTKAAQDN-VDFLTLHFFLGCYLTELSLLDYSCVQFLPSVVAASVFLARFTILPKVHPWNLALQ
CycA3;nt;2  EISPPHVEDFCYITTDNTYTKKDVVVKMEADLVKTLNFEMGNPTVKTFMGLRRFSVRAQED-YKTPNLQLEFLGYYLAELSLLDYSCVKIVVPSLLAAAVFLSRFTLQPKLHPWSVGLE
CycA3;nt;3  EISPPHVEDFCYITTDNTYTKKDVVVKMEADLKTLNFEMGNPTVKTFMGLRRFTVRAQ ED-CKNSNLKLEFLGCYLAELSLLDYNCVKFVPSLVAAAVFLSRFTLQPKLHPWSVGLE

Cons cycA3  EISPP.VEDFC.ITDNTYT..EVVKMEADVL.SL.FEMGNPT.KTFLR.F..VAQED----L..EFLG.YL.ELSLLDY.CV.FLPSLVAASVFLARPT..P..HPW..ALQ
```

## Fig. 6 (Contd.)

```
                                             end cyclin core>
CycA1;bn     HYTQYEAMELRGCVMDLQRLCSNAH-VSTLPAVRDKYSQHKYKFVAKKFCPSIIP-PDFFFKNSLY
CycA1;gm     HYTLYQPSDLCVCVKDLHRLCCNSP-NSNLPAIREKYSQHKYKYVAKKYCPPSIP-PEFFQN
CycA1;nt;1   HYTLYQPSDLRDCVVALHSLCCNNN-NSSLPAIREKYSQHKYKFVAKKYCPPTIP-VEFFQNISC
CycA1;os;2   FYTQYKPSDLCNCAKGLHRLFLVDP-GGNLRAVREKYS-HKYKFVAKKYSPPSIP-AEFFEDPSSYKPD
CycA1;os;1   HYTQYKSSELSDCVKALHRLFSVGP-GSNLPAIREKYTQHKYKFVAKKPCPPSIP-TEFFRDATC
CycA1;zm;1   HYTQYKPSKLSECVKALHRLCSVGS-GSNLPAIREKYTQHKYKFVAKKQCPPQIPSLEFFRDATCL

Cons cycA1   HYT.Y.PSDL..CV.LHRL......-.S.LPAIREKYSQHKYKFVAKK.CPP.IP-.EFF.N.......

CycA2;at;1   HYTRYETSALKNAVLAMEDLQLNTS-GSTLIAIRTKYNQQKFKRVATLTSP--ERVNTLFSR
CycA2;at;2   HYTRYEVAELKNTVLAMEDLQLNTS-GCTLAATREKYNQPKFKSVAKLTSP--KRVTLLFSR
CycA2;at;3   HYTTYKPSDLKASVHALQDLQLNTK-GCPLSAIRHKYRQEKYKSVAVLTSP--KLLDTLF
CycA2;bn     HYTTYKASDLKASVHALQDLQLNTK-GCPLSAIRHKYQEKFKSVAVLMSP--KLLDTLF
CycA2;gm     HYTKYKASDLKTVVLALQDLQLNTK-GCFLNAVREKYKQQKFNCVANL-SP--KSVQSLFQNQV
CycA2;ms     HYASYKASDLKATVLALQDLQLNSNDDCPLTTIRKKYTQDKLNCVAALSSP--QLLETLF
CycA2;nt     HYTRYKVSELRTTVFALQELQMNTS-GCTLNAIREKYRQPKFKSVATLAAS--KPVQSLF
CycA2;ps;1   HYASYEASDLKDAVLALQDLQLNSN-DCPLTSIRTKYTQDKLKCVATLSSP--KLLETMFSR

Cons cycA2   HYT.Y..SDLK..V.ALQDLQLNT.-GC.L.AIR.KY.QEKFK.VA.L.SP--K.LETLF...

CycA3;am     SNTGYKPEDLKECVGILHDLQLSRK-GSSLVAIRDKYNQHKFKCVSALSSPSEVPES-FFEAIKDD
CycA3;dc     KCSGYKSKDLKECVLILHDLQMGRR-GGSLSAVRDKYKHKFKCVS-TLSPAPEIPESIFNDV
CycA3;gm     ECSGYKPAELKECVLILHDLYLSRR-AASFKAVREKYKHQKFKCVAANLPTPPYV-PSCYFEDQ
CycA3;nt;1   QCTGYKPSELKDCVLVIHELQSGRR-AASVQAVRRKYMDHKYKCVAALHPP--DIPACFFDDA
CycA3;nt;2   QYSGYKAADLKECILILHDLQLSRR-QGSLAAVRDKYKQHKFKCVSSLTSP-VEIPASFFEDMRQL
CycA3;nt;3   QNSGYRA

Cons cycA3   ..SGYKP.DLKECVLILHDLQ..RR-GGS..AVRDKY..HKFKCVS.L..P....P..FFED.....
```

The sequences are described in Table 1. Dashes represent gaps introduced to maximize the alignment within a given group and, in the case of the cyclin core, between the various groups. The destruction box is shaded. The residues which are similar in all sequences of one group are in bold. The consensus (cons) line shows the residues which are similar in at least 80% of the sequences. The asterisks show the residues identical in all sequences of one group. Putative bipartite NLS, located close to the destruction box, are single (bipartite NLS) or double (clusters of five basic residues) underlined.

other animal cyclins, as can be deduced from their incorrect phylogenetic position (Figs. 3 and 4), owing to the long-branch attraction artifact [68].

The tree of eukaryotes as inferred from rRNA sequences is {Dictyostelium [plants (fungi, metazoans)]} [71]. Protein sequences often confirm the close relationship between fungi and metazoans [72], but suggest that Dictyostelium could be closer to this group than plants [73]. As a result, one expects the cyclins of fungi to be close to those of metazoans. As observed in Fig. 3, this is the case for cyclin B, but not for cyclin A. Two hypotheses can explain this discrepancy: (i) duplication of the gene coding for cyclins A and B occurred early in the evolution of eukaryotes and cyclin A has been lost in fungi; (ii) duplication of the gene coding for cyclins A and B occurred independently in the lineages leading to plants and to animals, and the grouping of plant CycA and animal cyclin A is an artifact of tree reconstruction. However, these cyclins seem to evolve slowly (see above), preventing the long-branch attraction phenomenon to occur. We thus favour hypothesis (i), but the cyclin sequences of other eukaryotic taxa are necessary to settle this question.

Cyclin F appears different from cyclin A homologues. This particular, protein lacks sequence features that are present in both cyclin A and B classes, and it also shows a much lower degree of similarity than to the other cyclin homologues [74]. These results indicate that cyclin F evolved by a separate gene duplication event of cyclin A and B. Owing to the availability of only a small number of cyclin F sequences, we cannot yet conclude whether this cyclin originates from cyclin A or

B, or whether it is also present in plants. Among cyclins, the B-types are uniformly distributed in animals, fungi and plants. Cyclin B3 has been found in nematodes and in chicken, which indicates that it is probably the ancestral B cyclin [75], and which implies that a plant homologue may exist.

The evolutionary rates of cyclins have varied throughout eukaryotic evolution, suggesting important changes in the functional constraints acting on these proteins. Therefore, from an evolutionary point of view, cyclins have shown a remarkable adaptative plasticity, allowing successive speciations either to select new features or to keep the old ones.

## Sequence analysis of CycA and CycB plant cyclins

All the reports about A/B-type plant cyclins have emphasized the fact that they displayed specific sequence characteristics that prevented a clear-cut assignment to a given class of cyclins [16,17,76]. The primary sequences of CycA and CycB plant cyclins reported in Table 1 have been aligned (Figs. 6 and 7). These cyclins display many features typical of animal and yeast A- and B-type cyclins. They have a variable N-terminal domain, approximately 180 amino acid long, at the beginning of which a nine-amino-acid motif closely resembles the destruction box of mitotic cyclins. This domain is followed by a strikingly conserved domain of approximately 270 amino acids which is the most homologous to the cyclin core of cyclins A and B and extends nearly to the C-terminal end of the sequences (Fig. 8).

## The loops connecting helices in the plant cyclin core of CycA and CycB vary in size according to the classification in classes and groups

From the alignment of plant and animal cyclin core sequences (Fig. 2) the 12 helices observed in the crystal structure of bovine and human cyclin A [48,49] can be easily positioned by homology in the sequence of plant CycA and CycB cyclins. Remarkably, the reverse process, i.e. the prediction of secondary and tertiary structure from a careful alignment of cyclin sequences, has been shown to be feasible with a good fitness to the observed structure [53].

As expected from this structural role, none of the helix sequences encompasses a gap in the sequence alignment. Although gaps occur quite often in the sequences connecting helices, which represent loops in the tertiary structure, the length of the connecting loops are very conserved inside the first cyclin fold, together with the loop connecting the two folds. In contrast, neither the inter-helix loops in the second fold, nor the loops connecting each cyclin fold to the N-terminal and to the C-terminal helices, are conserved in size.

Note that loop size appears to be a specific character shared by every member of a group, with the remarkable exception of CycB1 and of CycA2 in one case (Table 3). This can be used as a fingerprint for their classification into classes and groups. The very variable length of connecting loops of CycB1 cyclins is puzzling, indicating, on the one hand, that this group is not structurally

**Table 3    Comparative size of the inter-helix loops of the cyclin core**

The size is indicated as the number of residues in each of the 11 sequences of the loops separating the 12 helices of the cyclin core, positioned by sequence homology, as shown in Figs. 2 and 8. The loop sequences are named according to their flanking helices (e.g. 1–2: loop between the helix H1 and the helix H2). The columns give the sizes for either mammalian cyclin A, according to its three-dimensional structure [48,49] and for the plant cyclin A and B groups aligned as shown in Fig. 2.

| | Mammals | Higher Plants | | | | |
|---|---|---|---|---|---|---|
| Loop | A | CycA1 | CycA2 | CycA3 | CycB1 | CycB2 |
| Nter - 1 | 14 | 15 | 15 | 17 | 13–14 | 14 |
| 1–2 | 3 | 3 | 3 | 3 | 3 | 3 |
| 2–3 | 7 | 7 | 7 | 7 | 7 | 7 |
| 3–4 | 6 | 6 | 6 | 6 | 6 | 6 |
| 4–5 | 6 | 6 | 6 | 6 | 6–7 | 6 |
| 5–1' | 9 | 9 | 9 | 9 | 9 | 9 |
| 1'–2' | 8 | 12 | 12 | 12 | 7–12 | 7 |
| 2'–3' | 9 | 8 | 8 | 8 | 8–9 | 8 |
| 3'–4' | 5 | 7 | 7 | 7 | 6 | 6 |
| 4'–5' | 7 | 7 | 7 | 7 | 7–9 | 7 |
| 5'–Cter | 8 | 8 | 8-9 | 8 | 7–12 | 8 |

homogeneous, and on the other, that it could have evolved faster than the other groups to perform plant-specific functions, an assumption which seems to be confirmed by phylogenetic analysis. The compactness of the cyclin domain is controlled by the tight packing of the cyclin fold helices, associated with small-size residues at their crossing [49]. As can be seen in our alignment (Fig. 2), the 10 residues identified as positioned at such crossings [49] on helices H2, H3, H2' and H3' are indeed, also in plant cyclins, mostly small (G, A, S) or medium sized residues (V, T, C). There are obviously some variations in the size of the residues occupying these 10 positions according to the groups, suggesting that CycA and CycB plant cyclins may vary in the compactness of the cyclin core according to the group they belong to.

## The cyclin core of CycA and CycB plant cyclins share conserved sequence features with their homologues from yeast and animals

The consensus sequences of the cyclin core of the five groups of plant cyclins CycA and CycB are reported in Fig. 8, together with those of animal cyclin A, and of animal and yeast cyclin B. In this Figure, the residues are indicated when they are identical or conservatively substituted in 80% at least of the cyclins of a given group. Consensus sequences available for A- and B-type cyclins were incomplete or covered only a part of the cyclin core [44,47]. Hence, we established the consensus

# Fig. 7    Alignment of plant CycB cyclins

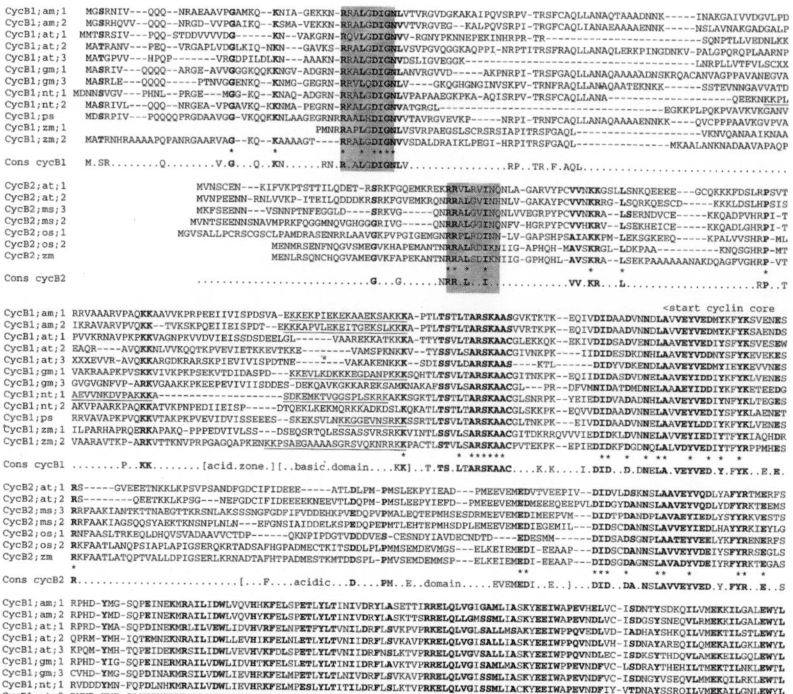

```
CycB1;am;1   MGSRNIV---QQQ--NRAEAAVPGAMKQ--KNIA-GEKKN-RRALGDIGNLVTVRGVDGKAKAIPQVSRPV-TRSFCAQLLANAQTAAADNNK-----INAKGAIVVDGVLPD
CycB1;am;2   MGSRHQVV--QQQ--NRGD-VVPGAIKQ--KSNA-VEKKN-RKALGDIGNVVTVRGVEG--KALPQVSRPI-TRGFCAQLIANAEAAAAENNK-----NSLAVNAKGADGALP
CycB1;at;1   MMTSRSIV---PQQ-STDDVVVVVQG----NK--VAKGRN-RQVLGDIGNVV-RGNYPKNNEPEKINRHPR-TR------------------------SQNPTLLVEDNLKK
CycB1;at;2   MATRANV---PEQ--VRGAPLVDGLKIQ-NKN---GAVKS-RRALGDIGNLVSVPGVQGQKAQPPI-NRPITRSFRAQLLANAQLERKPINGDNKV-PALGSPQGPLANRPT
CycB1;at;3   MATGPVV---HPQP--------VRGDPILDLKN---AAAKN-RRALGDIGNVVDSLIGVEGGK------------------------------------LRRFLLVTFVLGCXX
CycB1;gm;1   MASRIV---QQQQ--ARGE-AVVGGGKQQKKNGV-ADGRN-RKALGDIGNLANVRGVVD-----AKPNRPI-TRSPGAQLLANAQAAAAADNSKRQACANVAGPPAVANEGVA
CycB1;gm;3   MASRLE--QQQQ----PTNVGGENKQ---KNMG-GEGRN-RRVLGDIGNLV----GKQGHGNGINVSKPV-TRNFRAQLLANAQAATKENKK-----SSTEVNNGAVVATD
CycB1;nt;1   MDNNBVGV---PHNL--PRGE---MGG-KQ--KNAQ-ADGRN-RRALGDIGNLVPAPAAGKPKA-AQISRPV-TRSFCAQLLANA-----------------QEERKNKKPL
CycB1;nt;2   MASRIVL---QQQ--NRGEA-VPGAIKQ--KNMA-PEGRN-RKALGDIGNVVATGRGL------------------------------------EGKKPLPQKPVAVKVKGANV
CycB1;ps     MDSRPIV---PQQQPRGDAAVGG-VKQQKKNLAAGEGRNRRAALGDIGNVVTAVRGVEVKP----NRPI-TRNFCAQLLANAQAAAAENNK----QVCPFPAAVKGVPVA
CycB1;zm;1                                          PMNRRAPLGDIGNLVSVRPAEGSLSCRSRSIAPITRSFGAQL---------------------KKVQANAAIKNAA
CycB1;zm;2   MATRNHRAAAAPQPANRGAARVAG--KQ--KAAAAGT----RRALGDIGNVVSDALDRAIKLPEGI-HPITRSFGAQL------------------MKAALANKNADAAVAPAQP

Cons cycB1   M.SR.......Q........V.Q...........RN.R.ALGDIGNLV............RP..TR.F.AQL.
```

```
CycB2;at;1                    MVNSCEN---KIFVKPTSTTILQDET-RSRKFGQEMKKEKRRVLRVTINQNLA-GARVFPCVVNIKKGSLLSNKQEEEE---GCQKKKFDSLRPSVT
CycB2;at;2                    MVNPEENN-RNLVVKP-ITRILQDDDKRSRKFGVEMKRQNRRALGVININNLV-GAKAYPCVVHKKRG-LSQRKQESCD-----KKKLDSLRPSIS
CycB2;ms;3                    MKFSEENN---VSNNPTNFEGGLD----SRKVG------QNRRALGVINQNLVVEGRPYPCVVNKRA--LSERNDVCE-----KKQADPVHRPI-T
CycB2;ms;2                    MVNTSEENNSNAVMPRKFQGGMMQVGHGGGRIVG-----QNRRALGDIGNLVMRGHRPYPCVVHKRV--LSEKHEICE-----KKQADLGHRPI-T
CycB2;os;1                    MGVSALLPCRSCGSCLPAMDRASENRRLAAVGKPVPGIGEMGNRRPLRDIIN-LV-GAPSHPSAIAKKPM-LEKSGKREQ-----KPALVVSHRP-ML
CycB2;os;2                    MENMRSENFNQGVSMEGVKHAPEMANTNRRALRDKKNIIG-APHQH-MAVSKRGL-LDKPAA------------KNQSGHRP-MT
CycB2;zm                      MENLRSQNCHQGVAMEGVKFAPEKANTNRRALGDIKNIIG-GPHQHL-AVSKRA--LSEKPAAAAAANAKDQAGFVGHRP-VT

Cons cycB2   ........................G...G.....NRR.L.I....VV.KR....L...........RP..T
```

```
                                                                                                        <start cyclin core
CycB1;am;1   RRVAAARVPAQKKAAVVKPRPEEIIVISPDSVA-EKKEKPIEKEKAAEKSAKKKA-PTLFSTLTARSKAASGVKTKTK--EQIVDIDAADVNNDLAAVEVVEDHYKSVENES
CycB1;am;2   IKRAVARVPVQKK--TVKSKPQEIIEISPDT--EKKKAPVLEKEITGEKSLKKKA-PTLFSTLTARSKAASVVRTKPK--EQIVDIDAADVNNDLAAVEVVEDHKFYKSAENDS
CycB1;at;1   PVVKRNAVPKPKKVAGNPKVVDVIEISSDSDEELGL------VAAREKKATKKKA-TTYFSVLTARSKAACGLEKKQK--EKIVDIDSADVENDLAAVEYVEDIYSYKSVESEN
CycB1;at;2   EAQR----AVQKKNLVVKQQTPKVPEVIETKKEVTKKE-------VAMSPKNKKV--TYSSVLSARSKAACGIVNKPK----IIDIDESDKDNHLAAVEYVDDNYSYFKSVEKES
CycB1;at;3   XXXEVVR-AVQKKARGDKRARSKPIEVIVISPDTNE-----VAKAKENKKK--SDISSVLDARSKAAS----KTL------DIDYVDKENDLAAVEYVDDNYIEYKEVVNES
CycB1;gm;1   VAKRAAPKPVSRKVIVKPKPSEKVTDIIDASPD---IXXDKKEDKKGDANPKKKSQHTLSVLTARSKAACGITNKPK--EQIIDIDADVDNELAAVEYVIDDIYKTLVEENES
CycB1;gm;3   GVGVGNFVP-ARKVGAAAKPKEEPEVIVIISDDES-DEKQAVKGKKAREKSAMKNAKAYSVLGARSKAAS--SDKEMKTVGGSPLGSKRKAKSSRTLSFVLTARSKAACGLSNKPK--PR--DFVMNDATDMDNELAAAIIDDIYKYKETEEDG
CycB1;nt;1   AEVVNKDVPAKKKA------------------------SDKEMKTVGGSPLGSKRKAKSSRTLSFVLTARSKAACGLSNKPK--YEIEDIDVADADNHLAAVEYVINFVKLTEGKS
CycB1;nt;2   AKVPAARKPAQKKATVKPNPEDIIEISP-----DTQEKLKEKMQRKKADKDSLKQKATLFSTLTARSKAACGLSKKPK--EQVVDIDAADVNNDLAAVEYVEDIHFFYRSADRS
CycB1;ps     RRVAVAPKPVQKKVTAKPKPVEVIDVISSEEES--SKEKSVLNKKGGEVNSRKKSSRTLSFTLTARSKAACGLSNKPK--IDTIEDIDADNHLAAVEYVEDIYTFYKLTEDPS
CycB1;zm;1   ILPARHAPRQERKKAPAKQ-PPPEDVIVLSS---DSEQSRTQLESSASSVRSRKKVINTLSSVLSARSKAACGITDKRRQVVVIEDIDKLDVNNELAAVEYVIEDITFFKIAQHDR
CycB1;zm;2   VAARAVTKP-ARKVTTKNVPRPGAGQKPKNRKKPSABGAAAASGRSVQKNRRKKPACTLSVLSARSKAACTVEKPK--IDID..D.DNELA.VEYVED.Y.FYR..E.K.

Cons cycB1   ........P..KK........[acid.zone.][..basic.domain....KK]..T.TS.LTARSKAAC.....K.K...I.DID..D.DNELA.VEYVED.Y.FYK..E.
```

```
CycB1;am;1   RS-----GVEEETNKKLKPSVPSANDFGDCIFIDEEE-----ATLDLPM-PMSLEKPYIEAD---PMEEVEMEDVTVEEPIV--DIDVLDSKNSLAAVEYVQDLYAFYRTMERFS
CycB1;am;2   RS---------QEETKKLKPSG---NEFGDCIFIDEEEKNEEVTTLDQPM-PMSLERPYIEFD--PMEEEVEMEDMEEEQEEPVLDIDGYDANNSLAAVEYVQDLYDFYRTMEKES
CycB1;at;1   RRFAAKIANTKTTNAEGTTKRSNLAKSSSNGPGDFIFVDDEHKPVEDQPVPMALEQTEPMHSESDRMKEEVEMEDIMEKPVM--DIDTFDANDFLAAVEYVEDLYSYRKVMSTS
CycB1;at;2   RRFAAKIAGSQQSYAEKTKNSNPLNLN----KPQHSIAIIDELKSPEDQPEPMTLEHTEPMHSDPLDMEEVEMEDISGRMIL--DIDCDANNSLAVVEYIDLHAYFRKIEYLG
CycB1;at;3   RNFAASLTRKEQLDHQVSVADAAVVCTDP------QKNPIPDGTVDDDVES-CESNDYIAVDECNDTD---EDSGMM----DIDSADSDMFLAATEYVRDLYSFFRKNEKFS
CycB1;gm;1   RKFAATLANQPSIAPLAPIGSERQKRTADSAFHGPADMECTKITSDDLFLPMMSEMDKVMGS---ELKEIEMEDI-EEAAP---DIDSCDANNSLAVVEYIDLYSFFRKSELGS
CycB1;zm;2   RKFAATLATQPTVALLDPIGSERLKRNADTAFHTPADMESTKMTDDSPL-PMVSEMDEMMSP---ELKEIEMEDI-EEAA----DIDSGDANSLAVADYVEDIYRFYRKTEGAS

Cons cycB1   R........................[...F....acidic...D....PM..E..domain.....EVEMEDI..E..].D.DID..DA.NSLAVVEYVED.Y.FYR..E.S
```

```
CycB1;am;1   RPHD-YMG-SQPEINEKMRAILIDMLLVQVHHKFELSPETLYLTINIVDRYLASETTIRRELQLVGIGAMLIASKYEEIWAPEVHELVC-ISDNTYSDKQILVMEKKILGALKWYL
CycB1;am;2   RPHD-YMD-SQPEINEKMRAILIDMLVQVHYKFELSPETLYLTINIVDRYLASKTTSRRELQLLGMSSMLIASKYEEIWPPQVNDLVC-ISDGSYSNGEVLRMEKKILGALKWYL
CycB1;at;1   RPRD-YMA-SQPDINEKMRILIVEWLIDVHVRFELNPETFYLTVNILDRFLSVKPVPRKELQLVGLSALLMSAKYEEIWPPQVEDLVD-IADHAYSHKQILVMETILSTLEWYL
CycB1;at;2   QPRM-YMH-IQTEMNEKMRSILIDMLIEVHFKFELSPETLYLTVNIIDRFLSSLLIASKYEEIWAPEVNDFVC-ISDNAYAREQILQMEEKAILGKLKWYL
CycB1;at;3   KPQM-YMH-TQTEINEKMRSILIDMLVEVHVKFELSPETFYLTVNTIDRFHSLKTVPRKELQLVGVSALLIASKYEEIWPPDVNDLVC-ISDKSYTHDQVLQAMEKILGQLEWYL
CycB1;gm;1   RPHD-YIG-SQPEINERMRAILVDMLIDVHVKFELSLEPLYLTVNIIDRFLAVKPVPRKELQLGVGLSAMLASAKYEEIWPEVNDFVC-ISDRAYTHEEILLTMEKTILSLLEWYL
CycB1;gm;3   CVHD-YMG-SQPDINAKMRSILVDWLIEVHRKFELMPETLYLTINIVDRYLAVKPVPRKELQLVGISSMLICSKYEEIWPPEVNDFVC-ISDNGYVSYDQVLVMEKQILRKLEWYL
CycB1;nt;1   RVDDDYMN-FQPDLNHKMRAILVDML IEVHRKFELMPESLYLTITILDRFLSLKTVPRKELQLVGISSMLIACKYEEIWAPEVNDFIY-VTDNAYSSRQILVMEKILGNLRWYL
CycB1;nt;2   RVHD-YMD-SQPEINDMRAVILIDMLVEVHQKFENSPETLYLTAVKTTSRRELQLLGI SAMLIASKYEEIWAPEVNDFIC-ISDNGYVLSRQILVMEKGILGNLEWYL
CycB1;ps     RPHD-YMD-SQPEINDKMRAIDMLVEVHVKFEFLPRALYLTINITDRFLAISLVSRRELQLVGISAMLMASKYEIWPPEVNDFVCHSSDRAYTHEQILIMEKTILGKLEWYL
CycB1;zm;1   RHCD-YID-TQVEINDKMRRAILDMLIEVHLRFALNPETLYLMYTIDCYLSLQVTPRRELQLVGVSAMLLICKYEEIWAPEVNDFIL-ISDSAYSRDQILNNNLEWNL
CycB1;zm;2   RPID-YMGQTSPELSPRMRSILADMLIESHRRFLQMPETLYLTIYIVDRYLSLQPTPRRELQLVGVSAALIACRYEEIWAPEVNDVIH-IADGAFNRSQILAAAKKILANSMEWNL

Cons cycB1   R..D-YM.-SQPEIN.KMRAILDMLIEVH.KFEL.PETLYLTINIVDRFL.KQ.V.RKKLQLVGL..AMLIACKYEEVWAPEVNDFV..QIL.MEK.IL..LEW.L
```

```
CycB2;at;1   CVPLDYMA--QQFDISDKMRAILIDWLIEVHDKFDLMNETLFLTVNLIDRFLSKQNVMRKKLQLVGLVALLLACKYEEVSVPVVEDLVL-ISDKAYTRNDVLEMKKTMLSTLQFNI
CycB2;at;2   CVQPDYMS--QQGDINERMRAILIDWLIEVHDKFELMNETLFLTVNLIDRFLSKQAVARKKLQLVGLVALLLACKYEEVSVPVVEDLVL-ISDKAYTRNDVLEMKKIMLSTLQFNM
CycB2;ms;3   CVSPNYMA--QQFDINERMRAILIDWLVEVHDKFDLMHETLFLTVNLIDRFLAKQSVIRKKLQLVGLVAMLLACKYEEVSVPVVEDLVL-IIDRAYTRKDTVLEMKKIMLSTLQFNM
CycB2;ms;2   CVSPTYMD--EQLDLNERMRAILIDWLIEVHDKFDLMQETLFLTVNLIDRFLAKQNVVRKKLQLVGLVAMLLACKYEEVSVPVVEDLVL-IADRAYTRKDILEMKKLMLNTLQFNM
CycB2;os;1   CVPVDYMM--QQIDLNERMRAILIDWLVEVHVHFKFELMQETLFLTVNIIDRFLSKSKQNVVPRKKLQLVGVAMLLACKYEEVAVPVVEDLVL-ICDRAYTRKGQILEMEKLILNTLQFNM
CycB2;os;2   CVSPNYMM--QQTDLNERMRAILIDWLVEVHKLELLESTLFLTVNIIDRFLARENVVRKKLQLVGVAMLLACKYEEVSVPVVEDLVL-ICDRAYTRTDVLEMERMIVNTLQFDM
CycB2;zm     CVPTNYMS--SQTDINEKMRAILIDWLIEVHVHYKLELLESTLFLTVNIIDRFLARENVVRKKLQLVGVAMLLACKYEEVSVPVVEDLVL-ICDRAYTRADILEMERRIVNTLNFNM

Cons cycB2   CVP..YM..Q.D.NEKMR.ILIDWLIEVH.KFELM.ETLFLTVN.IDRFL.KQ.V.RKKLQLVGL..AMLLACKYEEVSVPVV.EDL.I.DRAYTR.DILEMEK..L.TL.FN.
```

```
CycB1;am;1   TVPTPYVFLVRFIKAS------MTDSDVENMVYFLAELGHMNYA-TLIYCPSMIAAASVYAARCTLNKA-PFWNETLQLHTGFSEP---QLMDCARLLVAFPKMAGDQ-----KLK
CycB1;am;2   TVPTPYVFLVRFIKAS-----LPDSDVEKMMVHYFLAELGHMNYATI1MVCPSMVAYTG----QLMDCAKLLIDFHGGSTDQ-----KLQ
CycB1;at;1   TVPTHYVFLARFIKAS-----IADEKMENMVHYLAELGWMHYDTMIMFSPSMVAASAIYAARSSLRQV-PIWTSTLKHHTGYSET---QLMDCAKLLAYQQWKQGERGSKSSTKG
CycB1;at;2   TVPTPYVFLVRYIKAA-----TPSDNQEMENMTFFAELGLMNYT-DLTFCPSMQAASAVYTARCSLNKS-PAWTDTLQLFHTGYTES--EINDCSKLLAFLHSCRCGES----RLR
CycB1;at;3   TVPTPYVFLVRFIKAA-----LPDSEINNMTVFAELGLD-SLMFCPSMLAASAVYTARCCLNKS-PWTIDTLKFHTGYSES--QLMDCSKLLAFFHSKAGES-------RLR
CycB1;gm;1   TVPTPYVFLVRDYKAA-----VPDQELDMNAFLSRLGLMNYA-TLMYCPSMVAAAVAARSCTLNKS-PFWNETLNLHTGYVDSE-QLMDCARLLVGFYSTLENG------KLR
CycB1;gm;3   TVPTYVHFLVRDYKAA-----TPSDKEMENMVFFLAELGLMHYFTVILYRPSLIAASAVYTARCTLGS-PFWTNTLKHHYTGYSEP--QLRDCAKIMANLHAAAAPGS-----RLR
CycB1;nt;1   TVPTQYVFLVRFIKAA-----NSDPEMENMVHFLAELGHMYKITISYRPSMLAASVVAARSTELKT-PLQL-GHHTGYSED--QLMECAKILVSYHLDAAES------RLR
CycB1;nt;2   TVPTPYVFLVRFLKAA-----GSDQKLENLVHFLAELGHMHHA-TIYSYRPSMLAAAYVAARHTLNRT-PFWNETLKLHTGFSES---QLIECARLLVSYQSAAATH----RLR
CycB1;ps     TVPTPYVFLVRFLKAAA-----SVSLPSSDLALENMAHFLSELGHMHYA-TLMYSPSMMAAAAVYAARCTLKKS-PVWDETLTMHTGYSEE---ELMGCARLLVSFHSASGSG----KLR
CycB1;zm;1   TVPTVYMFLVRFLKAA-----TLGNIVEKDMENMVFFAELALMQYG-LVTRLPSLVAAVVSRKKLKTLT-PIWDTLKHHTGFRESEAELIECTRCWSAHTRPHADS-----KLR
CycB1;zm;2   TVPTVHFLLRFAKAA------GSADEQLQHTINFFGELAIMDYG-MVMTNPSTAACAVYAARLTLGRS-PLWTETLKHHTG-PQRAADTGRAKTLVGSHAASASPDA-----KLR

Cons cycB1   TVPT.Y.FLVRF.KAS-------..D....ENMV.F.AELG.MHY.-...Y.PSM.ASAVYAAR..TL.KS-P.W.ETL..HTGYSE.--.QL.DCAKLLV..........KLR
```

```
CycB2;at;1   SLPTQYPPLKRFLKAAQ----ADKKCEVLASFLIELALVEYE-MLRPPSLLAATSVYTAQCTLDGS-RKWNSTCEFHCHYSED---QLMECSRKLVSLHQRAAGTN-----LT
CycB2;at;2   SLPTQYPFLKRFLKAAQ----SDKKLMILASFLIELALVEYE-MVRYPPSLLAATSYYTAQCTMHCS-QLLERCRRMVRLMQAGTGR----LT
CycB2;ms;3   SVPTAYVFMRRFLKAAQ----ADRKLELLAFFLIELSLVEYE-MLKFSPSLLAAAVVTAQCT-MYGV-KQWSKTCEWHTNYSED---QLLECSILMVDFHCHKKAGTGK----LT
CycB2;ms;2   SLPTAYVFMRRFLKAAQ----ADKKLELVAFFLVDLSLVEYE-MLKFPPSLVAAAAVYTAQCT-VGF-KHWNKTCEWHTNYSED---QLLECSMLMVGFHQKAQGA-----LT
CycB2;os;1   SVPTFYVFMRRFLKAAQ----SDKQLQLSFFIELSLVEYQ-MLKYRPSLSAAAVVTAQCALTRC-QQWTKTCELHSRYTGE---QLRECRSMKMVDFHKQKAGAGK----LT
CycB2;os;2   SVPTPYCFMRRFLKAAQ----SDKKLELMSFFIILSLVEYE-MLKFQPSMLAAAAIYTAQCTINGF-KSWNKCCELHTKYSEE---QLMECKSMMVGFHQKAGHGK-----LT
CycB2;zm     SVPTPYCFMRRFLKAAQ----SEKKLELLSFFMIRLSLVEYE-MLQFCPSMLAAAAIYTAQCTINGF-KSWNKCCELHTRYSEE---HLMVCSRRMMVELHQRAAGHGK----LT

Cons cycB2   SVPT.Y.FMRRFLKAAQ----SDKKLEL.A.F.IELSLVEYE-MLKF.PS.LAA.AVYTAQCT..G..-.W....CE.H..YSED--.QL.ECS..MV...HQKAG.GK.....LT
```

## Fig. 7 (Contd.)

```
              end cyclin core>
CycB1;am;1   SIYRKKYSNLERGAVA-LLSPAKSVFVFLIELLMNAIEKIQCSLFPTSEIWCRF
CycB1;am;2   GIYRKYSRLEKGAVA-LL-PQPLLA
CycB1;at;1   ALRKKYSKDERFAVA-LIPPAKALLTGTESA
CycB1;at;2   AVYKKYSKAENGGVA-MVSPAKSLISAAADWKKPVSS
CycB1;at;3   GVLRKYSKLGRGAVA-LISPAKSLMSSAP
CycB1;gm;1   VVYRKYSDPQKGAVA-VLPPAKFLLPEGSASQHS
CycB1;gm;3   AVYKKFSNSDLSAVA-LLSPAKDLSALS
CycB1;nt;1   AIYRKFSSPDRGAVA-FFPPARNLLPTTTTDAASLVLEEFYFLVWEAFLGSFYCCDLSSLD
CycB1;nt;2   VIYKKYSSPERGVVS-LLTPAKSLLAASSSSVLSEQADLRKSTEAAATSSSKMVVVGCQRCHMYVMVTEADPRCPQCKSTTTRKMT
CycB1;ps     GVYKKYADPQKGAVA-VLPPAKTLCHLPQLVLKFVIEINKGSCLSF
CycB1;zm;1   AVYKKYSEQFRTRA-RVRPPAAAVEINETKITCC
CycB1;zm;2   AVYQKYATEQFGRVA-LHPPAPAALPDLV
                 *                              *
Cons cycB1   .VYKKYS..E.G VA-...PA....................................

CycB2;at;1   GVYRKYSTSKFGYIA-KCEAAHFLVSESHHSF
CycB2;at;2   GVHRKYSSSKFGYIATKYEAAHFLVSDSH
CycB2;ms;3   GAHRKYCTSKFSYTA-KCEPASFLLENEL
CycB2;ms;2   GVHRKYGSAKFSFTA-KCEPACFLLENKNQP
CycB2;os;1   GVHRKYSTFKFGCAA-KTEPALFLLESGAGGYNLQKQPC
CycB2;os;2   GVHRKYSTFRYGCPA-KSEPAVFLLKSVAL
CycB2;zm     GVHRKYNTSRYSYAA-KSEPATFLLDA
                 * ***            * * * * **
Cons cycB2   GVHRKY.T.KFG..A-K.EPA.FLL......
```

*For explanations, see Fig. 6. The place of the domains in the N-terminal part, which are discussed in the text, is indicated in brackets on the consensus (cons) line. Putative bipartite NLS, located upstream from the conserved LTARSKAAC motif of CycB1 cyclins, are underlined.*

sequences of A- and B-type cyclins by using the same criteria as described above, from the sequences of nine animal A-type cyclins, and of 16 animal and eight yeast B-type cyclins.

A total number of 30 residues are highly conserved in the cyclin core of all A- and B-type cyclins from animal and yeast (indicated by asterisks in Fig. 8). Most of them, 72%, are located within the cyclin box. The plant cyclins CycA and CycB share 90–97% of them, and notably they have always the five residues, $R_{52}$, $D_{81}$, $L_{96}$, $K_{107}$ and $E_{136}$ (indicated in bold asterisks in Fig. 8) which mutational analysis has shown to be strictly required for cyclin activity [50–52]. It was shown in the three-dimensional structure of bovine cyclin A [48] that $R_{52}$–$D_{81}$ forms a buried ionic pair in an hydrophobic environment. The absolute conservation of that pair and of the hydrophobic character of each individual residue positioned in the environment by our alignment (Fig. 2) is very striking.

Using phylogenetic analysis, the relationship between CycA and CycB plant cyclins with yeast and animal cyclins A and B, respectively, is only partly confirmed at the level of consensus sequence similarities. The cyclin core of CycA cyclins is approximately 37% similar to that of animal A-type cyclins, and approximately 26% to that of animal and yeast B-type cyclins. However, the cyclin core of CycB cyclins has the same overall homology with A- and B-type cyclins, approximately 30% (Table 2). Similarly, the classification of plant cyclins as either A or B using conserved residues in each group is quite difficult: the cyclin core of CycA plant cyclins (i.e. CycA1, CycA2, CycA3) has only a mean number of 10 (i.e. 11, 10, 8) of the 27 residues specific to animal cyclin A, but it has also three (i.e. 4, 4, 2) of the 10 residues specific to cyclin B (Fig. 8). The plant cyclins, CycB1 and CycB2, have six of the nine residues which are specific to cyclin B in yeast and animal, but they also share three and five, respectively, of the 27 residues which are specific to animal cyclin A (Fig. 8).

## Fig. 8     The consensus sequence of the cyclin core of CycA and CycB plant cyclins

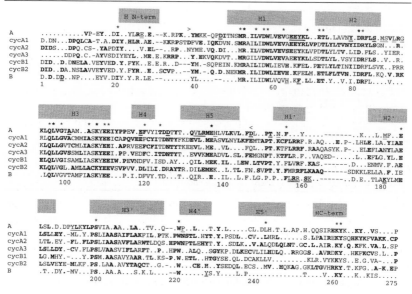

The consensus sequences of CycA and CycB plant cyclins indicated in Figs. 6 and 7 have been aligned together with the consensus sequences of animal A-type cyclins (A, top) and of yeast and animal B-type cyclins (B, bottom). The residues are indicated when they are conserved in at least 80% of the sequences. Residues in bold are identical in all the sequences of a given group. The dots represent non-conserved residues. The dashes indicate gaps required for maximum alignment. The residues specific to animal A-type cyclins (i.e. absent in >90% of B-type cyclins) are underlined in the A line. The residues specific to animal and yeast B-type cyclins (i.e. absent in >90% of A-type cyclins) are underlined twice in the B line. The residues which are identical in both A- and B-type cyclins are indicated by asterisks above the alignment and bold asterisks indicate the residues which have been shown to be essential for cyclin activity. The boxes drawn above the sequences indicate the helix domains of bovine cyclin A [48]. The consensus sequence of animal A-type cyclins has been established from the alignment of nine cyclin A sequences (SwissProt Accession numbers, otherwise indicated: Spisula solidissima, P04962; Patella vulgata, P24861; Xenopus laevis, P18606; Homo sapiens, P20248; Bos taurus, P30274; Mesocricetus auratus, P37881; Drosophila melanogaster, P14785; Mustela vison, Genbank Accession number U00595; Mus musculus, PIR Accession number S37280). The consensus sequence of B-type cyclins comes from the alignment of 16 animal cyclin B sequences (Arbacia punctulata, P07818; Asterina pectinifera, P18063; Drosophila melanogaster, P20439; Martasterias glacialis, P15206; Patella vulgata, P24862; Spisula solidissima, P13952; Cricetulus longicaudatus, Q08301; Mesocricetus auratus, P37882 and P37883; Homo sapiens, P14635; Mus musculus, P24860 and P30276; Rattus norvegicus, P30277; Xenopus laevis, P13350 and P13351; Gallus gallus, P29332) and of eight yeast B-type cyclins (Saccharomyces cerevisiae CLB1-6 P24868, P24869, P24870, P24871, P30283, and P32943; Shizosaccharomyces pombe CDC13 P10815 and cig2 P36630).

A major difference between plant cyclins and other A- and B-type cyclins lies in the typical animal and yeast cyclin B signature, $F_{155}$LRR-SK. Both CycA and CycB plant cyclins possess an altered form of this motif, lacking the serine and lysine residues in CycA, and the serine in CycB. Therefore, this motif cannot be considered as a typical hallmark of all B-type cyclins. Moreover, its conservation is not related, at least in plant cyclins, to the phosphorylation of the serine residue [44]. Plant CycB cyclins also lack the other most typical hallmark of animal and yeast B-type cyclins, the $A_{179}$KYL motif, and notably they do not have the basic residue typical of B-type cyclins (Fig. 8).

From the comparison of plant cyclins with other A- and B-type cyclins, the best conserved difference between the two types of cyclins (A and B) at the level of consensus motifs is found in the helix H1 of the first domain of the cyclin fold. All A-type cyclins, including plant CycA cyclins, have the typical $L_{59}$VEV-EEY signature, whereas all B-type cyclins have the typical cyclin B signature $(HQ)_{63}$-(KRQ)(FL), the plant CycB having the same motif H-KF as yeast B-type cyclins (Fig. 8). A second difference between A- and B-type cyclins lies close to the conserved motif $K_{260}$Y at the end of the cyclin core [44]: this motif is located downstream from an acid residue in A-type cyclins, including 65% of plant CycA cyclins, whereas it is preceded by a conserved basic residue in B-type cyclins, including all plant CycB cyclins, and in only 10% of the plant CycA cyclins.

The distinction between the three CycA groups and between the two CycB groups can be made by overall homology analysis of the cyclin core in addition to structural features of the N-terminal domain (see below). Compared with other plant cyclins, CycB1 cyclins are characterized by a conserved glutamic acid in position 93 of the cyclin box (Fig. 8), a position where cyclins A, B, E, and F have a basic residue (Fig. 2).

## Specific destruction boxes characterize each plant cyclin group

Plant cyclins, CycA and CycB, have in their N-terminal region a nine-amino-acid motif. This is closely homologous to the destruction motif characterized in animal and yeast A- and B-type cyclins, and associated with the dramatically regulated proteolysis of these cyclins during mitosis [54]. This motif is shaded in Figs. 6 and 7. It is followed by a stretch rich in basic residues, which is supposed to be a ubiquitin-binding site [54]. The consensus sequence of the destruction box varies according to the class and to the group (Table 4). These differences can be explained in functional terms, as cyclins have to be destroyed at various times during the cell cycle. The first essential residue, arginine, is systematically conserved [54,77]. The leucine residue in position 4 is also strongly conserved. The destruction box is perfectly conserved over nine residues in animal cyclin B1, RTALGDIGN, whereas it is poorly conserved in yeast B-type cyclins (R---L---V-N) and in cyclin A (R--LGV---).

Plant CycB1 cyclins have the most conserved destruction box among plant cyclins, with six conserved residues RRALGDIGN (Table 4). This motif is the most similar to that of vertebrate cyclin B1. They share an original feature with

CycB2 plant cyclins compared with yeast and animal cyclins: a conserved basic residue in the second position. The four first residues, and the seventh, of the destruction box of plant CycB2 cyclins are the same as in CycB1 (Table 4), but the two groups differ in the second half of their destruction motif; in CycB1 the sixth residue is aspartic acid and the last one is a conserved asparagine, two features typical of yeast and animal B-type cyclins [54]. However, in CycB2 the sixth residue is a valine in 50% of the cases, and there is no conserved asparagine at the end, two features reported to characterize animal A-type cyclins [54].

The presence of a conserved D/E/N in the sixth position, and of a conserved asparagine residue at the end of the destruction box, makes CycA plant cyclins closer to B-type cyclins than to animal A-type cyclins [54]. CycA3 plant cyclins have a highly conserved destruction box with seven conserved amino acids (RVLLGEI-N). The three subgroups of CycA plant cyclins have different consensus sequences in positions 2, 3, 5, 6 and 7 of their destruction box, which further allows their distinction (Table 4). Hence the regulation of the proteolysis may not be the same for all of the CycA plant cyclins.

## The variable N-terminus displays features specific for each group of plant cyclins

In the domain upstream of the cyclin core, but outside the destruction box, the plant CycA1, CycA2 and CycA3 have 17%, 11% and 6% of conserved residues, respectively. For CycB1 and CycB2, the conserved residues are 26% and 17%, respectively (Figs. 6 and 7). This is much less than the 73–83% conservation values for the cyclin core (Table 2). The residues which are conserved in the N-terminus of plant cyclins are mostly spread all along this region. However, conserved motifs are also present in the N-terminus of CycB1, CycB2, CycA1, and CycA2 plant cyclins.

The cyclins, CycB1 and CycB2, share a small conserved motif downstream from the destruction box. The consensus sequence of this motif, RP--TR, is the same for the two groups (Fig. 7). A large, strikingly conserved, motif located just upstream from the cyclin core, is characteristic of CycB1 cyclins (Fig. 7). This 12-amino-acid-long motif, which has no homologues in sequence databases, contains a stretch of six conserved residues, TS(V/T)LTARSKAAC. This motif is located downstream from a highly basic domain, itself following a small acidic domain of 15 residues, with no conserved motif.

CycB2 cyclins have a characteristically large conserved acidic domain upstream from the cyclin core, in the place of the basic domain mentioned above for CycB1 cyclins (Fig. 7). Acidic residues (D, E) represent 34% of the 51 residues of this domain, instead of a mean of 16% in whole CycB2 cyclins. A conserved motif, EVEMEDI, is present at the end of this domain in a region also rich in methionine residues and can serve to identify CycB2 plant cyclins.

CycA1 plant cyclins, like CycB1 and CycB2, have a conserved domain located upstream from the cyclin core, with 55% conservation over 45 residues, but without any strictly conserved motif (Fig. 6). CycA2 cyclins do not have a

**Table 4**  The destruction motif of cyclins A and B

The number of occurrences of each residue is indicated in subscript, except when the residue is conserved in all sequences. The $n$ sequences considered for each group are those mentioned in Figs. 6 and 7 for plant cyclins, and those mentioned in Fig. 8 for yeast and animal cyclins.

| Cyclin | (n) | 1 | 2 | 3 | 4 | 5 | 6 | 7 | 8 | 9 |
|---|---|---|---|---|---|---|---|---|---|---|
| A animals | 7 | R | $A_3S_2$ $T_1V_1$ | $A_4G_1$ $I_1V_1$ | L | $G_5$ $A_2$ | $V_5T_1$ $L_1$ | $I_4$ $L_3$ | $T_2R_2Q_1$ $K_1G_1$ | $N_2S_2A_1$ $D_1G_1$ |
| CycA1 plants | 5 | R | $A_2P_2V_1$ | $A_2P_2S_1$ | $L_4V_1$ | $G_2S_2T_1$ | $N_4S_1$ | $I_3L_2$ | $G_1S_2T_2$ | N |
| CycA2 plants | 7 | R | A | $V_5I_2$ | L | $H_1K_3S_1$ $Q_1G_1$ | D | $V_4I_2$ $M_1$ | $T_3K_1S_1$ | $N_6L_1$ |
| CycA3 plants | 6 | R | V | V | L | G | E | $I_4L_2$ | $P_1Q_2R_1$ $S_1T_1$ | N |
| B1 animals | 6 | R | T | A | L | G | D | I | G | N |
| B2 animals | 4 | R | A | $V_3A_1$ | L | $E_3G_1$ | E | I | G | N |
| B3 animals | 1 | R | S | A | F | G | D | I | T | N |
| B invertebrates | 6 | R | $A_3G_1S_1$ | $A_5T_1$ | L | $G_5E_1$ | $D_4N_2$ | $I_4V_1L_1$ | $S_3G_2Q_1$ | N |
| B yeasts | 8 | R | $T_2V_2G_1$ $L_1A_1H_1$ | $A_4V_1K_1$ $L_1I_1$ | $L_7$ $T_1$ | $G_2S_2D_2$ $N_1Q_1$ | $D_4N_2$ $V_1R_1$ | $V_6P_1$ $D_1$ | $T_4S_3V_1$ | $N_6S_1T_1$ |
| CycB1 plants | 12 | R | $R_7A_2K_2$ $Q_1$ | $A_9P_1V_2$ | L | $G_{10}H_1$ $Q_1$ | D | I | G | N |
| CycB2 plants | 7 | R | R | $A_5P_1V_1$ | L | $G_3R_3S_1$ | $V_3D_3G_1$ | I | $N_5K_2$ | $H_1N_3Q_3$ |

Position of the residue in the destruction motif

**Table 5          Organization of the N-terminal domain of plant cyclins**

The data are the length of the sequences upstream from the destruction box (only for full-length cyclins) and between the destruction box and the cyclin core. For each group, the *n* sequences which have been analysed are those which are presented in Figs. 6 and 7.

| Cyclins | No. of residues [average $\pm$ S.D. (*n*)] | | | |
|---------|-------------------------|------|---------------------------|------|
|         | **N-end to destruction box** | | **Destruction box to cyclin core** | |
| CycA1   | $42 \pm 4$   | (3)  | $147 \pm 35$ | (5)  |
| CycA2   | $59 \pm 15$  | (7)  | $102 \pm 24$ | (7)  |
| CycA3   | $38 \pm 10$  | (5)  | $43 \pm 6$   | (6)  |
| CycB1   | $31 \pm 2$   | (11) | $123 \pm 11$ | (11) |
| CycB2   | $33 \pm 6$   | (7)  | $112 \pm 6$  | (7)  |

conserved motif immediately upstream from the cyclin core, but instead a small motif with three conserved residues, E---R-TRARA, close to their N-terminal end (Fig. 6). CycA3 cyclins have no conserved motif in their N-terminal domain, except for the destruction box. Note that the length of the N-terminal domain of CycA cyclins is more variable than that of CycB cyclins. This is mostly owing to variation in length of the sequence located between the destruction box and the cyclin core (Table 5). This sequence is the longest in CycA1 cyclins (approximately 147 residues) and the shortest in CycA3 cyclins (approximately 43 residues) and about the same length (approximately 110 residues) in CycA2, CycB1 and CycB2 cyclins (Table 5.)

The region upstream from the cyclin core of CycB1 and CycA plant cyclins contains domains rich in basic residues, especially around the destruction box (Figs. 6 and 7). As a consequence, plant CycB1 cyclins and the three groups of CycA cyclins have a 'calculated isoelectric point (pI)' in the basic range, above 7.7, similar to that for animal B-type cyclins, especially cyclin B2 (Table 6). In contrast, plant cyclins CycB2 are relatively acidic, with a calculated pI very close to 5.5. This is also a property of animal A-type cyclins, which have a mean calculated pI of 5.6 (Table 6). The pIs for yeast B-type cyclins span a broad range of values between 4.8 and 8.3, with a mean of 6.5. The large dispersion of pI values for yeast cyclins may be related, at least partly, to the various functions these latter cyclins are committed to during the yeast cell cycle. In this respect, the more homogeneous pI values in the other cyclins A or B groups, including the plant cyclins groups, would be indicative of homogeneous functional groups.

The N-terminus of CycB1 and CycA cyclins has clusters of basic residues which feature putative nuclear localization sites (NLS). In plants, NLS are grouped into three main categories. The first comprises clusters of five basic residues; the second is characterized by hydrophobic regions with one or more basic residues. The third is a bipartite NLS made of a combination of a first region of two basic residues and a second region of at least three out of five basic residues, the two regions being separated by a spacer of approximately 10 (>4) residues [78]. Putative

**Table 6        The electric charge of cyclins A and B**

The isoelectric point (pI) of cyclins was calculated theoretically from amino acid sequences translated from full-length cDNAs. For plant cyclins, the *n* sequences considered in each subgroup are those presented in Figs. 6 and 7. The six sequences of animal cyclin A are from *Drosophila melanogaster, Homo sapiens, Mesocricetus auratus, Spisula solidissima, Patella vulgata* and *Xenopus laevis* (see Accession numbers in Fig. 2). The six animal cyclin B1 sequences are from *Cricetulus longicaudatus, Mesocricetus auratus, Homo sapiens, Rattus norvegicus* and *Xenopus laevis*. The four animal cyclin B2 sequences are from *Mesocricetus auratus, Gallus gallus, Mus musculus* and *Xenopus laevis*. The six animal cyclin B sequences are from *Asterina pectinifera, Martasterias glacialis, Patella vulgata, Arbacia punctulata* and *Drosophila melanogaster*. The eight yeast cyclins B are *Saccharomyces cerevisiae* CLB1-6 and *S. pombe* cdc13 and cig2.

| Cyclin | *n* | pI (average ± SD) | pI range |
|---|---|---|---|
| A animals | 6 | 5.6 ± 0.4 | 5.1–6.1 |
| CycA1 plants | 3 | 8.1 ± 0.1 | 8.0–8.2 |
| CycA2 plants | 7 | 8.3 ± 0.9 | 6.6–9.3 |
| CycA3 plants | 3 | 7.7 ± 0.9 | 6.8–8.6 |
| B1 vertebrates | 6 | 7.5 ± 0.8 | 6.8–8.8 |
| B2 vertebrates | 4 | 8.9 ± 0.1 | 8.7–9.0 |
| B3 animals | 1 | 8.2 | - |
| B invertebrates | 6 | 8.1 ± 1.1 | 6.3–9.6 |
| B yeasts | 8 | 6.5 ± 1.4 | 4.8–8.3 |
| CycB1 plants | 11 | 8.6 ± 0.8 | 7.1–9.7 |
| CycB2 plants | 7 | 5.5 ± 0.2 | 5.2–5.7 |

bipartite NLS are found in 13 of the 18 CycA cyclins, in the region surrounding the destruction box (Fig. 6, underlined) and in six of 12 CycB1 cyclins, in the region just upstream from the LTARSKAAC motif (Fig. 7, underlined). In these motifs, the spacer region has 5–19 residues. Motifs resembling bipartite NLS are found in nearly all the CycA and CycB1 cyclins, in the regions mentioned above. In addition, three CycA cyclins have, in the same region as the putative bipartite NLS, clusters of five basic residues, which represent another type of putative NLS (Fig. 6, CycA2;bn, CycA3;dc and CycA3;am;1, double underline). For CycA and CycB1 cyclins, there is no functional evidence, as yet, to suggest that these clusters are NLS or that they are located in the nucleus. However, these putative signals are located in the same conserved region within each group of cyclins, but they are absent from CycB2 plant cyclins.

From sequence analysis, no definite conclusion can be drawn as to the cellular compartmentation of plant cyclins. The conserved domains, different for each group and present in the variable N-terminus of CycA1, CycA2, CycB1 and CycB2 cyclins, could be involved in the specific recognition of other proteins, possibly associated with the compartmentation of plant cyclins. In the same way, the region upstream from the cyclin core of vertebrate cyclins B1 and B2 has a cytoplasmic retention signal, a conserved domain which is responsible for anchoring these cyclins to cytoplasmic components at interphase [79]. The putative NLS in CycB1 and CycA plant cyclins suggest that, at least during a part of the cell cycle, these cyclins are located in the nucleus. Such signals are not found in cyclin A and B from other eukaryotes, although animal cyclin A is present in the nucleus throughout its lifetime [22,30–32] and animal cyclin B relocates from the cytoplasm into the nucleus during mitosis [31,80–82].

## Conclusion: a rational classification of plant cyclins

The analysis of 51 plant cyclin sequences related to the cyclin A/B family shows that they cluster into two classes that we name CycA and CycB, with three CycA groups and two CycB groups. These groups and subgroups are very clearly defined by a number of features located both in the variable N-terminus of plant cyclins and in the cyclin core. A phylogenetic analysis from the alignment of the cyclin core tends to indicate that CycA plant cyclins are more related to A-type than to B-type animal cyclins, and that CycB plant cyclins are more related to yeast or animal B-type cyclins than to animal A-type cyclins. This classification is, however, not very robust and it is only partially visible at the level of the known consensus of animal A- and B-type cyclins (Fig. 8). In the same way, the fact that the number of residues specific to B-type cyclins (9) is lower than that of those specific to A-type cyclins (27) derives from the narrower phylogenetic distribution of cyclin A. In this analysis we found a common signature for separating every eukaryotic cyclin A from cyclin B, which resides in helix H1, the most strongly conserved part in cyclin sequences (Fig. 8).

The occurrence of modified consensus sequences in the cyclin core and the presence of putative NLS in CycA and CycB1 further substantiate the separate evolution of plant cyclins as compared with the other eukaryotic A- and B-type cyclins, in connection with the specificity of the regulation of the plant cell cycle [8,9,83–85]. However, functionally, plant cyclins clearly belong to the cyclin A/B family, since they are effective inducers of meiotic maturation of immature *Xenopus* oocytes [9,14,17]. As also demonstrated for animal A- and B-type cyclins, N-terminal truncated forms of plant CycA and CycB cyclins can rescue *CLN* mutants of *S. cerevisiae* [86,87].

No standardized system of nomenclature has yet been adopted for plant cyclins, despite some proposals [17,18,77], so that the names of plant cyclins used so far do not allow us to appreciate their relationship one to another (Table 1, old names). This constitutes a major drawback to the functional analysis of plant cyclins. As a result of sequence analysis, we have introduced in this chapter a new, standardized nomenclature of plant cyclins, which is presented in Table 1, and which we propose to adopt in further studies. The low sequence similarity of plant cyclins with other A- and B-type cyclins, as shown by phylogenetic analysis, led us to name the two groups of plant cyclins CycA and CycB, respectively. The groups are identified by numbers 1, 2 or 3.

Several examples have already been found of multiple cyclins from a given species in the same group, and we have seen that this could be a general situation among plants, which differs from animals. The nomenclature rules of plant gene names suggest that we treat each group as a multigene family [88], with the genes encoding the different members being indicated by individual member number. Thus the correct designation of the three *Arabidopsis thaliana* (abbreviation at) CycA2 genes from Table 1 should be : CycA2;at;1, CycA2;at;2 and CycA2;at;3. The member numbers do not belong to the gene name, such that CycA2;at;1 is not more similar to CycA2;ps;1 than to CycA2;ps;2. It may be expected to find in the future, when the number of sequences has become sufficient, a further level of structural subclassification within some groups, or to have evidence for functional differences among the cyclins from one group, which would be conserved between species. In such cases, the nomenclature proposed here could be kept and a lower case letter added after the plant-wide gene name to indicate these differences. A similar nomenclature is also proposed for the plant homologues to animal cyclins D, which are to be called CycD plant cyclins, and which are distributed into three subgroups, CycD1, CycD2 and CycD3 (Table 1).

An interesting question arising from this analysis is the extent to which the structural classification of plant cyclins will meet a functional one. As a preliminary answer, it must be strongly emphasized that the proposed nomenclature by no means signifies that CycA and CycB cyclins are true functional homologues of yeast or animal A- and B-type cyclins, respectively. Such a parallel would even be ambiguous since animal A-type cyclins are known to perform different functions in S and $G_2$/M, and yeast use various B-type cyclins all along their cell cycle. The few data available so far show that the cyclins from the two groups CycA and CycB are differentially regulated at the transcriptional level during the cell cycle, and that they display significant differences in their expression: CycA RNAs are expressed in S and $G_2$ phases; they may accumulate to high levels after blocking with aphidicolin or hydroxyurea and they are not found in mitotic cells [16,55,85–87,89]. An alfalfa cyclin, CycA2;ms, was found to be permanently expressed during the cell cycle, although at a lower level in $G_1$, and to be potentially involved in $G_0$/$G_1$ transition [87]. The cyclins CycB are only expressed in $G_2$ and M phases [15,55,61,62,85–87,89–91]. Interestingly, a few studies have already shown that the cyclins of different CycA groups [76,77,97] or CycB groups [16] display slightly different characteristics of expression when assayed in a same biological system. However, the data remain too scarce to assess their validity for the whole group. It is expected that the accurate definition of structural groups of plant cyclins will help clarify the specific functions they perform in cycling plant cells.

*The authors thank J. Hashimoto, T. Jacobs, and M. Sauter, for communicating unpublished sequence data, H. Logan for correcting English, C. Germonprez for preparation of Figures, L. Van Wiemeersch and P. Déhais for help in computing, and M. De Cock for final preparation of the manuscript. This work was supported by grants from the Belgian Programme on Interuniversity Poles of Attraction (Prime Minister's Office, Science Policy Programming, #38), the Vlaams Actieprogramma Biotechnologie (ETC 002), the Körber Stiftung, the Fonds voor Geneeskundig Wetenschappelijk Onderzoek (G.0121.96), and the EC-Concerted Action (BIO4-CT95-0247) in the frame of the BIOTECH Program. A.S. is indebted to the European Union for a Human Capital and Mobility post-doctoral fellowship (ERBCHBICT-941824). J.P.R., D.I. and P.R. belong to the Institut National de la Recherche Agronomique (France).*

## References

1.    Pines, J. (1995) Biochem. J. **308**, 697–711
2.    Nigg, E.A. (1995) Bioessays **17**, 471–480
3.    Doree, M. and Galas, S. (1994) FASEB J. **8**, 1114–1121
4.    Evans, T., Rosenthal, E.T., Youngblom, J., Distel, D. and Hunt, T. (1983) Cell **33**, 389–396
5.    Grana, X. and Reddy, E.P. (1995) Oncogene **11**, 211–219
6.    Sherr, C.J. and Roberts, J.M. (1995) Genes Dev. **9**, 1149–1163
7.    Norbury, C. and Nurse, P. (1992) Annu. Rev. Biochem. **61**, 441–470
8.    Francis, D. and Halford, N.G. (1995) Physiol. Plant. **93**, 365–374
9.    Ferreira, P., Hemerly, A., Van Montagu, M. and Inzé, D. (1994) Plant Mol. Biol. **26**, 1289–1303
10.   Jacobs, T.W. (1995) Annu. Rev. Plant Physiol. Plant Mol. Biol. **46**, 317–339
11.   Fitch, I., Dahmann, C., Surana, U., Amon, A., Nasmyth, K., Goetsch, L., Byers, B. and Futcher, B. (1992) Mol. Biol. Cell **3**, 805–818
12.   Schwob, E. and Nasmyth, K. (1993) Genes Dev. **7**, 1160–1175
13.   Fisher, D. and Nurse, P. (1995) Semin. Cell Biol. **6**, 73–78
14.   Hata, S., Kouchi, H., Suzuka, I. and Ishii, T. (1991) EMBO J. **10**, 2681–2688
15.   Hirt, H., Mink, M., Bögre, L., Györgyey, J., Jonak, C., Gartner, A., Dudits, D. and Heberle-Bors, E. (1992) Plant Cell **4**, 1531–1538
16.   Ferreira, P., Hemerly, A., Engler, J.D., Bergounioux, C., Burssens, S., Van Montagu, M., Engler, G. and Inzé, D. (1994) Proc. Natl. Acad. Sci. U.S.A. **91**, 11313–11317
17.   Renaudin, J.P., Colasanti, J., Rime, H., Yuan, Z. and Sundaresan, V. (1994) Proc. Natl. Acad. Sci. U.S.A. **91**, 7375–7379
18.   Soni, R., Carmichael, J.P., Shah, Z.H. and Murray, J.A.H. (1995) Plant Cell **7**, 85–103
19.   Dahl, M., Meskiene, I., Bogre, L., Ha, D.T.C., Swoboda, I., Hubmann, R., Hirt, H. and Heberle-Bors, E. (1995) Plant Cell **7**, 1847–1857
20.   Minshull, J., Blow, J.J. and Hunt, T. (1989) Cell **56**, 947–956
21.   Swenson, K.I., Farrell, K.M. and Ruderman, J.V. (1986) Cell **47**, 861–670
22.   Pagano, M., Pepperkok, R., Verde, F., Ansorge, W. and Draetta, G. (1992) EMBO J. **11**, 961–971
23.   Lehner, C. and O'Farrell, P. (1990) Cell **56**, 957–968
24.   Pines, J. and Hunter, T. (1990) Nature (London) **346**, 760–763
25.   Stewart, E., Kobayashi, H., Harrison, D. and Hunt, T. (1994) EMBO J. **13**, 584–594
26.   Peeper, D.S., Parker, L.L., Ewen, M.E., Toebes, M., Hall, F.L., Xu, M., Zantema, A., van der Eb, A.J. and Piwnica-Worms, H. (1993) EMBO J. **12**, 1947–1954
27.   King, R.W., Jackson, P.K. and Kirschner, M.W. (1994) Cell **79**, 563–571
28.   Devault, A., Fesquet, D., Cavadore, J.C., Garrigues, A.M., Labbé, J.C., Lorca, T., Picard, A., Philippe, M. and Dorée, M. (1992) J. Cell Biol. **118**, 1109–1120

29. Girard, F., Strausfeld, U., Fernandez, A. and Lamb, N.J.C. (1991) Cell **67**, 1169–1179
30. Zindy, F., Lamas, E., Chenivesse, X., Sobczak, J., Wang, J., Fesquet, D., Henglein, B. and Bréchot, C. (1992) Biochem. Biophys. Res. Commun. **182**, 1144–1154
31. Pines, J. and Hunter, T. (1991) J. Cell Biol. **115**, 1–17
32. Maridor, G., Gallant, P., Golsteyn, R. and Nigg, E.A. (1993) J. Cell Sci. **106**, 535–544
33. Sobczak-Thepot, J., Harper, F., Florentin, Y., Zindy, F., Brechot, C. and Puvion, E. (1993) Exp. Cell Res. **206**, 43–48
34. Lees, E., Faha, B., Dulic, V., Reed, S. and Harlow, E. (1992) Genes Dev. **6**, 1874–1885
35. Shirodkar, S., Ewen, M., DeCaprio, J., Morgan, D., Livingston, D.M. and Chittenden, T. (1992) Cell **68**, 157–166
36. Devoto, S.H., Mudryj, M., Pines, J., Hunter, T. and Nevins, J.R. (1992) Cell **68**, 167–176
37. Li, L.J., Naeve, G.S. and Lee, A.S. (1993) Proc. Natl. Acad. Sci. U.S.A. **90**, 3554–3558
38. Xu, M., Sheppard, K.A., Peng, C.Y., Yee, A.S. and Piwnica-Worms, H. (1994) Mol. Cell. Biol. **14**, 8420–8431
39. Kitagawa, M., Higashi, H., Suzukitakahashi, I., Segawa, K., Hanks, S.K., Taya, Y., Nishimura, S. and Okuyama, A. (1995) Oncogene **10**, 229–236
40. Obara-Ishihara, T. and Okayama, H. (1994) EMBO J. **13**, 1863–1872
41. Conolly, T. and Beach, D. (1994) Mol. Cell. Biol. **14**, 768–776
42. Bueno, A., Richardson, H., Reed, S.I. and Russell, P. (1991) Cell **66**, 149–159
43. Epstein, C.R. and Cross, F. (1992) Genes Dev. **6**, 1695–1717
44. Nugent, J.H.A., Alfa, C.E., Young, T. and Hyams, J.S. (1991) J. Cell Sci. **99**, 669–674
45. Bai, C., Richman, R. and Elledge, S.J. (1994) EMBO J. **13**, 6087–6098
46. Kraus, B., Pohlschmidt, M., Leung, A.L.S., Germino, G.G., Snarey, A., Schneider, M.C., Reeders, S.T. and Frischauf, A.M. (1994) Genomics **24**, 27–33
47. O'Farrell, P. and Leopold, P. (1991) Cold Spring Harbor Symp. Quant. Biol. **56**, 83–92
48. Brown, N.R., Noble, M.E.M., Endicott, J.A., Garman, E.F., Wakatsuki, S., Mitchell, E., Rasmussen, B., Hunt, T. and Johnson, L.N. (1995) Structure **3**, 1235–1247
49. Jeffrey, P.D., Russo, A.A., Polyak, K., Gibbs, E., Hurwitz, J., Massagué, J. and Pavletich, N.P. (1995) Nature (London) **376**, 313–320
50. Kobayashi, H., Stewart, E., Poon, R., Adamczewski, J.P., Gannon, J. and Hunt, T. (1992) Mol. Biol. Cell **3**, 1279–1294
51. Lees, E.M. and Harlow, E. (1993) Mol. Cell. Biol. **13**, 1194–1201
52. Zheng, X.F. and Ruderman, J.V. (1993) Cell **75**, 155–164
53. Bazan, J.F. (1996) Proteins **24**, 1–17
54. Glotzer, M., Murray, A.W. and Kirschner, M.W. (1991) Nature (London) **349**, 132–138
55. Swofford, D.L. (1990) PAUP: Phylogenetic analysis using parsimony, version 3.0. Computer program distributed by the Illinois National History Survey, Champaign, Illinois
56. Felsenstein, J. (1993) PHYLIP (Phylogeny Inference Package), version 3.5c. Computer program distributed by the author, Department of Genetics, University of Washington, Seattle.
57. Philippe, H. (1993) Nucleic Acids Res. **21**, 5264–5272
58. Saitou, N. and Nei, M. (1987) Mol. Biol. Evol. **4**, 406–425
59. Felsenstein, J. (1985) Evolution **39**, 783–791
60. Sauter, M., Mekhedov, S.L. and Kende, H. (1995) Plant J. **7**, 623–632
61. Fobert P.R., Coen, E.S., Murphy, G.J.P. and Doonan, J.H. (1994) EMBO J. **13**, 616–624
62. Swofford, D.L. and Olsen, G.J. (1990) in Molecular Systematics (Hillis, D.M. and Moritz, C., eds.), pp. 411–501, Sinauer Associates Inc., Sunderland, MA
63. Philippe, H., Chenuil, A. and Adoutte, A. (1994) Development **120**, Suppl., 15–25
64. Fitch, W.M. and Markowitz, E. (1970) Biochem. Genet. **4**, 579–593
65. Fitch, W.M. (1971) Biochem. Genet. **5**, 231–241
66. Philippe, H. and Adoutte, A. (1996) in Aspect of the Genesis and Maintenance of Biological Diversity (Clobert, J. and Barbault, R., eds.), pp. 41–59, Oxford University Press, Oxford
67. Philippe, H., Sörhannus, U., Baroin, A., Perasso, R., Gasse, F. and Adoutte, A. (1994) J. Evol. Biol. **7**, 247–265
68. Felsenstein, J. (1978) Syst. Zool. **27**, 401–410
69. Conway Morris, S. (1993) Nature (London) **361**, 219–225

70.  Herendeen, P. and Crane, P.R. (1995) in Monocotyledons: Systematics and Evolution (Rudall, P., Cribb, P.J., Cutler, D.F. and Humphries, C.J., eds.), pp. 1–21, Royal Botanic Gardens, Kew
71.  Schlegel, M. (1994) Trends Ecol. Evol. **9**, 330–335
72.  Baldauf, S.L. and Palmer, J.D. (1993) Proc. Natl. Acad. Sci. U.S.A. **90**, 11558–11562
73.  Kuma, K.-I., Nikoh, N., Iwabe, N. and Miyata, T. (1995) J. Mol. Evol. **41**, 238–246
74.  Bai, C., Richman, R. and Elledge, S.J. (1994) EMBO J. **13**, 6087–6098
75.  Kreutzer, M.A., Richards, J.P., De Silva-Udawatta, M.N., Temenak, J.J., Knoblich, J.A., Lehner, C.F. and Bennett, K.L. (1995) J. Cell Sci. **108**, 2415–2424
76.  Kouchi, H., Sekine, M. and Hata, S. (1995) Plant Cell **7**, 1143–1155
77.  Lorca, T., Devault, A., Colas, P., VanLoon, A., Fesquet, D., Lazaro, J.B. and Dorée, M. (1992) FEBS Lett. **306**, 90–93
78.  Raikhel, N. (1992) Plant Physiol. **100**, 1627–1632
79.  Pines, J. and Hunter, T. (1994) EMBO J. **13**, 3772–3781
80.  Gallant, P. and Nigg, E.A. (1992) J. Cell Biol. **7**, 213–224
81.  Ookata, K., Hisanaga, S., Okano, T., Tachibana, K. and Kishimoto, T. (1992) EMBO J. **11**, 1763–1772
82.  Jackman, M., Firth, M. and Pines, J. (1995) EMBO J. **14**, 1646–1654
83.  Furuya, M. (1984) Annu. Rev. Plant Physiol. **35**, 349–373
84.  Jacobs, T. (1992) Dev. Biol. **153**, 1–15
85.  Shaul, O., Van Montagu, M. and Inzé, D. (1996) Crit. Rev. Plant Sci., **15**, 97–112
86.  Setiady, Y.Y., Sekine, M., Hariguchi, N., Yamamoto, T., Kouchi, H. and Shinmyo, A. (1995) Plant J. **8**, 949–957
87.  Meskiene, I., Bogre, L., Dahl, M., Pirck, M., Ha, D.T.C., Swoboda, I., Heberle-Bors, E., Ammerer, G. and Hirt, H. (1995) Plant Cell **7**, 759–771
88.  Price, C.A., Reardon, E.M. and Lonsdale, D.M. (1996) Plant Mol. Biol. **30**, 225–227
89.  Shaul, O., Mironov, V., Burssens, S., Van Montagu, M. and Inzé, D. (1996) Proc. Natl. Acad. Sci. U.S.A. **93**, 4868–4872
90.  Savouré, A., Feher, A., Kalo, P., Petrovics, G., Csanadi, G., Szecsi, J., Kiss, G., Brown, S., Kondorosi, A. and Kondorosi, E. (1995) Plant Mol. Biol. **27**, 1059–1070
91.  Hemerly, A., Bergounioux, C., Van Montagu, M., Inzé, D. and Ferreira, P. (1992) Proc. Natl. Acad. Sci. U.S.A. **89**, 3295–3299
92.  Szarka, S., Fitch, M., Schaerer, S. and Moloney, M. (1995) Plant Mol. Biol. **27**, 263–275
93.  Day, I.S., Reddy, A.S.N. and Golovkin, M. (1996) Plant Mol. Biol. **30**, 565–575
94.  Qin, L.X., Richard, L., Perennes, C., Gadal, P. and Bergounioux, C. (1995) Plant Physiol. **108**, 425–426
95.  Logemann, E., Wu, S.C., Schröder, J., Schmelzer, E., Somssich, I.E. and Hahlbrock, K. (1995) Plant J. **8**, 865–876
96.  Renaudin, J.P., Doonan, J.H., Freeman, D., Hashimoto, J., Hirt, H., Inzé, D., Jacobs, T., Kouchi, H., Rouzé, P., Sauter, M., Savouré, A., Sorrell, D.A., Sundaresan, V. and Murray, J.A.H. (1996) Plant Mol. Biol. **32**, 1003–1018
97.  Reichheld, J.P., Chaubet, N., Shen, W.H., Renaudin , J.P. and Gigot, C. (1996) Proc. Natl. Acad. U.S.A. **93**, 13819–13824

# 5

# Plant D cyclins and retinoblastoma protein homologues

**James A.H. Murray\*, Donna Freeman, Judith Greenwood, Rachael Huntley, Joe Makkerh, Catherine Riou-Khamlichi, David A. Sorrell, Claire Cockcroft, Jeremy P. Carmichael†, Rajeev Soni‡ and Zahid H. Shah§**

Institute of Biotechnology, University of Cambridge, Tennis Court Road, Cambridge, CB2 1QT, UK

## Introduction

All eukaryotic cells undergo the same sequential series of events when they divide, and the term 'cell cycle' reflects the ordered nature and universality of these observations. The benefit of the eukaryotic cell cycle is that DNA replication (S) and cell division (M) are temporally separated by 'gap' phases ($G_1$ and $G_2$) in the sequence $G_1$–S–$G_2$–M. This arrangement allows entry to the critical processes of DNA replication and mitosis to be precisely controlled, and it is not surprising that underlying the cytological events of the cell cycle is an ordered series of temporally and spatially organized molecular and cellular processes which define the direction and order of the cycle.

In 1987, Lee and Nurse showed that human cells contain a homologue of the yeast Cdc2 kinase [1], and the following 10 years have emphasized the conservation of cell cycle controls between yeasts and mammals, particularly in regulating the entry to mitosis. It is therefore perhaps not surprising that plants contain many of the same conserved regulators, and are therefore likely to use similar mechanisms to control mitosis.

The yeast Cdc2 kinase has become the paradigm for a family of related proteins, the cyclin-dependent kinases (Cdks), which are characterized by a requirement for the binding of a regulatory subunit, known as a cyclin, for catalytic activity to be obtained [2]. The current view of the cell cycle in all eukaryotic organisms is essentially that it is driven forward by the sequential activation and destruction of Cdk kinase activities, affording Cdks and cyclins central roles in regulating cell cycle commitment and progression [3]. Regulation is in part

*\*To whom correspondence should be addressed.*
*Present addresses: †Biotechnology and Innovation Centre, School of Applied Biological and Chemical Sciences, University of Ulster, Colraine, BT52 1SA; ‡Lidak Pharmaceuticals, 11077 N. Torrey Pines Road, La Jolla, CA 92037, U.S.A.; §Institute of Medical Technology, University of Tampere, P.O. Box 607, Tampere, FIN-33101, Finland*

conferred by the synthesis and destruction of the Cdk and cyclin components, and further controls may be imposed by the binding of inhibitory proteins to the individual components or to the intact complex [4,5] and by activating and inhibitory phosphorylations on the Cdk molecule. In particular, all known Cdks are activated by phosphorylation of a conserved threonine (the 'T-loop threonine') at around position T160, which in many cases is known to be essential for Cdk–cyclin complexes to exhibit kinase activity.

In this chapter, we focus on molecules which regulate the exit from $G_1$ and entry to S phase in plants. This period includes not only the point of commitment to cell division, but may also represent the time during which differentiation decisions are made. Although there are parallels between yeast and mammalian controls operating during this stage in the cell cycle, not all the processes and proteins involved are functionally equivalent between the two systems. We discuss recent work suggesting that the proteins involved in plant $G_1$/S controls are closer to those of mammals than they are to their equivalents in fungi. Proliferative and differentiation decisions may therefore be made in analogous ways in all higher eukaryotes, both animal and plant.

## When to start and stop a cycle

The term 'cell cycle' has unsatisfactory aspects, since it tends to imply a view of cell division as an ever-rolling and repeatedly reiterated cycle of sequential events. While this is perhaps true of exponentially growing cell suspension cultures, it may not have much biological relevance *in vivo*. In all but single-celled organisms, cells constitute the component parts of organs, and have defined spatial relationships with their neighbours in creating higher-order structures. Cells must 'know' when they are required to divide, and when division must cease or be modified, to allow differentiation into specialized cell types to occur (see D. Francis, Chapter 9).

This 'knowledge' requires the integration of information from several sources. The extracellular environment contributes information on nutrient availability and environmental conditions. The developmental and positional context of a cell within higher-order structures is clearly of vital importance in determining when continuing divisions are appropriate. These various types of information must be transduced into the intracellular environment, and integrated with further controls which sense factors that preclude cell division, such as DNA damage.

These various factors may in principle impinge on the cell cycle at any point. To a certain extent this is true, since checkpoints operate at various stages to ensure completion of one phase before the next is initiated [6,7]. In general, however, checkpoints represent safe resting points for cells which have failed to execute a process, rather than the primary decision to divide unaltered, to differentiate or to cease division. By analogy with mammalian cells and yeast, this primary control point may be predicted to operate during $G_1$ in plants. Indeed, there is classical evidence suggesting that the $G_1$ control point is more stringent in cell suspension cultures deprived of nutrients [8,9], and cultures enter stationary phase with cells in $G_1$.

In intact plants, many differentiated plant cells also have a $G_1$ (2C) DNA content, indicating exit from the cell cycle from the $G_1$ state. However, significant proportions of cells cease division with a 4C DNA content, and can therefore be interpreted as undergoing $G_2$ arrest (the proportion of such cells is species- and tissue-specific). The status of these 4C cells is complicated by the frequency of endoreduplication in differentiated plant cells [9,10], and such 4C cells may therefore represent cells in $G_1$ that have undergone an endoreduplication event, rather than $G_2$ arrest [9]. This question remains largely unresolved, but the $G_1/G_2$ alternatives may in principle be distinguished by whether or not such cells proceed through mitosis without an intervening S phase. Nevertheless, it is likely that in most cells the most important decision to divide or differentiate operates in late $G_1$ before S-phase onset.

There is, therefore, circumstantial and correlative evidence suggesting that $G_1$ control is of primary importance in proliferation and differentiation decisions in plants, and our studies have focused on the molecular basis of regulation operating at this point. Here, we outline the controls that operate in mammals and yeast at this period of the cell cycle, and review the isolation and analysis of genes likely to have a role in plants. Finally, we discuss our current view of regulatory mechanisms in $G_1/S$, and the necessity for these to be integrated into our understanding of meristem function.

## How to start and stop a mammalian or yeast cycle

In this section we briefly outline the current view of how the $G_1/S$ transition and the commitment to cell division are controlled in yeast and mammalian cells. For more detailed discussions, refer to recent reviews [2,3,11,12].

### The control of $G_1/S$ transition in budding yeast

Among the yeasts, the controls that operate during $G_1$ and into early S phase are best understood in budding yeast *Saccharomyces cerevisiae*. After a point defined as START, cells become committed to a round of division, and cells can no longer respond to signals promoting alternative pathways, such as mating in haploids or sporulation (meiosis) in diploids. In the next section, we will see that this theme of a commitment point, before which different cell fates are possible, but after which the cell must progress into a round of division, is also relevant to mammalian cells.

The molecular event that underlies START is the onset of a positive feedback loop that leads to activation of CDC28, the budding yeast Cdk that is involved in both $G_1/S$ and $G_2/M$ controls. There are three principal $G_1$ cyclins associated with CDC28 during $G_1$, called Cln1, Cln2 and Cln3 [13]. Before START,

**Fig. I**           **G$_1$/S control in budding yeast and mammals**

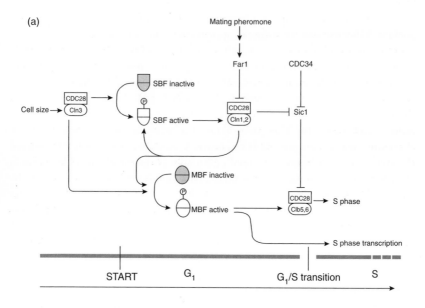

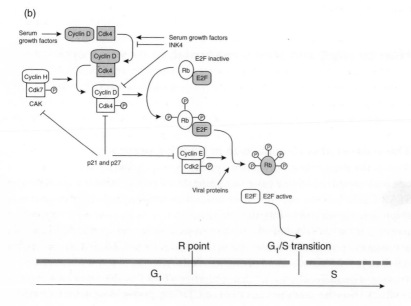

Cln3 is probably expressed in a cell-size-dependent manner [14], and is relatively constitutive, whereas Cln1 and Cln2 are strongly regulated by transcription [15]. Cln3 expression leads to a low level of CDC28 kinase activity, and is responsible for activating expression of the two cyclins Cln1 and Cln2 at START. These act in a positive feedback loop, activating their own transcription, by phosphorylation of the transcription factor complex SBF (Fig. 1$a$). This makes START irreversible [16,17]. START does not correspond to the onset of S phase, since this requires the expression of further genes under the control of a second transcription factor complex called MBF, which shares a subunit with SBF [18] (see P. Sabelli et al., Chapter 12).

During late (i.e. post-START) $G_1$, two further cyclins, Clb5 and Clb6, are expressed, which are required for S phase (Fig. 1$a$). Clb5 and Clb6 expression are also activated in a positive feedback loop, this time involving MBF [19,20]. However, the CDC28–Clb5 complex is initially inactive because it is bound to the Cdk inhibitor, Sic1 [21]. At S-phase entry, Sic1 is phosphorylated, triggering its destruction by the ubiquitin pathway, a process dependent on CDC34 [22]. It is noteworthy that Sic1 may be the only essential substrate of Cln kinases, since deletion of Sic1 rescues the otherwise lethal deletion of all three *CLN* genes [22].

---

**Fig. I (Contd.)**

*(a) $G_1$/S control in budding yeast. The two principal signals for division are cell size, to which Cln3 levels respond, and the presence of mating pheromones, which induce the Cdk inhibitor Far1 through a MAP kinase cascade (not shown). At START, transcription of the Cln1 and Cln2 cyclins is activated, and these operate with the transcription factor SBF in a positive feedback loop, whereby SBF is activated by phosphorylation, and Cln1 and Cln2 transcription is stimulated by SBF. Clb5 and Clb6 cyclins are then transcribed, and operate in a similar feedback loop with the transcription factor MBF. The cdk inhibitor Sic1 blocks the kinase activity of cdc28–clb5/6 until Sic1 is destroyed, a process which requires Cdc34, a ubiquitin ligase. Inactive forms of proteins and complexes are stippled, active forms are unshaded. The activation of Cdk–cyclin complexes by T-loop threonine phosphorylation is not shown, since CAK activity is not well understood in yeast. See text for further details. Adapted from [2,3]. (b) $G_1$/S control in mammals. Cyclin D transcription is activated by the presence of serum growth factors. Cyclin D binding to CDK4 is inhibited by the Cdk inhibitor INK4, and requires serum growth factors, since there is no activation in unstimulated cells that constitutively express cyclin D and CDK4 [85]. Phosphorylation of the T-loop threonine in CDK4 is required for kinase activity, which is carried out by CAK, itself composed of cdk7 and cyclin H. It is unclear whether CAK assembly or activity is regulated, although it is believed to be inhibited by p21 and p27, Cdk inhibitors which also inhibit all $G_1$ Cdks. Active cyclin D-CDK4 phosphorylates Rb, which during early $G_1$ binds and inactivates E2F. Cyclin E-CDK2 activity rises rapidly in late $G_1$ and contributes to hyperphosphorylation of Rb, which releases E2F, causing S phase to initiate. Viral proteins such as SV40 T antigen and adenovirus E1A cause premature E2F release and drive cells into S phase. Not shown is that cyclin E-CDK2 also requires T-loop phosphorylation by CAK for activity. Inactive forms of proteins and complexes are stippled, active forms are unshaded. See text for further details. Adapted from [2,3,5].*

Several key themes can be distilled from this pathway of events: first, the essential involvement of Cdk activity; secondly, the primary regulation at the level of cyclin transcription, particularly in START itself; thirdly, the important involvement of transcription factors, which are directly under the control of Cdk–cyclin complexes; fourthly, the key significance of Cdk inhibitors. Finally the separation of the commitment point (START) from S-phase entry. These themes are reiterated in mammalian cells.

## Mammalian $G_1$/S controls

In mammalian cells, the controls operating in $G_1$ are analogous rather than homologous to those of yeast (Fig 1b). Classical studies have defined a point of commitment called the restriction point or R [23], after which cells no longer require the presence of serum growth factors for progression to DNA replication and completion of the cycle. Partly because of the reduced synchrony levels that can be obtained in mammalian cell cultures compared with yeast, the molecular definition of R is somewhat hazy, although a consensus view is that it corresponds to the activation of cyclin-D-associated kinase activity. This suggests that D-type cyclins in mammals play an analogous role to the Clns in yeast.

Three D-type cyclins have been identified in mammals, originally by complementation of yeast strains engineered to be conditional for *CLN* gene function [24,25], or as mRNAs upregulated by growth factor addition [26]. Cyclin D1 was also identified as the oncogene *PRAD1* [27]. D-type cyclins are believed to act as sensors of the presence of growth factors [28], a proposition based on their short half-lives and on the absolute dependence of their transcription on the presence of serum growth factors [26]. D-type cyclins are strongly upregulated when serum-starved or quiescent cells are re-stimulated by serum addition, and on removal of serum growth factors their mRNA levels decline very rapidly (Fig. 1b).

The principal substrate of cyclin D–CDK4 kinase activity is the product of the retinoblastoma-susceptibility (Rb) gene, or more accurately the family of Rb-related proteins, which currently has three members: Rb, p107 and p130 [29]. Rb itself was first identified as a tumour-suppressor gene inactivated in familial childhood retinoblastomas, and was also found to be a cellular target of the transforming proteins of DNA tumour viruses. The Rb family of proteins have related, but distinct, functions in controlling $G_1$ exit. Targeted mutagenesis of the genes in transgenic mice suggests some degree of functional overlap may occur, since double mutants lacking two genes of the Rb family have a more extreme embryonic lethal phenotype than any single mutant.

Rb proteins are phosphorylated by cyclin D–CDK4 kinase at R. CDK4 is targeted to Rb as a consequence of a specific interaction between D-type cyclins and the 'pocket domain' of Rb. The binding of cyclin D to Rb depends on the motif LxCxE (single amino-acid code; x = any amino-acid) found at the N-terminus of cyclin D [30,31]. The same sequence motif is found in transforming proteins encoded by viruses such as SV40 (T antigen), adenovirus (E1A) and papillomavirus

(HPV16 E7), which also bind Rb and abrogate its growth-suppressing properties as part of the process of cellular transformation [32,33].

Rb, sometimes caricatured as 'the Guardian of the Cell', is therefore a negative regulator of cell proliferation, whereas cyclin D and CDK4 promote cell division by phosphorylation of the Rb proteins. During most of $G_1$, Rb is in its active, unphosphorylated state, and ensures cells remain in $G_1$ by sequestering a family of transcription factors known as E2F, actually heterodimers of E2F and DP family members [29,34]. E2F activity is required for expression of S-phase genes, such as DNA and precursor biosynthetic enzymes. Overexpression of E2F will activate DNA synthesis, and it is the only protein whose overexpression is in itself sufficient to drive cells into S phase [35] (see P. Sabelli et al., Chapter 12).

In the normal course of events, hyperphosphorylation of Rb on multiple sites in late $G_1$ leads to Rb inactivation, release of active E2F and hence the triggering of S phase. This process is begun by cyclin D–CDK4, but somewhat later than cyclin D–CDK4 activation is a burst of synthesis of cyclin E during late $G_1$ and early S phase [36]. Cyclin E associates with another Cdk, CDK2, and the peak of cyclin E–CDK2 kinase activity coincides with the onset of S phase [36]. Since Rb is also a substrate of cyclin E–CDK2 [37], the phosphorylation of Rb, started by cyclin D–CDK4, may be completed by cyclin E–CDK2 [29], and cyclin E–CDK2 is therefore ultimately responsible for triggering DNA synthesis (Fig. 1$b$).

The binding of cyclin D proteins to their CDK partners, principally CDK4, is inhibited by the Cdk inhibitor INK4, which is classified as a tumour-suppressor gene, and acts upstream of R [5]. Two other principal types of inhibitor operate during $G_1$, which inhibit both CDK4 and CDK2 kinase activity [5]. p21 inhibitor is induced by p53 in response to DNA damage, whereas p27 provides the means by which transforming growth factor β (TGFβ) exerts its growth inhibitory effects. p21 and p27 also inhibit the activity of CAK [5], which is required for T-loop phosphorylation, and hence catalytic activity of cyclin D–CDK4 and cyclin E–CDK2. CAK is itself composed of CDK7 and cyclin H, but is unclear if the synthesis of these subunits is regulated.

The general themes of yeast $G_1$/S control (see 'The control of $G_1$/S transition in budding yeast', pp. 101–104) also apply in mammalian cells, although the details are different. Nevertheless, parallels exist between mammalian and yeast $G_1$ controls, particularly in the activation of S phase by the transcription factors E2F in mammalian cells and MBF in yeast. Both of these are heterodimeric complexes, but there is no sequence similarity between the mammalian and yeast proteins, so their evolutionary relationship is unclear.

## Signal integration and the regulation of pathways of differentiation and proliferation

There are a number of lines of evidence supporting the idea that D-type cyclins, and their immediate downstream target Rb, play an important role in the switch between proliferation and differentiation [2,3]. First, D-type cyclins can act as oncogenes, presumably because their constitutive expression can bypass the normal requirement for correct growth factor signals for cell proliferation [27,38,39]. There is also evidence that constitutive cyclin D expression prevents the correct response to differentiation signals [40], and that CDK4 needs to be downregulated for cell differentiation to occur [41,42]. In tissue culture, overexpression of Rb blocks cell division, and Rb is normally highly expressed in cells *in vivo* as they undergo differentiation into specialized cell types [43]. Removal of Rb function by targeted mutagenesis in transgenic mice results in lethality relatively late in embryonic development, and such mutant embryos exhibit a failure of terminal differentiation in both haematopoietic and neuronal lineages [44,45]. Finally, in differentiating muscle cells, the factor MyoD induces both muscle-specific gene expression and the Cdk inhibitor p21, which acts to block cyclin D and cyclin-E-associated kinase activity and hence $G_1$ progression [46–48].

These various lines of evidence point to D-type cyclins and Rb acting as integrators of both positive and negative signals for cell proliferation [28,29]. Put another way, D-type cyclins are an important mechanism by which the information from external signals is transduced into mammalian cell cycle control. These signals may be proliferative and promote the commitment to division (R point transit). On the other hand, they may stimulate differentiation, probably by the dual mechanism of slowing $G_1$ progression, accompanied by the concomitant activation of specific sets of genes that cause alternative pathways to be followed.

## Isolation of plant D-type cyclins

The above discussion suggests that in yeasts and mammals the $G_1/S$ control point is of primary importance for understanding both cellular proliferation and differentiation. These issues are also of significance in plants, where the majority of post-embryonic division activity is concentrated in the meristems, particularly at the primary root and shoot apices. The regulation of meristem activity and of the continued pattern of organogenesis that characterizes plant growth, as well as the molecular differences between meristematic cells and nearby differentiated progeny, are currently areas of significant research effort. Moreover, many plant cells can de-differentiate in response to wounding, pathogen attack or exogenous application of plant hormones, suggesting more plastic controls over cell cycle exit and re-entry than are normally found in animals. Understanding the molecular basis of this reversibility is therefore of considerable interest, and may have an impact on areas such as regeneration and transformation of plant cultures, and the ability to manipulate plant growth characteristics.

We were therefore interested in investigating whether plant controls operating in $G_1$ were more related to those of yeasts or of mammals. As a first step, we adopted a yeast-complementation strategy to isolate $G_1$ cyclin homologues, paralleling the approach which led to the identification of the mammalian cyclins D and E [24,25].

In the yeast strain BF305-15d 21, the *CLN1* and *CLN2* genes have been disrupted, and the *CLN3* gene placed under galactose regulation [25]. This strain is therefore dependent on galactose for growth, and in glucose-containing media undergoes $G_1$ arrest. Using a plasmid library of *Arabidopsis* cDNAs in a yeast expression vector [49] (kindly donated by F. Lacroute, Gif-sur-Yvette), more than $3 \times 10^6$ transformants of the yeast strain were screened for clones capable of converting BF305 to galactose-independent growth, as previously described in detail [50,51].

The cDNAs obtained fell into four groups, three of which showed greater homology to mammalian D-type cyclins than to any other protein in the database. However, their identity to the mammalian proteins was low (about 10%; Table 1), but further conserved features of the plant cyclins subsequently confirmed their identification (see below). We initially named the three groups of plant D cyclins isolated from *Arabidopsis* δ cyclins, to indicate their greater sequence similarity to mammalian D cyclins than to any other cyclin group [50]. The three groups of plant D cyclins have, however, recently been renamed CycD1, CycD2 and CycD3 as part of a wider reorganization of the nomenclature of plant cyclins (see J.P. Renaudin et al., Chapter 4, and [52]). Two independent cDNAs were isolated for CycD2 and more than five for both CycD1 and CycD3.

In addition to the three *Arabidopsis* CycD cyclins described above, a CycD3 cyclin was isolated by Hirt and co-workers from the legume alfalfa by

| Table 1 | Homology of plant CycD cyclins |
| --- | --- |

This table shows the percentage similarity between the positions indicated in the alignment of Fig. 3 (corresponding approximately to the cyclin core) among the cyclins from one given subgroup (calculated from the number of conserved residues in each consensus sequence of Fig. 4), or between the three subgroups of plant cyclins and animal D-type cyclins. The percentage similarity within a group is calculated by counting the number of positions at which there is a consensus shown in Fig. 3. A consensus is defined when at least 80% of sequences in the group have identical residues. The percentage similarity between groups is calculated by counting the positions at which the consensus for the two groups are identical. The data for CycD2 are based on comparison with the CycD2;At sequence with the other consensus sequences.

| | CycD1 | CycD2 | CycD3 |
| --- | --- | --- | --- |
| CycD1 | 45 | | |
| CycD2 | 24 | — | |
| CycD3 | 20 | 27 | 49 |
| Cyclin D | 9.4 | 14 | 12 |

complementation of a similar yeast strain [53] (see H. Hirt et al., Chapter 6). Using probes derived from the *Arabidopsis* genes, further CycD cDNAs belonging to the CycD1 and CycD3 groups have been isolated from *Antirrhinum* (J.H. Doonan, Chapter 10; Gaudin et al., unpublished work) and *Helianthus* (J.A.H. Murray et al., unpublished work). Interestingly, two CycD3 homologues have been found in *Antirrhinum* which show differential expression patterns in meristems (see J.H. Doonan, Chapter 10).

## Characteristics and sequence relationships of plant D cyclins

The existence of a number of sequences for plant D cyclins has allowed the relationships between groups and tentative characteristic features of the CycD1 and CycD3 groups to be defined. Although only a single CycD2 sequence (from *Arabidopsis*) has been reported to date, it appears to form a distinct group as previously reported. The currently available CycD plant cyclins therefore fall into the three groups shown in Fig. 2, an analysis which is supported by a number of algorithms [52]. Fig. 3 presents consensus sequences for cyclin D homologues from plants and animals.

### LxCxE motifs

A defining hallmark of all cyclin D proteins is the presence of an LxCxE Rb interaction motif near their N-terminus. In mammalian D cyclins, this motif is within a few amino acids of the initiation methionine. In CycD2, this is also the case, but in CycD1, the leucine residue lies variously between positions 12–22, and in CycD3 is located between positions 15–24. Within the CycD1 and CycD3 groups, it is apparent that the LxCxE motif is embedded in a wider region of conservation that is specific for each group. Thus, the consensus of the three CycD1 sequences in this region is: MSxS(C/f)S(D/n)(C/d)(F/m)(S/-)D**L**(L/f)**C**(G/c)**E**DS (L-C-E of Rb motif is shown in bold, and the other residues shown are either invariant or the only alternatives found. Minority residues are shown in lower case, and an insertion by -). In two of the three sequences, the initial M represents the N-terminal methionine of the protein. The five cloned members of the CycD3 group are less highly conserved in this region and show the consensus: Sxx(F/L)D(A/t)L(Y,l,f)C(E,n,d)E(E/q). It is unknown whether these differences might reflect different affinities for an Rb-like protein, or might indicate the presence of multiple Rb homologues in plants.

### Cyclin box

All cyclins contain a homologous region of about 100 amino acids known as the cyclin box, which is involved in interactions with the Cdk partner. The structure of this region in mitotic cyclins is discussed in more detail by J.P. Renaudin et al.,

## Fig. 2        Relationships of plant CycD cyclins

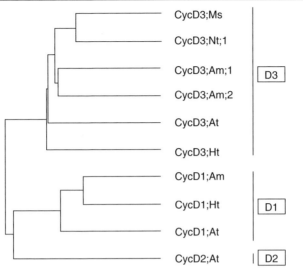

The relationship of cloned CycD cyclins was calculated by generating aligned sequences using the program PILEUP in the GCG package. These alignments were then manually edited to optimize the matches for known conserved residues. The graphical representation shown was calculated for these edited alignments by running PILEUP again with parameters chosen that prevent further changes to the relative alignments. The cDNAs cloned to date fall into the three groups originally defined in Arabidopsis, namely CycD1, CycD2 and CycD3 [50]. The nomenclature is in accordance with that used by Renaudin et al. [52]. The species abbreviations and EMBL accession numbers (where available) are Am: Antirrhinum majus (Gaudin et al., unpublished work); At: Arabidopsis thaliana (CycD1; At: X83369, CycD2; At: X83370, and CycD3; At: X83371, originally called δ1, δ2 and δ3, respectively) [50]; Ht: Helianthus tuberosus (D. Freeman and J..A.H. Murray, unpublished work); Ms: Medicago sativa (CycD3; Ms: X88864; originally called CycMs4) [53].

Chapter 4.) The corresponding region in D cyclins has relatively low homology to mitotic cyclins, but nine residues are invariant between the cyclin box of all cyclins of the A, B and D classes (animal and plant), including five residues shown experimentally to be essential for catalytic activity (Fig. 3). The structure of cyclin A has now been solved [54,55], and it is apparent that the cyclin box is embedded in a somewhat larger region of homology referred to as the cyclin core (see Figs. 1 and 8 in chapter 4 by J.P. Renaudin et al.). The approximate equivalent of the cyclin core is indicated on Fig. 3, and it was used to calculate the relatedness figures shown in Table 1.

The cyclin box is normally regarded as starting with the consensus sequence MRxIL in mitotic cyclins, which corresponds to helix H1 of the cyclin

**Fig. 3**              **The consensus sequences of plant and animal D cyclins**

```
        1                                           |< sequence used for % homology        100
CycD1   MS.S.S...DL.C.EDS.............................D..........D..SIA...E.ER...G........Q...........
CycD2   ......MAENLACGETSESWIIDNDDDDINYGGGFTNEIDYNHQLFAKDDNFGGNGSIPMMGSSSSSLSED.RIKEMLVREIEFCPGT..DYVKRLLSG
CycD3   .....S...DAL.C.EE...........................L...DL.........E.EEL..L..KE.EQ..............
CycD    ...........L.C.E...........................................E.....................
mamm D  ..........LLCCE............................RA.PD..LL...RVL...L..EE...P..SYF.CVQK.I.P.........
All D   ...........L.C.E...........................................E.....................

        101      *                          *                   *            *        * 200
CycD1   ..D.S.R..S.AWILKVQ..Y.FQPL.AYL.V.Y.DRF......P...GWP.QLL.VACLSLAAKMEE.L.PS..D.QV.G....FE...I.RME.LVL.
CycD2   DLDLSVRNQALDWILKVCAHYHFGHLCICLSMNYLDRFLTSYELPKDKDWAAQLLAVSCLSLASKMEETDVPHIVDLQVEDPKFVFEAKTIKRMELLVVT
CycD3   ..L...R.EAV.W.LKV...YGF..LTA.LA.NYLDRF....FQ.DKPWM.QL.AV.CLSLAAKVEETQVPLLLD.QVE..KYVFEAKTIQRMELL.LS
CycD    .......R.....W.LKV...Y.F..L.A.L...Y.DRF...........W..QL..V.CLSLAAK.EE..VP...D.QV...K..FE.K.I.RMELL...
mamm D  .....MR...A.WMLEVCEEQ.CEEEVFPLAMNYLDR.L...P..K.....LQLLGA.CM..ASK..ET..PLT.EKLCIYTD.....P..L...E.....
All D   .......R.....W.L.V............L...Y.DR..............QL...C...A.K..E...P...................E....
            ^                                          ^              ^^                ^  ^  ^

        201                                                                              300
CycD1   VLDWRLRS..PF....FF..KID.......G.L.S......ILS...E.S..EY.PS.IAAA..L..A..L...S.....H...E.WCDGL.K..I..C..
CycD2   TLNWRLQALTPFSFIDYFVDKISGHVSENLIYRSSR...FILNTTKAIEFLDFRPSEIAAAAAVSVSISGETECIDEEKALSSLIY..VKQERVKRCLN
CycD3   TLKWKM.PVTPISF.DHI.RRL.L.......F...CE...LS...D.RFV...PSV.A.ATM..VI...P......Y...L...L..K.K...CE.
CycD    .L.W.....TP..F.................L...................PS..A.A.................................K....C..
mamm D  KLKW.LAA...HDF...L...............KHAQTF.ALCAT.D..F...PPSM.A.GS..AA...GL.............LT..L.......DCL.
All D   .L.W.......F...................L...........PS..A...................................................C..
                                                   ^

        301      sequence for % homology >|                                             400
CycD1   L..................PK......RV.............S.....S..........S.S..K.RKL...S....................
CycD2   LMRSLTGEENVRGTSLSQEQARVAVRA..VPASPVGVLEATCL.SYR.SEERTVESCTNSSQSSPDNNNNNNNSKQEEEKTMRENKRVIHCLLQPKTTST
CycD3   LI.................KRK......P.SP.GVID....S...SN.SW........S...EP..KK.............
CycD    L.................K.................S...S.............S...........................
mamm D  ACQE.QIEA.L..SL....Q....................TPTDV..................................
All D   ..............................................................................
```

Alignments were generated as described in the legend to Fig. 2. Consensus sequences are presented for three CycD1 sequences, the single CycD2 sequence from Arabidopsis, and the five available CycD3 sequences. In these sequences, positions in bold indicate identical amino acids at this position between all members of the group. Positions showing at least 80% identity are shown in normal type. The complete sequence is shown of CycD2 for comparison. The CycD line shows positions at which at least 8/9 (normal type) or 9/9 CycD (bold type) sequences have identical residues. The 'mamm D' line shows positions at which at least 8/10 (normal type) or 10/10 (bold type) vertebrate cyclin D sequences have identical residues. The sequences for human, mouse and rat D1, D2 and D3, and chicken D2 were used. The 'all D' line shows positions at which all known cyclin D/CycD sequences have identical residues (bold), or more than 80% of sequences in each cyclin D/CycD group have identical residues (normal type). Underlined residues are also conserved (i.e. identical in ⩾80% of each group) in all A- and B-type cyclins from animals and yeast. Residues identical in every known animal, yeast and plant A-, B-, and D-type cyclins are indicated by ^, and residues shown to be essential for cyclin activity are indicated by asterisks above the sequence. The region (approximately equivalent to the cyclin core) used to calculate percentage homologies in Table 1 is also indicated. Dots indicate non-conserved positions or introduced gaps. Note that the corrected sequence of CycD2 from Arabidopsis is shown; two errors were located within the sequence originally submitted to the EMBL database (now corrected). The last three amino acids of CycD2 have been omitted.

fold. In D cyclins (plant and animal), however, it is only the invariant arginine that is conserved (MR in mammalian cyclin D; position 106 in Fig. 3). Further into the same helix, plant CycD cyclins have the characteristic W(I/M)LKV motif (position 113–117), which is also conserved in mammalian D-type cyclins, with the exception that the lysine is replaced by glutamate. In helix H2, the motif DRF (position 136–138) is conserved in many D and mitotic cyclins, the aspartate being a residue essential for cyclin function. The region corresponding to helix H3 (position 153–168) contains four of the nine residues invariant in all A, B and D cyclins, and is highly conserved in all CycD cyclin groups. The C-terminal end of helix H3 contains the conserved KYEE motif of mitotic cyclins, which is also present in CycD cyclins, except the tyrosine is replaced by a non-polar residue (M,V). All CycD3 sequences have a valine at this position, whereas all known CycD1 and CycD2 homologues have methionine.

The end of the cyclin box corresponds to helix H5 in the cyclin A structure, marked by MExx(I/V)L in mitotic cyclins. The corresponding consensus sequence in CycD cyclins is RMELLVL (positions 193–199), although only the ME residues are invariant. The cyclin core continues for approximately 120 amino acids over which there is scattered patches of homology between CycD cyclins (Fig. 3).

Overall, Table 1 shows that the percentage identity (defined as the proportion of positions at which 80% or more sequences have identical residues at a given position) is only 9–14% between plant CycD cyclins and the consensus sequence for mammalian D cyclins. Between plant CycD groups, percentage identity ranges from 20 to 27%, whereas within CycD groups it is 45% (CycD1 group) or 49% (CycD3).

## PEST sequences

A general feature of cyclins is that they are destroyed rapidly at certain points in the cell cycle, and the ability to activate this turnover at the correct time is central to cell cycle control. In the case of mitotic cyclins, a specific N-terminal sequence known as the destruction box targets the cyclins for specific ubiquitin-mediated destruction during mitosis [3]. The presence and types of destruction motifs in plant mitotic cyclins are discussed by J.P. Renaudin et al., Chapter 4.

$G_1$ cyclins in both yeast and mammals are short-lived proteins, and their rapid turnover depends on so-called PEST sequences found in the C-terminal regions of Clns, D-type cyclins and cyclin E. When these regions are removed, the cyclins are stabilized [58–60]. The PEST hypothesis [61,62] proposes that regions rich in proline, glutamate or aspartate, and serine or threonine, and located between positive-flanking residues, predict proteins which have a short half-life. Although biochemical data on the precise mechanism of PEST degradation are not available, it may also involve a ubiquitin-mediated pathway, and regions predicted as PEST motifs are indeed found in a number of proteins with short half-lives, and in the case of $G_1$ cyclins these coincide with regions shown experimentally to be required for rapid turnover.

PEST regions can be predicted with the program PESTFIND [61], which locates regions of potential PEST activity. We have examined plant CycD protein sequences with this program, and found that all contain potential PEST sequences (PESTFIND score $\geq 0$). These are located N-terminal or C-terminal to the cyclin box, or both. This suggests that PEST regions are a general feature of all $G_1$ cyclins, and predicts that plant CycD cyclins, like their animal homologues, will also be short-lived proteins.

# Regulation and expression of animal and plant D cyclins

Mammalian D-type cyclins show differential expression patterns, with most cells expressing D3 and either D1 or D2. During mouse embryogenesis, cyclin D1 is expressed particularly highly in the developing brain and retina [63], incidentally both tissues in which Rb protein is abundant [43]. There is possibly some redundancy between the D cyclins, as transgenic mice lacking cyclin D1 develop to term, although they show reduced body size and viability, neurological impairment, and a significant reduction in cell number in the retina. The mammary epithelium of adult females also fails to respond to steroid-induced proliferation during pregnancy [63].

During the cell cycle of cultured cells, the transcription of D-type cyclins is highly growth-factor-dependent, and declines very rapidly on withdrawal of serum as discussed above. In quiescent or resting cells, expression of cyclin D1 mRNA is low, and reaches a maximum 10 h after serum stimulation and 8 h before DNA synthesis begins [64]. However, in cycling cells, the abundance of cyclin D1 mRNA does not change significantly during the cell cycle [26,64], suggesting that the regulation of cyclin-D1-kinase activity is post-transcriptional. Indeed, D-type-cyclin protein levels are found to be high in $G_1$ cells, declining significantly by late S phase [26,64].

We have found that, like mammalian D-type cyclins, *Arabidopsis* CycD cyclins show tissue-specific variation in expression. CycD2 and CycD3 are transcribed into a single mRNA species, whereas CycD1 is represented by three distinct mRNA species [50]. We have not yet established the nature of these differences, which could be accounted by alternative 5' or 3' ends or splicing patterns.

RNA gel blot analysis (Fig. 4*a* and *b*) shows that CycD3 is highly expressed in roots and is present in lower amounts in flowers, young leaves and callus material. It is totally absent from stem tissue, even on long exposures of blots (Fig. 4*b*). The three transcripts of CycD1 show differential abundance in different tissues, with the longest transcript particularly prevalent in flowers, and to a lesser extent in roots and callus material, whereas in leaves the intermediate and shorter transcripts predominate (Fig. 4*c* and *d*).

We have examined the expression of CycD cyclins in two cell culture systems. In callus material cultured in liquid media, we have shown that CycD3 and CycD2 are expressed. Imposition and subsequent release of a $G_1$/S block with the ribonucleotide reductase inhibitor hydroxyurea led to strong induction of CycD3 slightly before the onset of S phase and histone H4 expression [50]. CycD3 expression was also nitrate dependent in this system, and accumulated in re-fed cells slightly in advance of histone H4 [66].

The cell cycle regulation of expression has also been examined for a number of *Arabidopsis* cyclins in a cell-suspension culture showing rapid growth and a well-dispersed characteristic [66]. This culture can be synchronized in $G_1$ by

## Fig. 4     Expression of CycD cyclins in *Arabidopsis* tissues

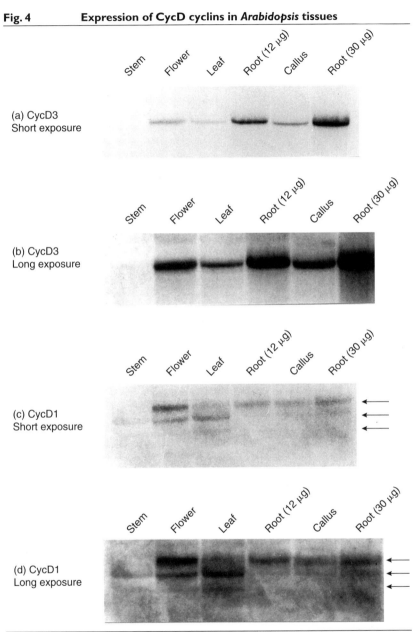

*Total RNA was extracted from the tissues indicated, and 30 μg was used for RNA-gel-blot analysis in all lanes, except lane 4, which contains 12 μg of RNA extracted from root [50]. Top panels: hybridized to the 0.7 kb HindIII–XbaI fragment from CycD3;At, exposed for 24 h (a) and 7 days (b). Lower panels: the same blot was reprobed with the 1.3 kb XbaI–SstI fragment of CycD1;At, exposed 24 h (c) and 7 days (d).*

**Fig. 5**        **Response of liquid-cultured *Arabidopsis* callus tissue to deprivation and re-addition of sucrose, auxin and cytokinin**

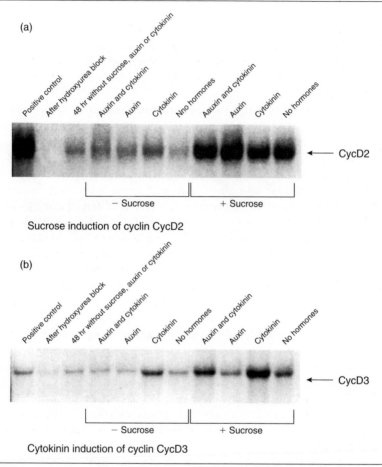

(a)

CycD2

Sucrose induction of cyclin CycD2

(b)

CycD3

Cytokinin induction of cyclin CycD3

*Callus cultures derived from excised seedling root segments were established in Murashige and Skoog (MS) culture medium (ICN Biomedicals) containing 3% (w/v) sucrose, and standard concentrations of the auxin 2,4-dichlorophenoxyacetic acid (2,4-D) and kinetin as described [50]. Cultures were washed and resuspended in MS medium lacking sucrose, 2,4-D and the cytokinin kinetin for 48 h. Samples were taken before (positive control) and after (48 h without sucrose, auxin or cytokinin) this treatment. The culture was then divided into eight aliquots, to which were added sucrose, auxin and cytokinin alone or in all possible combinations as indicated. After 4 h of continued incubation, RNA was prepared. As a reference, RNA extracted from cells after treatment with hydroxyurea (an inhibitor of ribonucleotide reductase that blocks S-phase entry) for 48 h was also included. RNA-gel blots were probed with CycD2 (a) or CycD3 probes (b). CycD2 was induced by sucrose re-addition regardless of the hormone combination present, whereas CycD3 was induced by cytokinin. This induction was stimulated by sucrose, but reduced by concomitant auxin addition.*

low concentrations of cycloheximide, which offers considerable advantages for studying $G_1$/S-related processes over drugs such as hydroxyurea and aphidicolin that probably arrest close to S-phase onset. These studies confirmed the original observations that CycD1 is undetectable or expressed at very low levels in liquid-cultured cells, and CycD2 mRNA levels are largely unaffected by block and release. CycD3 levels rose in late $G_1$, before the peak of S phase, and thereafter remained at a relatively constant level [66]. This mirrors the behaviour of cyclin D1 mRNA in mammalian cells [26,64].

A further important aspect of the regulation of mammalian D-type cyclins is their function as integrators of growth signals into cell cycle control. We have therefore examined whether *Arabidopsis* CycD cyclins are regulated in response to exogenous signals known to affect the growth of plant cells. Auxin and cytokinin are important plant hormones (growth regulators) that are required for the proliferation of most primary cell types in culture. Sucrose is the most important translocated carbon source and known to have a number of developmental effects. Liquid-cultured callus tissue was therefore subjected to withdrawal of these three substances for 48 h, being incubated for this time in minimal salts medium. The three components were then added back in all possible combinations, and RNA was prepared after a further 4 h to examine the immediate effect of re-addition on gene expression (Fig. 5). CycD3 was specifically induced by cytokinin (Fig. 5*b*), and CycD2 by sucrose (Fig. 5*a*), in both cases independently of the presence of other substances. The induction of CycD3 by cytokinin was also inhibited by auxin in this experiment, an interesting parallel to the often antagonistic physiological effects of these two hormones.

Further investigation of the cytokinin responsiveness of CycD3 has been carried out with suspension-cultured cells (Fig. 6). In both exponential and late log phase cells, we have found a rapid and specific response to cytokinin that is detectable after less than 2 h. The induction is most sensitive to zeatin at $10^{-6}$ M, although response are detected down to at least $10^{-8}$ M zeatin. There is no response to the purine adenine.

The data available on CycD expression in intact tissue are more limited. Expression of the two CycD3 homologues has been examined in sectioned material by hybridization *in situ* on *Antirrhinum* shoot apical and inflorescence meristems, and confirms that CycD3 mRNA does not accumulate in a cell-cycle-dependent manner, unlike CycB genes that have been studied (see J.H. Doonan, Chapter 10). Interestingly, there is a zone-specific expression of CycD3 genes within the meristem, with CycD3a (CycD3;1) mRNA limited to incipient and developing primordia. This not only adds weight to the concept of differential function within apical meristems (see J.H. Doonan, Chapter 10), but also suggests that changes in cell cycle control may be early events in the switch from a meristematic to a differentiated state (see CycD cyclins, cell proliferation and development, pp. 120–123).

**Fig. 6          Induction of CycD3 in suspension-cultured cells by cytokinins**

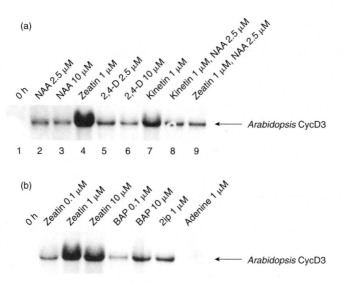

An Arabidopsis cell-suspension culture, dependent on exogenous auxin and cytokinin [66], was subcultured (1:10) and grown for 7 days to stationary phase. Cells were washed and resuspended in media containing sucrose, but lacking auxin and cytokinin for 24 h. A control sample was harvested (0 h), the substances indicated were added and incubation continued for 4 h. Cells were then harvested for RNA extraction and gel-blot analysis, using a CycD3 probe. (a) Auxins [naphthylacetic acid (NAA) and 2,4-D; lanes 2–3, 5–6 from left] increase CycD3 expression over the control level, but a super-induction occurs with the cytokinins zeatin and kinetin (lanes 4 and 7). The concomitant addition of NAA blocks super-induction (lanes 8 and 9). (b) Effects of different zeatin concentrations, and of the alternative cytokinins benzyladenine (BAP) and isopentenyladenine (2ip), and of adenine.

We have also generated transgenic plants in which the upstream region of the *Arabidopsis* CycD3 gene is fused to a β-glucuronidase marker gene. Preliminary results show staining in the shoot apical meristem of seedlings appearing 6–7 days after germination, but the total absence of staining in the primary root meristem. However, expression does appear later in the root, localized in putative sites of lateral root emergence, an observation that is supported by auxin-stimulated production of lateral roots. In older roots, expression appears to be more widespread, possibly associated with cambial activity, but is never observed in the primary root meristems at root tips.

The only other CycD expression pattern that has been studied is CycD3; Ms (CycMs4) from alfalfa, which also appears to be induced before the onset of DNA synthesis, and has exhibited high levels of expression in the root, with expression limited to the pericycle, endodermis and outer cortex [53] (see D. Dudits et al., Chapter 2).

In conclusion, there is clearly much still to be learned about the expression of CycD cyclins, but there is clear evidence for tissue- and meristem-specific expression, and cell cycle behaviour that mirrors that of mammalian D-type cyclins. It is clearly established that *Arabidopsis* CycD3 is regulated by cytokinin, in addition to other signals, such as nitrate, cell cycle progress, and, *in vivo*, positional information. CycD2 may be responsive to sucrose or carbon availability, but this observation requires further substantiation in suspension-cultured cells. Nevertheless, the weight of evidence implicates plant CycD cyclins as important integrators of proliferative signals in $G_1$.

# Retinoblastoma (Rb) homologues in plants

The LxCxE Rb-binding motif is a defining feature of mammalian D-type cyclins, that has not been found in any other cyclin class. The presence of an identical motif in plant CycD cyclins at the equivalent position, within proteins that are otherwise highly diverged (approx. 10% identity), suggests that this sequence has been specifically conserved. Moreover, the leucine is preceded by an acidic residue (D,E) at -1 or -2, a feature that is also found in the mammalian and viral proteins that interact with Rb. We therefore surmized the existence of Rb homologues in plants [50]. This proposal was subsequently further supported by the discovery of an LxCxE motif in a plant geminivirus replication protein. This motif is functional in that it confers binding in the yeast two-hybrid assay to p130, a human protein of the Rb family [67]. Moreover, in maize endosperm undergoing endoreduplication, a kinase activity was detected that binds both human E2F and adenovirus E1A proteins [68], an observation which could be explained by the presence in the extracts of a complex of a CycD-containing kinase bound to a plant Rb homologue, since Rb proteins might be predicted to bind both E2F and E1A.

## Interaction of plant cyclin D and human Rb

We have therefore directly examined the possibility that plant CycD cyclins can interact with mammalian Rb proteins. Using human Rb, expressed in *Escherichia coli* as a fusion protein with glutathione S-transferase (GST), we purified GST–Rb on glutathione-coupled agarose (Pharmacia), according to the manufacturer's instructions. CycD cyclins were expressed by *in vitro* transcription–translation (IVT) in a coupled T7-wheat-germ-lysate system (Promega TnT) and thereby labelled with [35S]methionine. After incubation of the translated protein with the GST–Rb agarose, specific binding of the IVT CycD cyclins was detected to Rb, but not to beads coupled only to the GST moiety, or to naked beads (Fig. 7). Moreover, when a point mutation in the pocket domain of Rb, which converts cysteine 706 to phenylalanine (C706F), was used as a GST fusion protein, a reduction of binding

**Fig. 7**          **Biochemical interaction of human Rb and plant CycD2 cyclin**
                     ***in vitro***

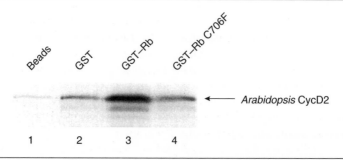

Fusion proteins of GST and human Rb, or human Rb with the C706F mutation, were expressed in E. coli
and purified on glutathione Sepharose beads. These beads were then incubated with ³⁵S-methionine-
labelled in vitro translated Arabidopsis CycD2, and washed four times according to standard procedures
[86]. Proteins bound to the beads were boiled and loaded on to SDS/PAGE gels, and visualized by
fluorography. Specific binding was observed of CycD2 to GST–Rb (lane 3), but not to GST–Rb C706F
(lane 4), or to GST with no fusion partner (lane 2), or uncoupled beads (lane 1).

was observed, suggesting that the interaction between plant CycD and Rb depends
on the presence of a functional pocket domain in Rb.

This represents the first biochemical evidence for interaction between
plant CycD proteins and Rb homologues, and suggests that these interactions are
highly conserved between mammals and plants. Moreover, it provides strong
supporting evidence for the existence of Rb homologues.

## Isolation of a maize Rb homologue

cDNA sequencing projects exist for several plant species, and provide a library of
randomly selected single-pass sequences (expressed sequence tags or ESTs) that can
be searched for homologues to genes of interest. We carried out a comprehensive
search of available plant ESTs from *Arabidopsis*, rice, maize and Loblolly pine for
plant Rb homologues. Because of the relatively low homology between known Rb
homologues, we used short peptide sequences derived from regions conserved
between the three human Rb homologues, Rb, p107 and p130, to search dBEST
using the program TBLASTN. This compares a peptide probe sequence against
translations of DNA sequence databases in all six possible reading frames. We also
searched the indexed database of *Arabidopsis* ESTs at the University of Minnesota
(http://lenti.med.umn.edu). A single EST of 531 bp from maize (Genbank accession
number T18395) was identified with suggestive homology to the conserved domain
B of the Rb pocket. We obtained this clone, and sequenced the remainder of the
cDNA that it contained.

Rb and its homologues are proteins composed of two principal domains of roughly equal sizes [69,70]. The N-terminal part of the protein has been found to bind a number of cellular proteins [70], but it is the C-terminal half which includes the so-called pocket domain that has been implicated in cell cycle regulation. The Rb pocket is the site of binding of LxCxE motif proteins such as D-type cyclins and transforming viral proteins, and also of E2F. The pocket contains two conserved domains called A and B, which are both required for binding. A third domain C may also play a role [70].

We found that T18395 encodes a truncated cDNA, which includes the C-terminal part of a maize Rb homologue that we call Rb1. The cDNA is truncated in the B domain, at a point corresponding to position 675 in human Rb sequence, about 35 residues into the B domain. Figs. 8 and 9 shows that the maize sequence encodes a protein with significant homology to mammalian Rb homologues, and to the recently identified *Drosophila* Rb homologue [71]. The conserved residues with mammalian Rb are concentrated in regions also conserved between different human Rb family members (Fig. 8), and notably maize Rb1 includes the critical conserved cysteine residue at the equivalent position to C706 in human Rb. It is not possible to define maize Rb1 as belonging to a specific branch of the Rb family (Fig. 8).

| Fig. 8 | **Relationship of partial maize Rb1 to animal Rb homologues** |
|---|---|

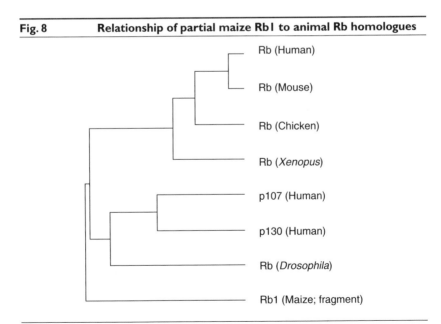

The protein sequences of human Rb, p107 and p130, mouse, chicken and Xenopus Rb, and the Drosophila Rb homologue were compared with the partial sequence of maize Rb1 using the GCG program PILEUP.

**Fig. 9**        **Alignment of partial maize Rb clone sequence with other Rb family proteins**

```
Rb human   K V Y R L A Y L R L N T L C E R L L S E H P E L E H I I W T L F Q H T L Q N E Y E L M R D R H L D Q   702
Rb mouse   K V Y R L A Y L R L N T L C A R L L S D H P E L E H I I W T L F Q H T L Q N E Y E L M R D R H L D Q   695
Rb chicken K V F R L A Y L R L H T L F F R L L S E H P D L E P L I W T L F Q H T L Q N E S E L M R D R H L D Q   694
Rb Xenopus K V Y L L A Y K R L S S L C S S L L S D H P E L E Q V I W T L L Q H T L Q Q E Y E L M R D R H L D Q   675
p107 human K V Y H L A S V R L R D L C L K - L D V S N E L R R K I W T C F E F L V H C P D L M K D R H L D Q   842
p130 human K V Y H L A A V R L R D L C A K - L D I S D E L R K K I W T C F E F S I I Q C P E L M M D R H L D Q   890
Rb Dros    K V Y L L G W L R I Q K L C S E - L S L C E K T P E S I W H I F E H S I T H E T E L M K D R H L D Q   636
Rbl maize  - - - - - - - - - - - - - - - - - - - - - - - Q T E R V Y N V F K Q I L E Q Q T T L F F N R H I D Q    27

Consensus  K V Y - L A - - R L - - L C - - - L S - - - E L - - - I W T - F - H T L - - E - E L M - D R H L D Q   950


Rb human   I M M C S M Y G I C K V K N I - D L K F K I I V T A Y K D L P H A V Q E T F K R V L I K - - - - - -   745
Rb mouse   I M M C S M Y G I C K V K N I - D L K F K I I V T A Y K D L P H A Q E T F K R V L I R - - - - - -   738
Rb chicken I M M C S M Y G I C K V K N V - D L R F K T I V S A Y K E L P N T N Q E T F K R V L I R - - - - - -   737
Rb Xenopus I M M C S M Y G I C K A K N I - D L R F K T I V T A Y K G L T N T N Q E T F K H V L I R - - - - - -   718
p107 human L L L C A F Y I M A K V T K E - E R T F Q E I M K S Y R N Q P Q A N S H V Y R S V L L K - - - - - -   885
p130 human L L M C A I Y V M A K V T K E - D K S F Q N I M R C Y R T Q P Q A R S Q V Y R S V L I K G K R K R R   939
Rb Dros    N I M C A I Y I Y I R V K R M E D P K F S D I M R A Y R N Q P Q A V N S V Y R E V F I - - - - - - -   679
Rbl maize  L I L C C L Y G V A K V C Q L - E L T F R E I L N N Y K R E A Q C K P E V F S S I Y I - - - - - - -    69

Consensus  - - M C - - Y G - - K V K - - - D L - F - - I - - A Y K - - - P - A - - E - F - - V L I - - - - - - -  1000


Rb human   - - - - - - - - - - - - - - - - - - - - - - - - - - - - - - - - - - - - - - - - - - - - - - - -   745
Rb mouse   - - - - - - - - - - - - - - - - - - - - - - - - - - - - - - - - - - - - - - - - - - - - - - - -   738
Rb chicken - - - - - - - - - - - - - - - - - - - - - - - - - - - - - - - - - - - - - - - - - - - - - - - -   737
Rb Xenopus - - - - - - - - - - - - - - - - - - - - - - - - - - - - - - - - - - - - - - - - - - - - - - - -   718
p107 human - - - - S I P R E V V A Y N K N I N D D F E M I D - - - - - - - - - - - - - - - - - C D L E D A T K   914
p130 human N S G S S D S R S H Q N S P T E L N K D R T S R D S S P V M R S S S T L P V P Q P S S A P P T P T R   989
Rb Dros    - - - - - - - - - - D I N E D - - - - - - - - - - - - - - - - - - - - - - - - - - - - - - - - - -   684
Rbl maize  - - - - - - - - - - - - - - - - - - - - - - - - - - - - - - - - - - - - - - - - - - - - - - - -    69

Consensus  - - - - - - - - - - - - - - - - - - - - - - - - - - - - - - - - - - - - - - - - - - - - - - - -  1050


Rb human   - - - - - - - - E E E Y D S I I V F Y N S V F M Q R L K T N I L Q Y A S T - - - R P P T L S P I P   783
Rb mouse   - - - - - - - - E E E F D S I I V F Y N S V F M Q R L K T N I L Q Y A S T - - - R P P T L S P I P   776
Rb chicken - - - - - - - - E E Q Y D S I I V F Y N L V F M Q K L K T N I L Q Y G S A - - - R P P T L S P I P   775
Rb Xenopus - - - - - - - - D G Q H D S I I V F Y N L V F M Q K L K S H I L Q Y G S A - - - R H P T L S P I P   756
p107 human T P D C S S G P V K E E R S D L I K F Y N T I Y V G R V K S F A L K Y D L A - - - N Q D H M M D A P   961
p130 human L T G A N S D M E E E E R G D L I Q F Y N N I Y I K Q I K T F A M K Y S Q A - - - N - - - - M D A P  1032
Rb Dros    - - - - - - - - G E P K V K D I I H F Y N H T Y V P L M R Q F V I D Y L N V - - - T P D V S G R A S   723
Rbl maize  G S T N R N G V L V S R H V G I I T F Y N E V F V P A A K P F L V S L I S S G T H P E D K K N A S G   119

Consensus  - - - - - - - - - E E - - - - I I - F Y N - V F - - - - K - - - L - Y - S - - - - - - - - - - - - P  1100


Rb human   H I P R S P Y - - K F P S - - - - S P L R I P G G - N - - I Y I S P L K S P Y K I S E G L P T P T K   824
Rb mouse   H I P R S P Y - - K F S S - - - - S P L R I P G G - N - - I Y I S P L K S P Y K I S E G L P T P T K   817
Rb chicken H I P R S P Y - - Q F S N - - - - S P R R V P A G N N - - I Y I S P L K S P Y K F S D G F H S P T K   817
Rb Xenopus H I P R S P Y - - R F G N - - - - S P - K V P G - - N - - I Y V S P L K T P Y K T A D G L L S P S K   795
p107 human P L - - S P F P H I K Q Q P G - - S P R R I - - S Q Q H S L Y I S P H K N - - - - - - - - - - G S G   995
p130 human P L - - S P Y P V R - - T G - - S P R R I Q L S Q N H P V Y I S P H K N - - - - - - - - - - E T M  1066
Rb Dros    D L Q L S P H P K E R A A - - - - Q P K K V - - T Q S H S L F V S Q M S K - - - - - - - - N E I   757
Rbl maize  Q I P G S P K P S P F P N L P D M S P K K V S A S H N - - V Y V S P L R Q T - - - - - - - K L D L L   160

Consensus  - I P - S P Y - - - F - - - - - - S P - R - - - - - N - - I Y I S P L K - - - - - - - - - - - - - -  1150
```

# CycD cyclins, cell proliferation and development

The involvement of $G_1$ controls in development, and particularly the roles of CycD cyclins and Rb homologues, is an area where significant advances are to be expected in the next few years. In particular, we need progress in understanding the interaction between proliferation and development, the reversibility of this switch, its relationship to meristem function, and how cell cycle decisions are regulated by meristem and organ identity genes.

## Fig. 9 (Contd.)

```
Rb human    M T P R S R I L V - S I G E S F - - - - G T S E K F Q K I N Q M V C N S D R V L K R S A E G S N P P    869
Rb mouse    M T P R S R I L V - S I G E S F - - - - G T S E K F Q K I N Q M V C N S D R V L K R S A E G G N P P    862
Rb chicken  M T P R S R I L V - S I G E T F - - - - G T S E K F Q K I N Q M V C N S E S H V K R S A E P S D A P    862
Rb Xenopus  M T P K T S F L I - S L G E T F - - - - R S P D R F Q K I N Q M L N S C E R P I K R S A D T G T T P    840
p107 human  L T P R S A L L Y - K F N G S - - - - - - P S K S L K D I N N M I R Q G E Q R T K - - - - - - - - -    1029
p130 human  L S P R E K I F Y - Y F S N S - - - - - - P S K R L R E I N S M I R T G E T P T K - - - - - - - - -    1100
Rb Dros     Q Q S P N Q M V Y - S F F R S - - - - - - P A K D L Q A M N E K V R G G - - - - - - - - - - - - - -    786
Rbl maize   L S P S S R S F Y A C I G E G T H A Y Q S P S K D L A A I N S R L N Y N G R K V N - - - - - - - - -    201

Consensus   - T P R S - - L - - S - G E S - - - - - - - S - - - Q - I N - M - - - - - - - - - K - - - - - - - - -    1200


Rb human    K P L K K L R F D I E G S D E A D G S K H L P G E S K F Q Q K L A E M T S T R T R M Q K Q K M N D S    919
Rb mouse    K P L K N V R F D I E G A D E A D G S K H L P A E S K F Q Q K L A E M T S T R T R M Q K Q R M N E S    912
Rb chicken  K P L K R L R F D I E G Q D E A D G G K H L P Q E S K F Q Q K L A E M T S T R T R M Q K Q K L N D G    912
Rb Xenopus  K P L K K L R F D S D G Q D E A D G S K H I Q G E S K F Q Q K L A E M T S T R T R M Q K Q K L E E S    890
p107 human  - - - K R V I A I D S D A E S P A K R V C Q E N D D V L L K R L Q D V V S E R A N H - - - - - - - - -    1068
p130 human  - - - K R G I L L E D G S E S P A K R I C P E N H S A L L R R L Q D V A N D R G S H - - - - - - - - -    1139
Rb Dros     - - - K R M L S F G D E P D - - - - - - - - - - - - - - - - - - - - - - - - - - - - - - - - - - - - -    797
Rbl maize   - - - S R L N F D M V S D S V V A G S L G Q I N G G S T S D P A A F S P L S K K R E T D T - - - - -    244

Consensus   - - - K R - - F D - - G - D - - - G - - - - - - - S - - - - - L A - - - S - R - - - - - - - - - - -    1250


Rb human    M D T S N K E E K    928
Rb mouse    K D V S N K E E K    921
Rb chicken  N D T S A N E E K    921
Rb Xenopus  L E S S Q Q E E K    899
p107 human  - - - - - - - - -    1068
p130 human  - - - - - - - - -    1139
Rb Dros     - - - - - - - - -    797
Rbl maize   - - - - - - - - -    244

Consensus   - - - - - - - - -    1259
```

*The alignment was created in the program PILEUP and displayed using the EGCG program PRETTYPLOT on the UK HGMP server. The numbers refer to amino acid positions in each protein. The region shown starts about 14 residues into the B region of the pocket domain. The B region ends at position 771 in human Rb. Domain C extends from 772 to the end of the human Rb protein, and contains a large number of mapped (S807, S811, T821, T826) and potential (S780, S788, S795) phosphorylation sites, but the role of the C domain is not as well understood as the A and B domains [70].*

However, some clues are already available that point to the likely importance of $G_1$ controls, perhaps even suggesting that they may regulate developmental events. Although it may seem intuitive that cell cycle regulation should normally be subservient to developmental controls, it is conceivable that changes in cell cycle regulation may drive downstream differentiation events. The discovery that one *Antirrhinum* CycD3 homologue is expressed only in incipient and developing primordia, and a second is down-regulated in boundary layers of cells that lie between proliferating regions (see J.H. Doonan, Chapter 10), opens the possibility that CycD cyclins may be intimately involved in meristem function and the control of proliferation and differentiation.

Further evidence points to a direct link between homoeotic and developmental genes, and those involved in cell cycle regulation. Mutations in the *Arabidopsis* gene *UFO* (unusual floral organs) induce flower abnormalities and variable homoeotic changes in petals and stamens [72]. An *Arabidopsis* protein interacting with the *UFO* gene product has recently been identified using a yeast two-hybrid screen, and found to be homologous to the mammalian protein p19 [73], previously identified as interacting with cyclin A in transformed cells [74]. It seems probable that further homoeotic and meristem-identity genes will be found to interact either directly or indirectly with cell cycle regulators and influence proliferation and differentiation.

This relationship between proliferation and differentiation is extremely subtle, and likely to be complex. The issues involved may be illustrated by the structure of the shoot apical meristem (SAM) of dicotyledonous plants such as *Arabidopsis* or *Antirrhinum*. The central zone of the SAM consists of a group of cells, whose size and morphology does not change during most of the post-embryonic development of the plant (see D. Francis, Chapter 9). This population of stem cells, arranged in three layers, represents the ultimate source of all aerial growth (see J.H. Doonan, Chapter 10). The meristem cells divide anticlinally (division plane perpendicular to the surface), so the progeny of central zone cells are pushed into the surrounding peripheral zone. In this area, leaf or floral primordia are specified and develop in a spiral pattern, becoming visible as small bump-like protrusions.

Cells in the peripheral zone are therefore undergoing decisions that will result in their ultimately forming determinate structures such as leaves or flowers, which consist of differentiated, non-proliferating cells. The fate of cells within the peripheral zone is therefore different to that of stem cells within the central zone, and they may therefore be regarded as having undergone a differentiation decision. However, the cells of the peripheral zone show few if any morphological differences to central zone cells, their rate of division is actually faster than in the central zone, and they produce progeny cells which will develop into a range of cell types (see D. Francis, Chapter 9). Nevertheless, in contrast to central zone cells, all progeny cells of the peripheral zone will ultimately undergo final specification and terminal differentiation.

From this description, it is clear that we must distinguish between at least four classes of differentiation events made by plant cells, which can be used to define populations of cells. The first differentiative decision represents the loss of totipotency, and corresponds to exit from the central zone of the SAM. Such cells can be conceptually distinguished from central zone stem cells as they are 'committed' in the sense that their ultimate terminal differentiation as non-dividing cells is assured. (They are not, however, determined in the sense that applies to organisms such as *Caenorhabditis elegans* for which the precise lineage of every cell can be traced, and, given the appropriate conditions, these cells, or derivatives from them, can proliferate and reorganize a meristem in culture). Such decisions may involve cell cycle control since the proliferation rate may increase as an immediate response. In *Antirrhinum*, the expression of CycD3a may be a molecular marker for these cells.

The second class of differentiation event occurs when cells start to become morphologically or physiologically distinct and recognizably on a pathway to a defined specialized cell type. This is likely to correspond to a reduction in division potential, and a loss of CycD3a expression. One may speculate that Rb-related proteins may be important in mediating this or, indeed, other differentiation events. A third class of decision is made in the terminal stage of differentiation, when a cell ceases all division. This represents an exit from the cell cycle, and such cells are predicted to lack CycD3 expression and have high levels of Rb-related proteins. A final type of decision is represented by programmed cell death, during processes such as xylogenesis.

## A working model for hormonal control of the plant cell cycle

The requirement for the plant hormones auxin and cytokinin in promoting the growth of cell cultures, and the re-entry into the division cycle of quiescent cells, has been known for many years [9,75–78]. It is likely that cultures not requiring one or other of these hormones exhibit endogenous synthesis. It is therefore likely that both auxin and cytokinin mediate effects on cell cycle regulation, and, in particular, control the re-entry into division of quiescent or differentiated tissue.

In studies of tobacco pith explants, a tissue dependent on both auxin and cytokinin for division, treatment with auxin alone is sufficient to induce expression of an immunoreactive Cdk protein detectable by protein-gel-blot analysis. However, this protein was not found to have catalytic activity unless cytokinin was also present [79]. We therefore propose that the plant Cdks may be induced by auxin, a conclusion also supported in soybean [80]. Given the response of cyclin CycD3 expression to cytokinin induction both in cell cultures (see Regulation and expression of animal and plant D cyclins pp. 14) and intact seedlings (our unpublished work), we propose that CycD3 may represent the cytokinin-limited component required for cell division. We are currently investigating this hypothesis experimentally.

It is likely, however, that auxin, and perhaps also cytokinin, may exert multiple effects on cell cycle progress that are mediated through different mechanisms, perhaps in different cell types (see D. Dudits et al., Chapter 2). The discovery that the *Arabidopsis* gene *AXR1*, the mutation of which leads to an auxin-resistant phenotype, encodes a protein with limited homology to ubiquitin-activating enzyme (E1) [81] is intriguing in this regard. Genes related to *AXR1* are found in mammals and yeast, and although the yeast gene is non-essential, its mutation enhances the lethality of temperature-sensitive alleles of *cdc34*. This an enzyme required for ubiquitin-mediated destruction of the Cdk inhibitor Sic1 and progression into S phase (see 'The control of $G_1$/S transition in budding yeast', pp. 3) [82] (see B. Plesse et al., Chapter 7). It is therefore an attractive, but speculative, possibility that an additional action of auxin in promoting cell division may be through triggering the destruction of Cdk inhibitor proteins, and suggests that multiple checkpoints may require auxin presence.

We have previously proposed that various signals, both stimulatory and inhibitory, are integrated into control of a START-like point in plant cells by modulating the activity of Cdks [83]. We present in Fig. 10 a refined model,

**Fig. 10**   **Model for control of the $G_1$/S transition in plants, indicating the possible points at which plant hormones and growth regulatory substances may act.**

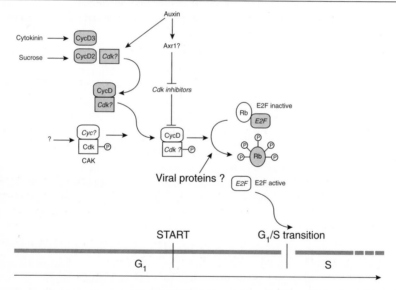

We propose that different CycD cyclins respond to specific signals and, in particular, that CycD3 is induced by cytokinin. Cdk homologues are known to be induced by auxin, and it is also possible that auxin may promote the destruction of (as yet unidentified) Cdk inhibitors through the activity of Axr1. A homologue of CDK7 called R2 has been cloned from rice [84], indicating the likely existence of CAK in plants, but its regulation is unknown. The presence of an Rb-binding motif in a geminivirus replication protein suggests that viral proteins in plants may drive cells into S phase directly by promoting release of the putative E2F from Rb. Proteins for which cDNAs have been cloned are shown in normal type; interactions, phosphorylations or proteins whose existence are speculative are indicated by italics. Thus CycD cyclins and an Rb homologue have been identified; several cdks are known, but it is not known with which cyclins they interact; the existence of E2F and cdk inhibitors is speculative. See text for further details.

incorporating recent discoveries of CycD and Rb proteins. The existence of a rice homologue (called R2) of human CDK7 [84], a component of Cdk-activating kinase (CAK), suggests that CAK is also a feature of plant cell cycle control.

## Conclusions

We have reviewed here the features of plant CycD cyclins, and described their interaction with mammalian Rb, and the identification of a maize Rb homologue.

Despite the relatively low homology of plant CycD cyclins to mammalian D cyclins, additional sequence features, such as the presence of potential PEST sequences and the N-terminal-binding motif for the Rb family of proteins, support their identification as plant homologues of mammalian D-type cyclins.

This is substantiated by functional analyses of plant CycD cyclins, which have revealed a number of important parallels with their mammalian homologues. First, CycD3 responds rapidly and specifically to induction by cytokinin, known to be required with auxin for induction of cell division in many quiescent cell types. CycD2 is similarly responsive to sucrose levels [50]. These transcriptional responses may be compared with the induction of mammalian cyclin D1 to growth factors. Secondly, the timing of activation of CycD3 after arresting cell cycle progression is very similar to mammalian cyclin D1. Thirdly, we have shown that plant CycD cyclins bind mammalian Rb protein *in vitro*, but not a point mutation of Rb that shows reduced binding to mammalian D cyclins. Finally, we describe a partial cDNA in maize that is highly homologous to the pocket domain of mammalian Rb proteins.

The existence of proteins and interactions controlling the $G_1$/S transition in plants with strong similarities and conserved interactions with mammalian components suggest that common themes will be found. The next few years will be exciting as we discover the extent to which these molecules are involved in proliferation and differentiation. We expect to see the convergence of studies on cell cycle regulators in plants with those on genes that control meristem function, organ identity and developmental processes. It will be these controls on the central pathway which will reveal features that explain the unique features of plant growth and development.

## References

1.   Lee, M.G. and Nurse, P. (1987) Nature (London) 327, 31–35
2.   Pines, J. (1995) Adv. Cancer Res. 66, 181–212
3.   Pines, J. (1995) Biochem. J. 308, 697–711
4.   Elledge, S.J. and Harper, J.W. (1994) Curr. Opin. Cell Biol. 6, 847–852
5.   Sherr, C.J. and Roberts, J.M. (1995) Genes Dev. 9, 1149–1163
6.   Hartwell, L.H. and Weinert, T.A. (1989) Science 246, 629–634
7.   Murray, A.W. (1992) Nature (London) 359, 599–604
8.   Gould, A.R., Everett, N.P., Wang, T.L. and Street, H.E. (1981) Protoplasma 106, 1–13

9.  Bayliss, M.W. (1985) in The Cell Division Cycle in Plants (Bryant, J.A. and Francis, D., eds.), pp. 157–177, Cambridge University Press, Cambridge
10. Galbraith, D.W., Harkins, K.R. and Knapp, S. (1991) Plant Physiol. **96**, 985–989
11. Reed, S.I. (1995) in Cell Cycle Control (Hutchinson, C. and Glover, D.M., eds.), pp. 40–62, IRL Press, Oxford
12. Lees, E.M. and Harlow, E. (1995) in Cell Cycle Control (Hutchinson, C. and Glover, D.M., eds.), pp. 228–263, IRL Press, Oxford
13. Richardson, H.E., Wittenberg, C., Cross, F. and Reed, S.I. (1989) Cell **59**, 1127–1133
14. Tyers, M., Tokiwa, G. and Futcher, B. (1993) EMBO J. **12**, 1955–1968
15. Wittenberg, C., Sugimoto, K. and Reed, S.I. (1990) Cell **62**, 225–237
16. Cross, F.R. and Tinkelenberg, A.H. (1991) Cell **65**, 875–883
17. Dirick, L. and Nasmyth, K. (1991) Nature (London) **351**, 754–757
18. Koch, C., Moll, T., Neuberg, M., Ahorn, H. and Nasmyth, K. (1993) Science **261**, 1551–1557
19. Schwob, E. and Nasmyth, K. (1993) Genes Dev. **7**, 1160–1175
20. McIntosh, E.M. (1993) Curr. Genet. **24**, 185–192
21. Schwob, E., Böhm, T., Mendenhall, M. and Nasmyth, K. (1994) Cell **79**, 233–244
22. Schneider, B.L., Yang, Q.-H. and Futcher, A.B. (1996) Science **272**, 560–562
23. Pardee, A.B. (1989) Science **246**, 603–608
24. Lew, D.J., Dulic, V. and Reed, S.I. (1991) Cell **66**, 1197–1206
25. Xiong, Y., Connolly, T., Futcher, B. and Beach, D. (1991) Cell **65**, 691–699
26. Matsushime, H., Roussel, M.F., Ashmun, R.A. and Sherr, C.J. (1991) Cell **65**, 701–713
27. Motokura, T., Bloom, T., Kim, H.G., Jüppner, H., Ruderman, J.V., Kronenberg, H.M. and Arnold, A. (1991) Nature (London) **350**, 512–515
28. Sherr, C.J. (1993) Cell **73**, 1059–1065
29. Weinberg, R.A. (1995) Cell **81**, 323–330
30. Xiong, Y., Zhang, H. and Beach, D. (1992) Cell **71**, 505–514
31. Dowdy, S.F., Hinds, P.W., Louie, K., Reed, S.I., Arnold, A. and Weinberg, R.A. (1993) Cell **73**, 499–511
32. Whyte, P., Buchkovich, K.J., Horowitz, J.M., Friend, S.H., Raybuck, M., Weinberg, R.A. and Harlow, E. (1988) Nature (London) **334**, 124–129
33. DeCaprio, J.A., Ludlow, J.W., Figge, J., Shew, J.-Y., Huang, C.-M., Lee, W.-H., Marsilio, E., Paucha, E. and Livingston, D.M. (1988) Cell **54**, 275–283
34. Shirodkar, S., Ewen, M., DeCaprio, J.A., Morgan, J., Livingston, D.M. and Chittenden, T. (1992) Cell **68**, 157–166
35. Johnson, D.G., Schwarz, J.K., Cress, W.D. and Nevins, J.R. (1993) Nature (London) **365**, 349–352
36. Dulic, V., Lees, E. and Reed, S.I. (1992) Science **257**, 1958–1961
37. Hinds, P.W., Mittnacht, S., Dulic, V., Arnold, A., Reed, S.I. and Weinberg, R.A. (1992) Cell **70**, 993–1006
38. Hinds, P.W., Dowdy, S.F., Eaton, E.N., Arnold, A. and Weinberg, R.A. (1994) Proc. Natl. Acad. Sci. U.S.A. **91**, 709–713
39. Lovec, H., Sewing, A., Lucibello, F.C., Müller, R. and Möröy, T. (1994) Oncogene **9**, 323–326
40. Kato, J. and Sherr, C.J. (1993) Proc. Natl. Acad. Sci. U.S.A. **90**, 11513–11517
41. Ewen, M.E., Sluss, H.K., Whitehouse, L.L. and Livingston, D.M. (1993) Cell **74**, 1009–1020
42. Kiyokawa, H., Richon, V.M., Rifkind, R.A. and Marks, P.A. (1994) Mol. Cell. Biol. **14**, 7195–7203
43. Szekely, L., Jiang, W.-Q., Bulic-Jakus, F., Rosén, A., Ringertz, N., Klein, G. and Wiman, K.G. (1992) Cell Growth Differ. **3**, 149–156
44. Lee, E.Y.-H.P., Chang, C.-Y., Hu, N., Wang, Y.-C.J., Lai, C.-C., Herrup, K., Lee, W.-H. and Bradley, A. (1992) Nature (London) **359**, 288–294
45. Lee, E.Y.-H.P., Hu, N., Yuan, S.-S.F., Cox, L.A., Bradley, A., Lee, W.-H. and Herrup, K. (1994) Genes Dev. **8**, 2008–2021
46. Skapek, S.X., Rhee, J., Spicer, D.B. and Lassar, A.B. (1995) Science **267**, 1022–1024
47. Halevy, O., Novitch, B.G., Spicer, D.B., Skapek, S.X., Rhee, J., Hannon, G.J., Beach, D. and Lassar, A.B. (1995) Science **267**, 1018–1021
48. Parker, S.B., Eichele, G., Zhang, P., Rawls, A., Sands, A.T., Bradley, A., Olson, E.N., Harper, J.W. and Elledge, S.J. (1995) Science **267**, 1024–1027

49. Minet, M., Dufour, M.-E. and Lacroute, F. (1992) Plant J. **2**, 417–422
50. Soni, R., Carmichael, J.P., Shah, Z.H. and Murray, J.A.H. (1995) Plant Cell **7**, 85–103
51. Murray, J.A.H. and Smith, A.G. (1996) in Plant Gene Isolation (Foster, G. and Twell, D., eds.), pp. 177–211, John Wiley and Sons, Chichester
52. Renaudin, J.-P., Doonan, J.H., Freeman, D., Hashimoto, J., Hirt, H., Inzé, D., Jacobs, T., Kouchi, H., Rouzé, P., Sauter, M., Savouré, A., Sorrell, D.A., Sundaresan, V. and Murray, J.A.H. (1996) Plant Mol. Biol. **32**, 1003–1018
53. Dahl, M., Meskiene, I., Bögre, L., Ha, D.T.C., Swoboda, I., Hubmann, R., Hirt, H. and Heberle-Bors, E. (1995) Plant Cell **7**, 1847–1857
54. Kobayashi, H., Stewart, E., Poon, R., Adamczewski, J.P., Gannon, J. and Hunt, T. (1992) Mol. Biol. Cell **3**, 1279–1294
55. Lees, E.M. and Harlow, E. (1993) Mol. Cell. Biol. **13**, 1194–1201
56. Jeffrey, P.D., Russo, A.A., Polyak, K., Gibbs, E., Hurwitz, J., Massagué, J. and Pavletich, N.P. (1995) Nature (London) **376**, 313–320
57. Brown, N.R., Noble, M.E.M., Endicott, J.A., Garman, E.F., Wakatsuki, S., Mitchell, E., Rasmussen, B., Hunt, T. and Johnson, L.N. (1995) Structure **3**, 1235–1247
58. Tyers, M., Tokiwa, G., Nash, R. and Futcher, B. (1992) EMBO J. **11**, 1773–1784
59. Cross, F.R. (1988) Mol. Cell. Biol. **8**, 4675–4684
60. Wittenberg, C. and Reed, S.I. (1988) Cell **54**, 1061–1072
61. Rogers, S., Wells, R. and Rechsteiner, M. (1986) Science **234**, 364–368
62. Rechsteiner, M. (1990) Seminars Cell Biol. **1**, 433–440
63. Sicinski, P., Donaher, J.L., Parker, S.B., Li, T., Fazeli, A., Gardner, H., Haslam, S.Z., Bronson, R.T., Elledge, S.J. and Weinberg, R.A. (1995) Cell **82**, 621–630
64. Sewing, A., Bürger, C., Brüsselbach, S., Schalk, C., Lucibello, F.C. and Müller, R. (1993) J. Cell Sci. **104**, 545–555
65. Lukas, J., Pagano, M., Staskova, Z., Draetta, G. and Bartek, J. (1994) Oncogene **9**, 707–718
66. Fuerst, R.A.U.L., Soni, R., Murray, J.A.H. and Lindsey, K. (1996) Plant Physiol. **112**, 1023–1033
67. Xie, Q., Suárez-López, P. and Gutiérrez, C. (1995) EMBO J. **14**, 4073–4082
68. Grafi, G. and Larkins, B.A. (1995) Science **269**, 1262–1264
69. Wiman, K.G. (1993) FASEB J. **7**, 841–845
70. Wang, J.Y.J., Knudsen, E.S. and Welch, P.J. (1994) Adv. Cancer Res. **64**, 25–85
71. Du, W., Vidal, M., Xie, J.-E. and Dyson, N. (1996) Genes Dev. **10**, 1206–1218
72. Levin, J.Z. and Meyerowitz, E.M. (1995) Plant Cell **142**, 529–548
73. Samach, A., Kohalmi, S.E., Crosby, W.L. and Haughn, G.W. (1996) Abstracts of the 7th International Conference on Arabidopsis Research, (Abstract) S54a
74. Zhang, H., Kobayashi, R., Galaktionov, K. and Beach, D. (1995) Cell **82**, 915–925
75. Skoog, F. and Miller, C.O. (1957) Symp. Soc. Exp. Biol. **11**, 118–131
76. Everett, E.R., Wang, T.L., Gould, A.R. and Street, H.E. (1981) Protoplasma **106**, 15–22
77. Wang, T.L., Everett, N.P., Gould, A.R. and Street, H.P. (1981) Protoplasma **106**, 23–35
78. Hanke, D.E. (1993) in Molecular and Cell Biology of the Plant Cell Cycle (Ormrod, J.C. and Francis, D., eds.), pp. 173–178, Kluwer, Dordrecht
79. John, P.C.L., Zhang, K., Dong, C., Diederich, L. and Wightman, F. (1993) Aust. J. Plant Physiol. **20**, 503–526
80. Miao, G.-H., Hong, Z. and Verma, D.P.S. (1993) Proc. Natl. Acad. Sci. U.S.A. **90**, 943–947
81. Leyser, H.M.O., Lincoln, C.A., Timpte, C., Lammer, D., Turner, J. and Estelle, M. (1993) Nature (London) **364**, 161–164
82. Cernac, A., Lammer, D., Lincoln, C., Timpte, C., Tuner, J. and Estelle, M. (1996) Abstracts of the 7th International Conference on Arabidopsis Research (Abstract), S24
83. Murray, J.A.H. (1994) Plant Mol. Biol. **26**, 1–3
84. Hata, S. (1991) FEBS Lett. **279**, 149–152
85. Matsushime, H., Quelle, D.E., Shurtleff, S.A., Shibuya, M., Sherr, C.J. and Kato, J.-Y. (1994) Mol. Cell. Biol. **14**, 2066–2076
86. Ausubel, F.M., Brent, R., Kingston, R.E., Moore, D.M., Seidman, J.G., Smith, J.A. and Struhl, K. (1987) Current Protocols in Molecular Biology, John Wiley and Sons, New York

# 6

# Functional isolation and analysis of plant cell cycle genes in yeast

**Heribert Hirt\*, Laszlo Bögre, Irute Meskiene, Marlis Dahl, Karin Zwerger and Erwin Heberle-Bors**

Institute of Microbiology and Genetics, Vienna Biocenter, Dr. Bohrgasse 9, 1030 Vienna, Austria

## Introduction

In recent years, a refined picture of eukaryotic cell cycle regulation has emerged. At the centre of this regulatory network are cyclin-dependent kinases (CDKs). In yeast, a single CDK provides the functions required for both the $G_1$/S and $G_2$/M transitions, but in animals and plants several related kinases have evolved (for reviews, see [1–3]).

CDKs are not active as monomers [4], but only become active when associated with a cyclin regulatory subunit. Cyclins are cell cycle stage-specific activators of CDKs, but also bind to other regulatory proteins such as retinoblastoma protein (Rb) in animal cells [5,6] and FAR1 in yeast [7]. Cyclin association also appears to be involved in alteration of CDK substrate specificity [8], availability for upstream regulators [9], and intracellular localization of the CDK complex [10]. The stage specificity of cyclins is ensured mainly by their oscillating appearance in specific cell cycle stages. This is accomplished by regulation of the expression [11] and specific degradation of the respective cyclins [12,13].

Cyclins can be grouped according to sequence similarities, which are believed to reflect different functional properties. Whereas A-type cyclins act in S and $G_2$ phases, B-type cyclins act at the $G_2$/M-phase transition, and D- and E-type cyclins function at the $G_1$/S-phase transition (for reviews, see [2,14]).

So far, four classes of plant cyclins have been isolated. Three classes have highest homologies to mammalian A- and B-type cyclins [15–22]. One class shows similarity to mammalian D-type cyclins [23,24]. Expression analysis of the majority of these cyclins in plant cells is restricted to cells in certain stages of the cell cycle, indicating phase-specific functions for different cyclins [15,18,19,22,23].

The investigation of the cell cycle regulatory components of plants has greatly benefited from the high degree of conservation of the cell cycle machinery in all eukaryotes. Sequence homology approaches, such as techniques based on polymerase chain reaction (PCR), were applied for the isolation of CDK

*To whom correspondence should be addressed.*

components and A- and B-type cyclins. Whereas these approaches can be used for the isolation of genes encoding highly conserved domains, the advanced molecular genetics of yeast also offers the possibility of isolating genes with a low degree of homology. This chapter takes the format of a conventional experimental paper to provide a succinct account of the methodologies and analyses that form part of yeast complementation strategies.

## Materials and methods

### Yeast strains and techniques

The parental strains CY2 (*MATα::sup40, HMLa, HMRa, ho-βgal, ura3, HIS4, ade6-0, ade2-1, can1-100, met, his3, leu2-3, 112, trp-1-1, swi6::TRP1, bar1::HISG*) and K2149 (*MATa, HLMa, HMRa, ho-βgal, ura3, HIS4, ade2-1, can1-100, met, his3, leu2-3, 112, trp1-1, bar1::HISG*) were provided by Kim Nasmyth and Christian Koch (Vienna, Austria). Strain K2149, which contained the plasmid containing a *GAL1* promoter–*HO* fusion gene and *URA3*, was mated with strain CY2 to produce the diploid strain GA2201, carrying the *MATa, MATa*, and silent a mating locus genes. GA2201 was used for transformation with the alfalfa cDNA expression library. Using a diploid strain considerably reduced the frequency of isolating recessive mutations in the yeast pheromone pathway. Exchange of all three copies of the MAT locus to the a-type eliminated the possibility of mating-type switching. Finally, the *bar* mutation prevented proteolysis of the pheromone by the secreted pheromone-specific protease.

The yeast strain K3413 (relevant genotype: *cln1::HisG, cln2::del, cln3::LEU2*, Yiplac204-*MET2-CLN2*) was kindly provided by A. Amon and Kim Nasmyth (Vienna, Austria). The strain was used for transformation with the alfalfa cDNA expression library. Yeast transformation was carried out according to Gietz et al. [25]. After transformation, cells were plated on uracil-free medium containing 2% (w/v) glucose and, after 2 days, replica-plated on uracil-free medium containing galactose. After 24 h, cells were replica-plated on to selective induction plates [2 mM methionine, 2% (w/v) galactose, without uracil]. Growth was assayed for 5 days. Methionine-resistant transformants were propagated on selective induction plates. Standard methods were used for culturing and manipulating yeast.

### Cloning and sequencing

Most molecular techniques were performed as described in [26]. The isolation of the truncated and full-length cDNA clones of the *cycMs3* and the *cycMs4* gene was as described in [22] and [23], respectively. For expression in yeast, the truncated and the full-length cDNA clones of the *cycMs4* genes were cloned into the yeast/ *Escherichia coli* shuttle vector pYEUra3, which contains the *URA3* gene as selectable marker and the GAL1/10 promoter conferring inducible expression on galactose-containing medium. The full-length and the truncated *cycMs2* gene were also cloned into the pYEUra3 vector. For this purpose, the full-length *cycMs2* open-reading frame was amplified by PCR with the following primers: 5′AC CTC GAG CAT ATG GTG AAT ACT TCT GAA GAG 3′ and 5′TCA CTC ACC GCG GCC GCA TGG CTG GTT CTT GTT CTC 3′. The truncated CycMs2 construct, which was missing the first 108 amino acids, was amplified with the primers 5′ GCC AGT CAT ATG AAT GAG TTT GGA AAC TTC 3′ and the same 3′ oligonucleotide as for the full-length CycMs2. Before insertion into pYEUra3, the PCR fragments were subcloned into the plasmid pTz19U (Pharmacia) and sequenced with a Pharmacia T7 Sequencing Kit. Sequence homology searching was performed at the National Center of Biological Information (Bethesda, MD) using the BLAST network service. Sequence comparisons were done with the program GAP from the Genetics Computer Group (Madison, WI). 'PEST' rich regions were identified by using the PCGENE program PESTFIND (Intelligenetics).

## Results

### Functional selection systems for the isolation of plant cyclin genes in yeast

Plant cyclins that contain a highly conserved central domain (cyclin box) were isolated from a variety of plant species by PCR techniques. In contrast, the very low similarities between $G_1$ cyclins have so far excluded the possibility of using a similar approach for the isolation of this class of cyclin genes. To identify plant cyclins that act in the $G_1$ phase of the cell cycle, two yeast selection systems were employed that are based on the functional interaction of a plant cyclin with the yeast Cdc28 protein kinase at the $G_1/S$ transition.

(i) One approach exploited the fact that haploid yeast cells prepare for mating by cessation of cell division owing to the binding of a secreted peptide pheromone to receptors on the cell surface. We assumed that expression of a plant $G_1$ cyclin in yeast would override the cell cycle block imposed by α-pheromone and allow continued growth. This technique led to the isolation of *cycMs3*, an A-type alfalfa cyclin gene.

(ii) Another approach was based on the fact that yeast cells that are deficient in all three $G_1$ cyclin genes, CLN1, CLN2 and CLN3, are arrested in the $G_1$ phase and cannot divide. We assumed that expression of a plant $G_1$ cyclin in yeast might replace the function of the CLN genes in the cell cycle. In this approach, we isolated the D-type alfalfa cyclin gene *cycMs4*.

### A selection of alfalfa genes that overcome pheromone-induced cell cycle arrest in yeast

To identify higher plant genes which might be involved in the transition through the $G_1$ phase of the cell cycle, we selected for plant genes, that when expressed in yeast, would override the α-factor arrest of *MATa* yeast cells and allow normally pheromone-sensitive yeast cells to form colonies on medium containing α-pheromone. To eliminate most spontaneous mutations which might result in α-pheromone resistance in the yeast host, we constructed a homozygous yeast diploid strain, GA2201. This strain is hypersensitive to pheromone because it lacks the major α-pheromone-degrading enzyme Bar1, and also contains *MATa* at both the silent and expressed mating type loci, and thus cannot become *MATα* by mating-type switching, which would lead to α-pheromone resistance. A cDNA library was prepared from an alfalfa-cell suspension culture in a galactose-inducible yeast expression vector and used to transform this library into the yeast strain GA2201 (Fig. 1). Eight transformants formed colonies on media containing both α-pheromone and galactose. In the absence of pheromone, GA2201 cells transformed with any of the eight alfalfa cDNA clones did not show any significant difference

---

**Fig. I          Functional selection scheme of alfalfa genes that suppress pheromone-induced cell cycle arrest**

---

Selection procedure

Transform GA2201 with pYEUra3 alfalfa cDNA expression
library (*URA3 GAL1-10*/alfalfa cDNA)

↓

Select for uracil prototrophy

↓

Replica plate on to galactose- and pheromone-containing
uracil-free medium

↓

Isolate pYEUra3 plasmids from colonies that divide and
'Shmoo' and re-transform GA2201

---

*GA2201 was transformed with an alfalfa cDNA expression library in the pYEUra3 vector containing the URA3 selection marker and the cDNAs under the control of the GAL1-10 promoter. Transformants were selected for uracil prototrophy. Colonies were replica-plated on to galactose- and pheromone-containing, uracil-free medium. The majority of colonies were arrested in the $G_1$ phase and could not proliferate. Only eight colonies showed growth under these conditions. To omit the possibility that chromosomal mutations were responsible for the pheromone resistance, plasmid DNA was isolated from these colonies and used for transformation of GA2201 cells. Galactose-inducible growth on pheromone-containing medium indicated that the expression of the cDNA from the pYEUra3 vector was responsible for cell division.*

compared with non-transformed cells (compare Fig. 2C and 2A, respectively). In contrast to non-transformed GA2201 cells, or GA2201 cells transformed with the empty yeast–*E.coli* shuttle vector, cells expressing the plant cDNAs continued to grow and divide in the presence of α-pheromone up to 50 µg/ml.

Normally, the addition of a pheromone has multiple effects on yeast: it results in the activation of a transcriptional programme with characteristic morphological changes (projection formation called 'shmooing') and arrest in the $G_1$ phase of the cell cycle, as shown for the untransformed GA2201 cells (Fig. 2B). Expression of two of the eight alfalfa cDNAs still allowed yeast cells to respond to the pheromone-exhibiting projection formation (shmooing), but no cell cycle arrest

---

**Fig. 2**      **Alfalfa *cycMs3* overrides pheromone-induced cell cycle arrest in yeast**

---

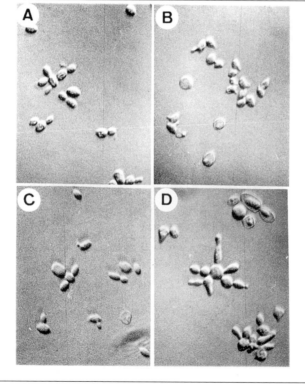

---

*(A) In the absence of α-pheromone, Saccharomyces cerevisiae GA2201 cells divide normally. (B) Four hours after addition of α-pheromone, S. cerevisiae GA2201 cells show elongated morphology ('shmoo' phenotype) and are $G_1$ arrested. (C) In the absence of α-pheromone, S. cerevisiae GA2201 cells containing ectopically expressed cycMs3 divide normally. (D) Four hours after addition of α-pheromone, S. cerevisiae GA2201 cells containing ectopically expressed cycMs3 show 'shmoo' phenotype, but still divide [22].*

(Fig. 2D). These data suggest that the products of these two alfalfa cDNAs specifically act on the yeast cell cycle machinery.

## cycMs3 encodes an A-type cyclin

Restriction mapping, hybridization, and partial sequencing of one of the two above described clones revealed strong homology to cyclins, and was therefore selected for further study. Because the translational start codon of the longest open-reading frame of 1143 nucleotides was not preceded by an in-frame stop codon, another cDNA library, that was derived from somatic alfalfa embryos, was hybridized with a radiolabelled cDNA fragment. The longest clone, termed *cycMs3* (for cyclin *Medicago sativa*), contained an open-reading frame that potentially encodes a 452-amino acid polypeptide with an estimated molecular mass of about 51 kDa [22].

Previously, we have isolated cDNA clones encoding two different alfalfa cyclins CycMs1 and CycMs2 [3,21]. Alignment of the three predicted alfalfa cyclins over the entire length (see Table 1) showed 72% identity between the mitotic cyclins CycMs1 and CycMs2 compared with 29% or 28% identity between mitotic cyclins and CycMs3. Considering the cyclin box only, the mitotic cyclins showed 84% identity compared with 41% and 42% with the respective sequence in the CycMs3 protein. CycMs1, CycMs2 and CycMs3 showed 33–45% identity to yeast B-type and mammalian A- and B-type cyclin boxes.

Besides the well-conserved cyclin box, A- and B-type cyclins contain a highly conserved destruction box motif near the N-terminus, which is required for ubiquitin-mediated degradation of these cyclins at the end of mitosis [12]. The consensus motif that is found in CycMs1 and CycMs2, and other plant mitotic cyclins, is the nonameric amino acid sequence RXA/VLGXIX(X)N. CycMs3 contains the sequence RAILQDVTN at the N-terminus which resembles the plant consensus, but only three out of the six consensus residues are identical (Fig. 3).

---

**Fig. 3**      **Schematic depiction of important features of the primary structure of the alfalfa CycMs1, CycMs2, CycMs3, and CycMs4 cyclins**

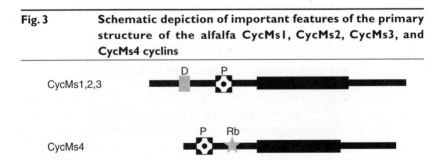

*CycMs1, CycMs2, CycMs3 and CycMs4 have a central cyclin box (closed bar) and 'PEST'-rich sequence elements (denoted by P above the box). CycMs1, CycMs2, and CycMs3 have a destruction box motif (indicated by D above the hatched box), whereas CycMs4 has an Rb-binding motif (denoted by Rb above star).*

'PEST'-rich regions, enriched in proline, glutamic acid, serine and threonine, are also found in the N-terminal region of the alfalfa CycMs1, CycMs2 and CycMs3 protein sequences (Fig. 3). The presence of a PEST-rich sequence renders proteins proteolytically unstable [27] and is found at the N- or C-terminus of many $G_1$ cyclins [28].

## Isolation of alfalfa genes that complement *CLN1*-, *CLN2*- and *CLN3*-deficient yeast cells

Deficiency of the *CLN1*, *CLN2*, and *CLN3* genes in yeast results in a $G_1$-phase arrest in 'START'. Since this effect is lethal, no colonies can form under these

---

**Fig. 4**     **Functional selection scheme of alfalfa genes complementing G1 yeast cyclins**

Selection procedure

Transform K3413 with pYEUra3 alfalfa cDNA expression library (*URA3 GAL1-10*/alfalfa cDNA)

$\downarrow$

Select for uracil prototrophy

$\downarrow$

Replica plate on to galactose- and methionine-containing uracil-free medium

$\downarrow$

Isolate pYEUra3 plasmids from colonies and re-transform K3413

---

*K3413 cells were transformed with an alfalfa cDNA expression library in the pYEUra3 vector, containing the URA3 selection marker, and the cDNAs under the control of the GAL1-10 promoter. Transformants were selected for uracil prototrophy. Colonies were replica-plated on to galactose- and methionine-containing, uracil-free medium. Owing to the repression of the CLN2 gene by methionine, the majority of colonies were arrested in the $G_1$ phase and could not proliferate. Only some colonies showed growth under these conditions. To omit the possibility that chromosomal mutations were responsible for the pheromone resistance, plasmid DNA was isolated from these colonies and used for transformation of K3413 cells. Galactose-inducible growth on methionine-containing medium indicated that the expression of the cDNA from the pYEUra3 vector was responsible for cell division.*

conditions. To select for alfalfa cyclins that can functionally substitute the yeast $G_1$ cyclins, the yeast strain K3413 [11] was used, which has deletions of the *CLN1, CLN2* and *CLN3* genes, but is conditionally viable on medium without methionine owing to the ectopic expression of *CLN2* under the control of the methionine-respressible *MET2* promoter (Fig. 4).

K3413 cells were tranformed with an alfalfa cDNA expression library which could be selectively expressed in medium containing galactose, but not glucose. From $10^6$ yeast transformants, 123 colones were able to grow on medium containing methionine (Fig. 4).

Isolation and analysis of the plasmids revealed two types of cDNA inserts. One type of insert, denoted as *cycMs4-1*, potentially encoded a novel alfalfa cyclin and was investigated further. To prove that expression of the *cycMs4-1* cDNA conferred the complementing activity, we introduced plasmid pYEUra3/cycMs4-1 (in which the *cycMs4-1* cDNA is under control of the galactose-inducible promoter *GAL1-10)* into the K3413 strain and compared it with K3413 cells that were transformed with the vector (Fig. 5). K3413 cells containing the pYEUra3 vector were viable only when *CLN2* was expressed under methionine-free conditions. When *CLN2* expression was repressed by addition of methionine, the yeast cells were not able to divide unless expression of *cycMs4-1* was induced by galactose (Fig. 5) indicating that *cycMs4-1* can substitute the function of the *CLN* genes in the yeast cell cycle.

---

**Fig. 5**      **Alfalfa cycMs4-1 complements the function of $G_1$ cyclins in yeast**

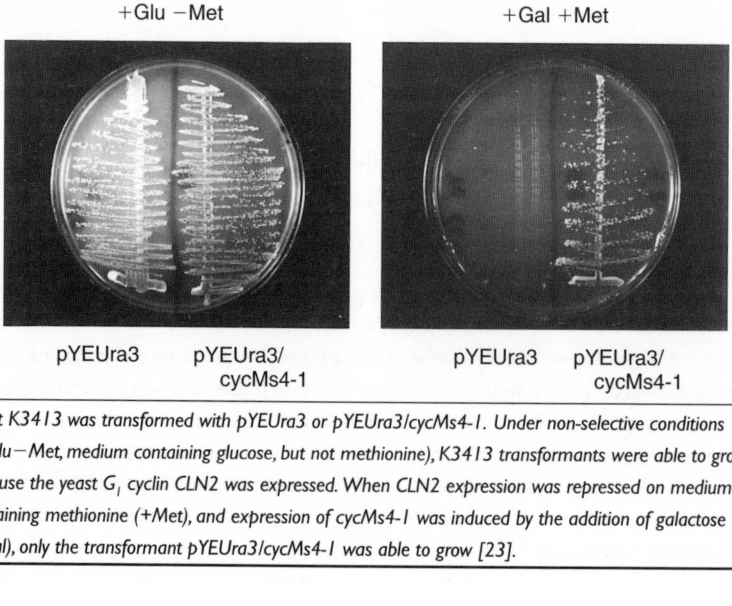

| +Glu −Met | +Gal +Met |
|:---:|:---:|

| pYEUra3     pYEUra3/ <br> cycMs4-1 | pYEUra3     pYEUra3/ <br> cycMs4-1 |
|:---:|:---:|

Yeast K3413 was transformed with pYEUra3 or pYEUra3/cycMs4-1. Under non-selective conditions (+Glu−Met, medium containing glucose, but not methionine), K3413 transformants were able to grow, because the yeast $G_1$ cyclin CLN2 was expressed. When CLN2 expression was repressed on medium containing methionine (+Met), and expression of cycMs4-1 was induced by the addition of galactose (+Gal), only the transformant pYEUra3/cycMs4-1 was able to grow [23].

### cycMs4 encodes a D-type cyclin

The pYEUra3/cycMs4-1 plasmid contained a 1000 bp insert. However, the translational start codon of the longest open-reading frame was not preceded by an in-frame stop codon. To obtain a full-length cDNA, another cDNA library, that was derived from somatic *M. sativa* embryos, was hybridized with a radiolabelled cDNA fragment. The longest clone, termed *cycMs4* (for cyclin *M. sativa*), contained an insert of 1857 bp. The open-reading frame of the previously isolated cDNA could be extended by 348 bp. Several in-frame stop codons upstream of the first possible ATG indicate that this clone represents a full-length cDNA sequence. The identified open-reading frame potentially encodes a 386 amino acid polypeptide with an estimated molecular mass of 44 kDa.

Alignment of the predicted CycMs4 protein sequence with the previously identified alfalfa CycMs1, CycMs2 and CycMs3 proteins over the entire length revealed only 26–29% identity to these cyclins (Table 1). Considering the cyclin box only, identity scores up to 42% were obtained with the CycMs4 protein. Whereas the A- and B-type alfalfa cyclins contain a destruction box motif, responsible for mitotic degradation in animals [12], no such motif could be found in CycMs4 (Fig. 3). Sequence comparison of CycMs4 with other plant cyclins showed highest identity (56%) to the *Arabidopsis* cyclin δ3 [24]. Comparison of the CycMs4, and the three δ-type *Arabidopsis* cyclin sequences isolated by Soni et al. [24], with current data banks showed highest identity to mammalian D-type cyclins [29,30]. Alignment of CycMs4 and human cyclin D1 revealed 33% identity over the entire length of the proteins. The predicted CycMs4 protein sequence has, like its related *Arabidopsis* δ-type cyclins, the typical structural elements of $G_1$ cyclins, a cyclin box in the central region, and 'PEST'-rich regions in the N- and C-terminal domains (Fig. 3). These features are typical for short-lived $G_1$ cyclins and suggest that a 'PEST'-dependent protein-degradation machinery must also exist in plants.

| Table 1 | Sequence comparison of the predicted protein sequences of the *cycMs1, cycMs2, cycMs3* and *cycMs4* genes | | |
|---|---|---|---|
| | **CycMs2** | **CycMs3** | **CycMs4** |
| CycMs1 | 72% | 29% | 26% |
| CycMs2 | | 28% | 27% |
| CycMs3 | | | 28% |

The identity scores were calculated over the entire length of the sequences.

Mammalian D-type cyclins interact directly with pRB [5,6] through the amino acid sequence motif LxCxE (where x is any amino acid). CycMs4 and all three *Arabidopsis* δ-type cyclins contain an Rb-binding motif in the N-terminal regions indicating that D-type cyclins might also interact with Rb homologues in plants (Fig. 3). Overall, the sequence similarity and the presence of all the hallmarks of mammalian D-type cyclins in these plant sequences suggest that the $G_1$ phase of the plant cell cycle may be regulated by a similar mechanism as the cell cycle of mammalian cells (see J.A.H. Murray et al., Chapter 5)

# Functional studies of plant cyclins in yeast: yeast pheromone-dependent cell cycle arrest is exclusively suppressed by cycMs3

One surprising result of our yeast selection approaches was the fact that we did not isolate the *cycMs1*, *cycMs2* or *cycMs4* gene in the pheromone-suppression system. To understand the basis of these differences, we investigated the properties of the different cyclins in more detail.

The CycMs1, CycMs2 and CycMs3 proteins have highly similar structures, containing a central cyclin box and a destruction motif at their N-termini. Moreover, all three proteins show the greatest similarity to A- and B-type animal cyclins. Why did not we then isolate the *cycMs1* and *cycMs2* genes in the pheromone-dependent selection system? One clue for solving this problem came from expression experiments with the full-length *cycMs3* gene in yeast. The original selection procedure for pheromone-resistant colony formation only yielded a truncated version of the *cycMs3* gene, which was used for the isolation of the entire coding region from another alfalfa libary [21]. When the full-length *cycMs3* gene was expressed in GA2201 cells on pheromone-containing medium, cell proliferation was considerably reduced compared with the truncated *cycMs3* construct (data not shown). This indicated that the truncation of the CycMs3 protein strongly affected the ability of CycMs3 to function in the $G_1$/S transition of the yeast cell cycle. The truncation deleted the N-terminal 73 amino acids of CycMs3, which contained the entire destruction motif, but left the cyclin box intact. We suspected that deletion of the destruction box was responsible for the ability to override cell cycle arrest.

To investigate the possibility that expression in yeast of any B-type cyclin would suffice to suppress the pheromone-induced arrest of the cell cycle, we expressed the full-length and a truncated version of the *cycMs2* gene from the episomal yeast–*E. coli* shuttle vector pYEUra3, which was used before to obtain pheromone-resistant cell divisions with the *cycMs3* gene. GA2201 cells that were transformed with pYEUra3/*cycMs2*, pYEUra2/*cycMs2*Δ, or pYEUra3/*cycMs3*Δ grew equally well on pheromone-free medium, but failed to grow on medium containing pheromone and glucose (Fig. 6). Since both the *cycMs2* and the *cycMs3* genes were expressed under control of the GAL1-10 promoter, which is only slightly active on glucose, the pheromone-induced arrest of the cell cycle was not expected to be suppressed under these conditions. However, when the cells were grown on medium containing pheromone and galactose, which resulted in 100-fold higher expression levels of the *cycMs2* and *cycMs3* constructs, only *pYEUra3/ cycMs3*Δ colonies could form (Fig. 6). This, therefore, disproved our theory that expression of any B-type cyclin would lead to pheromone-resistant cell division when the destruction box was missing.

Since *cycMs3* was found to complement a yeast mutant that was deficient in the function of all three *CLN* genes, we also tried to answer the question as to

**Fig. 6**     **Yeast pheromone-dependent cell cycle arrest is exclusively suppressed by CycMs3Δ.**

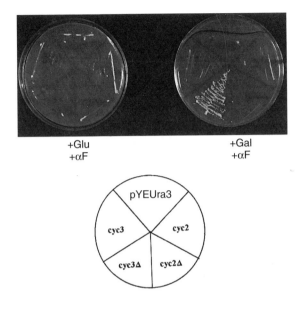

+Glu                    +Gal
+αF                     +αF

In the presence of α-pheromone (αF), GA2201 cells that were transformed with an empty pYEUra3 vector were arrested in the cell cycle and no colonies formed, irrespective of whether the medium contained glucose (+Glu) or galactose (+Gal). GA2201 cells that were transformed with pYEUra3/cycMs2, pYEUra3/cycMs2Δ, or pYEUra3/cycMs3 could not suppress the pheromone-dependent arrest of the cell cycle. In contrast, galactose-induced expression of cycMs3Δ led to colony formation in the presence of pheromone.

whether *cycMs4* would suppress the pheromone-dependent arrest of the cell cycle. However, expression of the *cycMs4* gene in GA2201 cells on pheromone-containing medium did not lead to colony formation (data not shown). This result stresses the functional difference between the different classes of cyclin genes, indicating that the expression of a protein containing the cyclin box alone is not sufficient to mediate pheromone-resistant cell division.

# Over-expression of cycMs2 leads to a lethal phenotype in yeast

During the yeast expression studies with the *cycMs2* constructs, we made an observation that clarified why this gene could never be identified in a selection procedure that is based on over-expression of plant cDNAs. We analysed GA2201 cells that were transformed with episomal expression vectors containing no gene construct, the full-length, or the truncated *cycMs2* and *cycMs3* gene constructs. On medium containing glucose, all transformants grew equally well (Fig. 7). When the expression of the plant cyclin constructs was induced on galactose-containing medium, GA2201 cells which were transformed with any of the *cycMs2* constructs could not proliferate and no colony formation was observed (Fig. 7). In contrast, expression of the *cycMs3* constructs did not arrest the cell cycle and colony formation was similar to GA2201 cells which were transformed with the empty pYEUra3 vector (Fig. 7). Although the basis for the lethal phenotype produced by the expression of the *cycMs2* gene constructs is currently unclear, the results clearly

---

**Fig. 7**         **Over-expression of the cycMs2 gene leads to a lethal phenotype in yeast.**

---

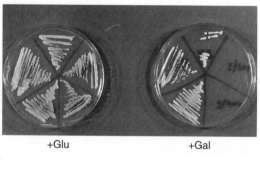

+Glu                       +Gal

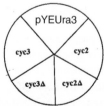

---

*GA2201 cells that were transformed with the empty pYEUra3 vector, the pYEUra3/cycMs3, or the pYEUra3/cycMs3Δ construct, formed colonies on glucose- (+Glu) or galactose- (+Gal) containing medium. GA2201 cells that were transformed with the pYEUra3/cycMs2, or pYEUra3/cycMs2Δ construct, however, only formed colonies on glucose-containing medium, indicating that galactose-induced over-expression of full-length and truncated CycMs2 is lethal for yeast.*

show that any over-expression strategy in yeast will fail to isolate the class of *cycMs2* cyclins.

## Discussion

The regulatory principles of cell cycle control, including its molecular components, are highly conserved in all eukaryotes. This finding made it possible to develop a variety of techniques to isolate and study the function of cell cycle factors from higher eukaryotes in the genetically tractable yeast. In this report, we show the usefulness of this concept for the isolation and functional studies of plant cyclins.

Any gene that is highly conserved can nowadays be isolated from any organism by PCR-based techniques with partially redundant primers. However, it is much more difficult to isolate genes that are conserved in function, but not necessarily in primary sequence. Ultimately, protein biochemistry may be necessary to purify and partially sequence the protein of interest. A possible solution to this problem may be to use any of a set of yeast techniques that have revolutionized molecular biology in the past 10 years. Complementation of yeast cells with cDNA expression libraries was the first success story in this respect and generated the first clones for human CDC2 [31]. We [23], and others [24], have used this approach to isolate the first D-type cyclin genes from plants. The limitation of the complementation technique is the availability of the appropriate yeast mutants.

The problem that many researchers are faced with today is that they have identified a certain gene which plays a role in a particular biochemical process, and they want to know more about the immediate regulation of this gene product. For this kind of problem, the so called two-hybrid system can provide an answer in the search for interacting partners [32]. In this approach, a bait is produced by the fusion of the gene of interest with a transcription factor that by itself is unable to drive the expression of a selection marker gene. For efficient gene expression to occur, the yeast transcription factor requires the association with a transcriptional activator that, however, is unable to recognize the transcription factor. The fusion of a cDNA library to the transcriptional activator can overcome this problem. If the appropriate factor can interact with the bait, the transcriptional activator and the transcription factor come into contact, resulting in expression of the selection marker gene.

What does one do if one wants to isolate components further upstream in a signalling cascade? Another yeast-based selection system can be of considerable help. Over-expression of wild-type yeast genes often rescues temperature-sensitive yeast mutants under restrictive conditions, thereby identifying components that affect the regulation of the biochemical process under investigation. By analogy, we have used this approach to identify genes that act on the yeast cell cycle machinery under pheromone-induced inhibitory conditions. The identification of a variety of upstream factors, which are involved in the signalling process from the plasma membrane to the nucleus, identified several novel plant factors that are under further investigation.

Many factors induce transcription of particular cell cycle genes at different phases, or during re-entry into the cell cycle. Is there a yeast technology available for identifying the transcription factors responsible for gene activation? Studies in our laboratory have shown that it is possible to activate a particular plant promoter in yeast when the correct transcription factor is expressed in the same cell [33]. We have developed a selection system in yeast that is based on the fusion of the heterologous promoter with a yeast selection marker gene. Expression of a cDNA expression library yielded novel transactivating factors to this particular promoter (J. Rüth et al., unpublished work). Deletion versions of the promoter also allowed the identification and study of the binding sites of the respective transcription factors, showing that this method can be very useful for the analysis of gene expression studies.

Although yeast techniques are available for almost any kind of problem, one word of caution must be added. Despite the similarities of many processes in yeast and higher eukaryotes, there are also many differences that most likely stem from the different lifestyles and the higher complexity of multicellular organisms. Therefore, it is clearly not sufficient to study yeast by itself or in a heterologous context. Yeast probably can provide a good idea of what might be going on, but it still has to be shown if the gene of the higher eukaryote functions similarly or in another way.

*The work was supported by a Lise Meitner fellowship to M.D. and grants (nos. S6004-BIO, P10020-MOB, and P10394) from the Austrian Science Foundation.*

## References

1.   Nasmyth, K. (1993) Curr. Opin. Cell Biol. **2**, 166–170
2.   Pines, J. (1993) Trends Biol. Sci. **18**, 195–197
3.   Hirt, J. and Heberle-Bors, E. (1994). Sem. Dev. Biol. **5**, 147–154
4.   Poon, R.Y.C., Yamashita, K., Adamczewski, J.P., Hunt, T. and Shuttleworth, J. (1993) EMBO J. **12**, 3123–3132
5.   Dowdy, S.F., Hinds, P.W., Louie, K., Reed, S.I., Arnold, A. and Weinberg, R.A. (1993) Cell **73**, 499–511
6.   Kato, J., Matsushime, H., Hiebert, S.W., Ewen, M.E. and Sherr, C.J. (1993) Genes Dev. **7**, 331–342
7.   Peter, M., Gartner, A., Horecka, J., Ammerer, G. and Herskowitz, I. (1993) Cell **73**, 747–769
8.   Peeper, S.D., Parker, L.L., Ewen, M.E., Toebes, M., Hall, F.L., Xu, M., Zantema, A., van der Eb, A.J. and Piwnica-Worms, H. (1993) EMBO J. **12**, 1947–1954
9.   Booher, R.N., Deshaies, R.J. and Kirshner, M.W. (1993) EMBO J. **12**, 3417–3426
10.  Maridor, G., Gallant, P., Golsteyn, R. and Nigg, E.A. (1993) J. Cell Sci. **106**, 535–544
11.  Amon, A., Tyers, M., Futcher, B. and Nasmyth, K. (1993) Cell **74**, 993–1007
12.  Glotzer, M., Murray, A.W. and Kirschner, M.W. (1991) Nature (London) **349**, 132–138
13.  Tyers, M., Tokiwa, G., Nash, R. and Futcher, B. (1992) EMBO J. **11**, 1773–1784
14.  Sherr, C.J. (1993) Cell **73**, 1059–1065

15. Hirt, H., Mink, M., Pfosser, M., Bögre, L., Györgyey, J., Jonak, C., Gartner, A., Dudits, D. and Heberle-Bors, E. (1992) Plant Cell **4**, 1531–1538
16. Hemerly, A., Bergounioux, C., Van Montagu, M., Inzé, D. and Ferreira, P. (1992) Proc. Natl. Acad. Sci. U.S.A. **89**, 3295–3299
17. Day, I.S. and Reddy, A.S.N. (1994) Biochim. Biophys. Acta **1218**, 115–118
18. Ferreira, P., Hemerly, A., Engler, J.D.A., Bergounioux, C., Burssens, S., Van Montagu, M. and Inzé. D. (1994) Proc. Natl. Acad. Sci. U.S.A. **91**, 11313–11317
19. Fobert, P.R., Coen, E.S., Murphy, G.J.P. and Doonan, J.H. (1994) EMBO J. **13**, 616–624
20. Hata, S., Kouchi, H., Suzuka, I. and Ishii, T. (1991) EMBO J. **10**, 2681–2688
21. Renaudin, J.-P., Colasanti, J., Rime, H., Yuan, Z. and Sundaresan, V. (1994) Proc. Natl. Acad. Sci. U.S.A. **91**, 7375–7379
22. Meskiene, I., Bögre, L., Dahl, M., Pirk, M., Ha, D.T.H., Swoboda, I., Heberle-Bors, E., Ammerer, G. and Hirt, H. (1995) Plant Cell **7**, 759–771
23. Dahl, M., Meskrene, L., Bögre, L., Ha, D.T.H., Swoboda, R., Hubman, R., Hirt, H. and Heberle-Bors, E. (1955) Plant Cell **7**, 1847–1887
24. Soni, R., Carmichael, J.P., Shah, Z.J. and Murray, J.A.H. (1995) Plant Cell **7**, 85–103
25. Gietz, D., St. Jean, A., Woods, R.A. and Schiestl, R.H. (1992) Nucleic Acids Res. **20**, 1425
26. Sambrook, J., Fritsch, E.F. and Maniatis, T. (1989) Molecular Cloning, Cold Spring Harbor Laboratory Press, Cold Spring Harbor, NY
27. Rogers, S., Wells, R. and Rechsteiner, M. (1986) Science **234**, 364–368
28. Tyers, M., Tokiwa, G. and Futcher, B. (1993) EMBO J. **12**, 1955–1968
29. Lew, D.J., Dulic, V. and Reed, S.I. (1991) Cell **66**, 1197–1206
30. Xiong, Y., Connolly, T., Futcher, B. and Beach, D. (1991) Cell **65**, 691–699
31. Lee, M.G. and Nurse, P. (1987) Nature (London) **327**, 31–35
32. Feilds, S. and Song, O.K. (1989) Nature (London) **340**, 245–246
33. Rüth, J., Schweyen, R.J. and Hirt, H. (1994) Plant Mol. Biol. **25**, 323–328

# 7

# The ubiquitin-dependent proteolytic pathway and cell cycle control

**Bertrand Plesse, Jacqueline Fleck\* and Pascal Genschik**

Institut de Biologie Moléculaire des Plantes du CNRS, Université Louis Pasteur, 12 rue du Général Zimmer, 67084 Strasbourg cedex, France

## Introduction

The transition from one phase of the cell cycle to another is accomplished through changes of activity of key regulatory proteins. Re-entry and correct progress through the cell cycle are thus under the control of a succession of events where steps of protein activation [1,2] alternate with steps of protein degradation. More and more evidence emphasizes that the programmed and co-ordinated destruction of most of the regulatory proteins involved in cell cycle progress, such as the cyclins, the cyclin-dependent kinase inhibitors (CKIs) and oncoproteins (for reviews see [3–6]), are under the control of a system which plays a central role in the cell's regulatory arsenal: the ubiquitin-dependent proteolytic pathway.

Ubiquitin-mediated proteolysis is highly conserved in eukaryotes. Although it is one of the best-characterized proteolysis processes, some of its various biological functions have been determined only recently [7,8]. A simplified view of the ubiquitin pathway is depicted in Fig.1. Degradation of a protein *via* the ubiquitin-dependent pathway is a two-step process: the protein is first tagged by the covalent attachment of ubiquitin, and then subsequently degraded. Conjugation of ubiquitin to the protein involves a cascade of three enzymes: E1, E2 and E3. The E1 (ubiquitin-activating) enzymes form a high-energy bond between a cysteine at its active site and the C-terminal glycine of ubiquitin, consuming the energy of ATP hydrolysis. Ubiquitin is then transferred to one of the multiple E2 (ubiquitin-conjugating or UBC) enzymes. At least twelve such E2 enzymes have been identified in yeast [8]. The specificity of the pathways resides, in part at least, in the multiple E2 enzymes and in their ability to form multimers necessary for the recognition of specific substrates. Finally, ubiquitin is linked either directly to a lysine from the substrate protein or in concert with a third class of enzymes E3 (ubiquitin protein ligases), which select the protein to be tagged and present it to the E2 to be ubiquitinated. The E3 itself can transfer ubiquitin to the target protein [9].

\*To whom correspondence should be addressed.

**Fig. I**          **The ubiquitin pathway**

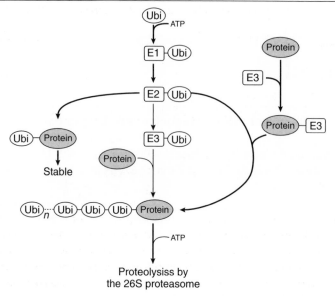

*The usual pathways are represented by large arrows leading to a stable monoubiquitinated protein (on the left) or to a multiubiquitinated protein (on the right) which will undergo proteolysis. The narrow arrow represents a third pathway in which E3 enzyme is the ubiquitin donor to the target protein. This possibility has been described only once. Ubi, ubiquitin; E1, ubiquitin-activating enzyme; E2, ubiquitin-conjugating enzyme; E3, ubiquitin protein ligase.*

E3 enzymes contribute most significantly to the specificity of the reaction. At this point, two different ways are opened: (i) the protein is multiubiquitinated and thus degraded by the 26S proteasome complex in an ATP-dependent manner. In this case, ubiquitin is removed from the substrate by an isopeptidase and is recycled; (ii) the protein is monoubiquitinated and thus stable.

In this review, we will discuss our current understanding of the key role ubiquitin plays in cell cycle progression.

## Ubiquitin and the cell cycle in animals and yeast

### Proteins degraded by the proteolytic ubiquitin-dependent pathway

The proteins involved in the cell cycle (see Fig.2), known to be degraded by the ubiquitin-mediated proteolysis pathway, are the different cyclins, the CKIs, some oncoproteins and tumour suppressors.

**Fig. 2**     **Schematic representation of the branchpoints between the ubiquitin-dependent proteolytic pathway and the cell cycle**

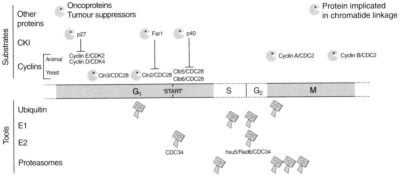

*Data from animal and yeast cell cycle studies have been compiled. At the top of the diagram (indicated by heads), are the known substrates of this pathway. The tools implicated in their degradation have not been reported. The closed bars represent an inhibiting effect. On the bottom (indicated by placards) are the tools which have been shown to be implicated in cell cycle control, but for which the substrates are not defined. The different phases of the cell cycle are not drawn to scale. CKI, cyclin-dependent kinase inhibitor; for explanation of E1 and E2, see legend of Fig.1.*

## S- and M-phase cyclins

The observation that cyclins disappeared at the end of mitosis [10], and that their stabilization arrests the cell cycle [11], illustrated the role of protein degradation in the control of the cell cycle. The demonstration by Glotzer et al. [12], that multi-ubiquitinated forms of a truncated cyclin were present in the course of its degradation, illustrated a link between ubiquitin and cyclin degradation. These observations were correlated by Hershko et al. [13], who stabilized cyclins by the use of a methylated ubiquitin unable to form polyubiquitin chains, a process that has been shown to be required for proteolysis. Recently, Mahaffey et al. [14] detected ubiquitin conjugates of full-length *Xenopus* cyclin B2 in mitotic extracts.

A highly conserved sequence, RxxLxxIxN, in the N-terminus of mitotic cyclins was important for this degradation [11]. This 'destruction box' is found in most of S- and M-phase cyclins (see J.P. Renaudin et al., chapter 4). When this box is removed or mutated, cyclins A and B are stabilized in cell-free systems [12,15–17], and addition of this box to an unrelated protein (protein A of *Staphylococcus aureus*) allowed the specific mitotic destruction of the chimerical protein [12]. Nevertheless, other observations [12] suggested that this 'destruction box', even if it is necessary, is not always sufficient to allow recognition of cyclin for proteolysis. Indeed, in *Xenopus* extracts, a truncated protein composed only of the N-terminal fragment of sea urchin cyclin B is able to stabilize endogenous cyclin and undergo mitosis-specific proteolysis [12], while an equivalent truncated protein consisting of the N-terminal fragment of cyclin B2 is able to stabilize cyclin in mitotic extracts, but is unfit for degradation [17].

An additional requirement for proteolysis may be the binding of the regulatory subunit (cyclin) to the catalytic subunit [cyclin-dependent kinase (CDK)]. Indeed, mutations affecting the binding of $Xenopus$ cyclins A and B2 to their CDK stabilizes these cyclins, even if the 'destruction box' is intact [18,19]. Nevertheless, this interaction is not required for $Xenopus$ cyclin B1 or for sea urchin cyclin B [12,18–20]. Signalling mitotic cyclin for destruction is thus not only dependent on a simple determinant, such as a 'destruction box', but also involves a more complex recognition code, including in some cases binding to the catalytic subunit. These multiple requirements for cyclin destruction may account for the precision and specificity of this event.

A further element in the complexity of degradation is that the timing of cyclin A and B destruction is different [21]. Both require the 'destruction box' and ubiquitination [22], but cyclin A is degraded in metaphase, whereas cyclin B is degraded during metaphase/anaphase transition. Moreover, cyclin B, but not cyclin A, is stabilized when spindle assembly is inhibited [23–25]. The spindle assembly checkpoint may prevent cyclin B destruction, preventing an exit from mitosis and incorrect segregation of the chromosomes.

Is earlier timing of cyclin A proteolysis due to an activation of a specific sub-pathway in the cyclin proteolysis or, simply, is cyclin A a better substrate for the general degradation pathway?

One way to answer this question was to identify the components responsible for the specificity and regulation of cyclin ubiquitination. This was first explored by Hershko et al. [26], who fractionated clam oocyte extracts and reconstituted cyclin B ubiquitination from three fractions. All fractions contained enzyme activities involved in the ubiquitin-dependent proteolysis. One contained an E1-like activity, the second one a novel species of the E2 enzymes (E2-C) and the third, in association with particulate material, an E3 activity. This last fraction contained a 1500 kDa complex with a cyclin-selective ligase called cyclosome [22]. Only the mitotic extracts naturally contain this E3 activity, whereas, in interphase extracts, cyclosome was inactive. Nevertheless, this activity can be restored by supplementing the extracts with CDC2 kinase.

The same approach was used by King et al. [27], who fractionated mitotic $Xenopus$ extracts and identified: (i) an E2 activity recognized by an antibody against the human homologue to the yeast E2, UBC4 and (ii) a large E3 complex. Nevertheless, the double mutant $ubc4ubc5$ failed to show cell cycle arrest [28]. This suggests that UBC4 is not the only E2 involved in cyclin ubiquitination and is not absolutely necessary. Seufert et al. [29], found a yeast temperature-sensitive mutant in another E2, $ubc9$, presenting an arrest at the $G_2$ or early M phase in the cell cycle with stabilized Clb5, an S-phase cyclin, and Clb2, an M-phase cyclin. The large E3 complex has different activities in interphase and mitosis. This complex contained CDC27 and CDC16, compounds found in mammalian cells, localized to the centrosome and mitotic spindle [30]. Injection of CDC27 antibody into HeLa cells caused a cell cycle arrest in metaphase. In yeast, these two proteins, together with CDC23, are involved in degradation of cyclin B [31]. In budding yeast, a genetic relation between *CDC23* and *UBC4* exists [31].

Although identification of the diverse components of cyclin ubiquitination and destruction is in progress, we cannot yet explain the difference of timing in cyclin A and B proteolysis.

Destruction of cyclin B at the metaphase/anaphase transition leads to the inactivation of the maturation-promoting factor (MPF), composed of cyclin B and CDK1. This inactivation is a negative feedback loop since MPF itself switches on the cyclin-destruction machinery. Indeed, cyclin A and B degradation may be triggered by adding an active CDK to interphase *Xenopus* egg extracts [32], and a stable truncated version of cyclin B can activate cyclin destruction in interphase extracts of Clam embryo [15]. Cyclin A added to these interphase extracts is unable to turn on cyclin degradation [15]. In fact, it delays both cyclin A and B proteolysis [33]. Cyclin A–CDK kinase activity preceding MPF activation, and the inability of cyclin A to turn on cyclin proteolysis, may be essential to prevent premature inactivation of MPF before the metaphase/anaphase transition. Nevertheless, even if MPF is able to turn on the cyclin-destruction machinery, there is a lag time between MPF activation and cyclin degradation. This lag suggests serial events from MPF activation to cyclin destruction, which could in part involve protein phosphorylation. In amphibian eggs, an okadaic-acid-sensitive phosphatase negatively controls the cyclin-degradation pathway [34]. Thus, this lag time between MPF activation and cyclin destruction may be another point of control before turning on the proteolysis machinery.

What is the effect of stabilizing mitotic cyclins? Obviously, it is to prevent MPF inactivation, and thus, chromosome decondensation, cytokinesis and spindle disassembly, i.e. exit from anaphase. More controversial is the stage of the cell cycle arrest observed. Whereas, in *Xenopus* extracts, expression of a mutated version of cyclin lacking the 'destruction box' induced arrest in metaphase [11–17], other studies have failed to show a correlation between MPF inactivation and sister chromatid separation, indicating that MPF inactivation is not required for chromosomes to enter anaphase in *Xenopus* extracts and yeast [20–35]. These results are in apparent agreement with some observations of Gallant and Nigg [36] who found that an N-terminal truncated form of avian cyclin was able to arrest HeLa cells in a state where chromosome separation had begun. Holloway et al. [20] further demonstrated that MPF inactivation is not required for sister chromatid separation, but that the 'destruction box'-dependent proteolysis machinery is nevertheless implicated in the degradation of a putative protein involved in chromatid linkages. This is confirmed because CDC23, CDC27 and CDC16, components of the recognition machinery for destruction of cyclins (part of a complex with an E3-like activity), are needed for transition from metaphase to anaphase in human and yeast cells [30,31]. The putative protein involved in chromatid linkages has not yet been identified, but some evidence indicates a link between two kinetochore proteins and the ubiquitin system. CENP-E, a putative kinetochore motor [37], has a destruction kinetic following that of cyclin B and possesses a sequence very similar to the 'destruction box' [38]. Moreover, a genetic and biochemical interaction between another kinetochore protein (Cbf2p/Ndc10p) and the E2 enzyme, CDC34, has been shown in yeast [39]. Over-expression of CDC34 suppresses the temperature-sensitive phenotype of the *ndc10-1* mutant and Cbf2p is multiubiquitinated and is a substrate of CDC34 *in vivo*.

In summary, it can be concluded that: (i) turning on the mitotic cyclins destruction pathway requires the MPF activity, and thus depends on the activity of a CDC–cyclin complex; (ii) turning off the machinery is dependent on another CDK activity. The destruction of mitotic cyclin is turned off in late $G_1$, when $G_1$ CDK is activated [40].

## $G_1$ cyclins

$G_1$ cyclins (Cln1, Cln2, Cln3 in budding yeast) regulate entry into the cell cycle, the 'START', by activating the CDC28 kinase activity [41,42]. So far, the mechanism of $G_1$ cyclin degradation has only been studied in yeast.

$G_1$ cyclins are very unstable proteins [43–45], but their stability is restored in some mutant cells. Cln2 and Cln3 are stabilized in *cdc34ts* mutant cells [46], while their multiubiquitination is lowered. Note that the *CDC34* gene codes for the yeast UBC enzyme, UBC3 [47], mutation of which leads to cell cycle defects; further addition of wild-type CDC34 to extracts restores ubiquitination [46–48]. Cln3 is also strongly stabilized in two other double mutants, one for two UBC enzymes *ubc4ubc5*, the other *pre1–1pre1–4* for two genes encoding proteins from the 20S complex known to be part of the 26S proteasome (the multicatalytic proteinase complex at the end of the ubiquitin pathway) [46].

A more detailed analysis of the structure of $G_1$ cyclins may explain their instability. Indeed, even if all $G_1$ cyclins lack the consensus sequence for the 'destruction box', they possess 'PEST' sequences, i. e. sequences rich in Pro, Glu, Ser and Thr, which have been postulated to be a signal for rapid proteolysis [49]. Such 'PEST' sequences are found in many unstable proteins, including mammalian C- and D-type cyclins and $G_1$ plant cyclins (see J.P. Renaudin et al., chapter 4). Interestingly, Cln3, Cln3.1 and Cln3.2 could be stabilized in yeast cells after removal of the C-terminal third part containing the different 'PEST'-rich sequences [44–50] and, in parallel, the fusion of the whole Cln2 'PEST'-rich C-terminal tail to an unrelated protein was sufficient to destabilize this protein [51]. Nevertheless, as a fragment of 37 amino acids encompassing only one 'PEST'-like motif fused to a heterologous protein could not promote its degradation, the 'PEST' motif is necessary, but not sufficient, for cyclin degradation [51]. A more detailed understanding of the mechanism responsible for $G_1$ cyclin instability has been provided by two studies on Cln2 [48] and Cln3 [46]. The minimal defined sequence able to destabilize a heterologous protein was composed of 85 amino acids of the

Cln3 C-terminal tail and encompasses two 'PEST' motifs, but also contains two regions required for efficient phosphorylation.

Phosphorylation seems to play also an important role in the signalling of cyclin degradation. A truncated version of the C-terminal tail of Cln2 abolished degradation and multiubiquitination of the cyclin, as well as its phosphorylation. Extracts from a *cdc28ts* mutant of *Saccharomyces cerevisiae* failed to restore efficient phosphorylation and multiubiquitination of Cln2. However, these defects can be partially rescued by complementation of these extracts with activated CDC28 and *in vitro* activated CDC28 promoted extensive phosphorylation of Cln2. However, the stabilization of Cln2 *in vivo* was only partial in the *cdc28ts* mutant [48]. On the other hand, a mutation on a potential phosphorylation site of the Ser/Thr kinase CDC28 (in the 85 amino acid sequence sufficient to drive specific proteolysis of a fusion protein) stabilized a chimeric protein composed of β-galactosidase and the C-terminal tail of Cln3 in the wild-type strain. This chimerical protein was stable when expressed in a *cdc28ts* mutant strain [46]. Phosphorylation of Cln2 was not disturbed by inefficient multiubiquitination in a *cdc34ts* mutant [48]. Thus, CDC28-dependent phosphorylation is part of the signal to target $G_1$ cyclin for proteolysis.

Together these results clearly implicate the ubiquitin-dependent proteolytic pathway in the control of the 'START' by regulating $G_1$ cyclin degradation.

## Cyclin-dependent kinase inhibitors

Another group of proteins implicated in cell cycle control, and which are broken down by the ubiquitin-dependent pathway, are the CKIs [52]. Progression through the cell cycle is regulated by the periodic and specific association of a catalytic kinase subunit and a regulatory subunit, to form an active kinase. Modulation of the kinase activity may be regulated at several levels: (i) accumulation of the catalytic subunit alone; (ii) phosphorylation or dephosphorylation of the complex; (iii) degradation of the regulatory subunit, or finally (iv) complexing the cyclin–CDK complex to a third element, a CKI. Three of these CKIs are degraded through the ubiquitin pathway: Far1, p40$^{sic1}$ and p27$^{kip1}$.

Far1 is an inhibitor of the $G_1$ cyclin Cln2 [53] in budding yeast. It is induced in response to the mating pheromone α-factor. Degradation of Far1 is CDC34-dependent [54]. p40 inhibits S-phase CDKs (CDC28 in association with Clb5 or Clb6), but not the $G_1$ CDKs [55]. Clb5 and Clb6 cyclins accumulate before 'START', but cannot drive any kinase activity when associated with CDC28 until after 'START' because of the association of p40. At this point, p40 is degraded, allowing Clb5/CDC28 and Clb6/CDC28 to fulfil their role in DNA replication. p40 is stabilized in a *cdc34ts* strain, possibly explaining the phenotype of these mutant cells which arrest in late $G_1$. This is confirmed in a *cdc34sic1* double mutant able to replicate DNA. However, this double mutant is lethal. It fails to undergo nuclear division and accumulates as large budded cells with a single nucleus, suggesting an essential function for CDC34 in S or $G_2$ phase, in addition to degradation of the proteins during the $G_1$ phase [55].

p27 is a mammalian CKI. It binds and inhibits cyclin E/CDK2 and cyclin D/CDK4 in response to two antimitogenic signals [56]: transforming growth factor β (TGFβ) and cell–cell contact. The phenotype of this inhibition is an arrest in $G_1$ phase. Many results have shown that p27 is degraded in a ubiquitin-dependent manner [57]: for example, (i) an inhibitor of the chymotrypsic site of the proteasome stabilized p27 *in vivo*, and allowed detection of multiubiquitinated forms of p27; (ii) a murine cell line expressing a mutant E1 enzyme accumulated stable p27; (iii) p27 is degraded in an ATP-dependent manner in rabbit reticulocyte lysate and this turnover is enhanced when UBC2 (RAD6) and UBC3 (CDC34) human homologues are supplied to the lysate; (iv) removal of the proteasome fraction from the lysate or addition of proteasome inhibitor stabilizes p27.

The ubiquitin degradation pathway is thus implicated in two new points in the cell cycle: (i) control of DNA replication, allowed if p40 is degraded, and (ii) response to external signals, through the destruction of two CKIs, Far1 and p27, both of which lie at the end of a transduction pathway sensitive to external changes.

## Oncoproteins and tumour suppressors

A large set of oncoproteins or tumour suppressors are degraded by the ubiquitin-mediated proteolytic pathway. Among them are N-myc, C-myc, C-fos, p53, E1A, C-mos and C-jun [58–60]. The ubiquitin-dependent degradation of p53 has been successfully reconstituted *in vitro* [61,62].

# Tools used for the ubiquitin-mediated proteolytic pathway in the cell cycle control

Many components of the ubiquitin-dependent proteolytic pathway are tools of cell cycle control: for example, ubiquitin, the E-type enzymes and the proteasomes (see Fig. 2). They have been identified, either by phenotypic effects of mutation or modification during the cell cycle. However, their natural corresponding substrates have not been identified.

## Ubiquitin

Expression of UbK48R, a ubiquitin where the Lys-48 responsible for multiubiquitination has been replaced by Arg in yeast cells, results in an inhibition of proteolysis and cell cycle arrest, mostly in $G_2/M$, but also in $G_1$ [63].

## Ubiquitin-activating enzymes

Several mammalian cell lines have been used to select for conditional, temperature-sensitive mutants with defects in cell cycle progression. Many of them bear a single lesion, a mutation in an E1 gene. Among them are the mouse Ts85 cells [64,65], TsA1S9 [66] and Ts20 from CHO cells [67], and several other mutants that belong to the same complementation group [68]. Most of these E1 mutants arrest at the

S/G$_2$ boundary at the non-permissive temperature, except TsA1S9 cells that appear to arrest earlier in the S phase of the cell cycle. A different cell cycle arrest of this cell line may be due to a different E1 mutation disturbing interaction with a specific S-phase E2 enzyme. Recently, the phenotype of the temperature-sensitive Chinese hamster Ts20 cells was rescued by transfection with cDNA encoding the human E1 enzyme [69]. Expression of a mouse E1 enzyme also rescued the Ts85 mutant cell line and, moreover, suppressed the accompanying abnormal integrity of the nucleolus, confering a role for ubiquitin in nucleolus dissolution [70].

Results concerning the localization of these E1 enzymes during the cell cycle are contradictory. Grenfell et al. [71] reported a cell cycle dependent localization in HeLa cells: (i) during G$_1$, they are both cytoskeletal and nuclear; (ii) in S phase, they become exclusively cytoskeletal and perinuclear; (iii) during G$_2$ phase, they are nuclear before the nuclear envelope breaks down and, finally, (iv) in mitosis, they are localized to both mitotic spindles and the cytosol, but are not associated with chromosomes. In contrast, Cook and Chock [72] reported that the E1 enzymes are associated with the condensed chromosomes in mitotic HeLa cells.

E1 has also been found to be phosphorylated [73]. This phosphorylation, which was at least in part dependent on the CDC2 protein kinase, reached a peak in G$_2$/M phase of the cell cycle [74].

Cell cycle dependent regulation of both localization and CDC2-dependent phosphorylation of the activating enzyme reinforce the relation between ubiquitin and the cell cycle.

## Ubiquitin-conjugating enzymes

Other mutants deficient in cell cycle progress are those bearing E2 mutations: a null *rad6* (ubc2) yeast mutant, a hus5 allele, abnormally sensitive to inhibitors of DNA synthesis and radiation, and a mutant bearing a G$_2$ checkpoint mutation.

The null *rad6* yeast mutant exhibits a Cdc temperature-sensitive defect in late S or G$_2$ [75]. Recently, a novel yeast UBC, hus5, has been isolated as a mutant deficient in the S-phase checkpoint [76]. This novel UBC is not directly involved in this checkpoint, but is required for efficient recovery from DNA damage or S-phase arrest. Since the *hus5* deletion mutant failed to go through mitosis, it may have a role in chromosome segregation. The last mutant is a yeast one with increased sensitivity to DNA damage due to a mutation in G$_2$ checkpoint which can be complemented by the human homologue of *CDC34* [77]: *cdc34ts* mutants in yeast usually arrest in G$_1$ phase. CDC34 cell cycle functions reside in a 73 amino acid C-terminal region which when fused to RAD6 confers on it these functions [78,79].

## Ubiquitin hydrolases

Two yeast de-ubiquitinating enzymes, DOA4 and UBP5, present highest homology with the human TRE-2 oncogene [80,81]. This oncogene has been identified as a de-ubiquitinating enzyme [80].

## Proteasomes

In yeast, several mutants of either the regulatory or the catalytic subunit of the 26S proteasome have been associated with defects in cell cycle progression. In *Saccharomyces cerevisiae,* one of these mutants is mutated on the *mts2* gene which codes for a protein 75% similar to an ATPase regulatory S4 subunit of human proteasome [82]. The phenotype of these temperature-sensitive *mts2* cells, which lose the ability to segregate chromosomes and accumulate polyubiquitinated proteins, may be rescued by expression of the human S4 subunit. Two other mutants defective in putative ATPase subunits of the 26S proteasome, *cim3-1* and *cim5-1*, have arrested cell division in $G_2$/metaphase at the restrictive temperature and fail to separate their chromosomes [83]. The N-end rule controlled proteolysis [84,85] is functional in both, whereas other ubiquitin-dependent degradation pathways are not. When *cim1* mutants were arrested in $G_1$ phase by treatment with α-pheromones, Clb2 and Clb3 levels greatly diminished, indicating that the cyclin degradation machinery was still functional. Thus, the $G_2$/M arrested phenotype of the *cim* mutants is probably due to a failure to degrade other proteins, possibly those involved in sister chromatid separation, a step necessary to enter anaphase. Another yeast mutant, *prg1,* exhibits a lesion in the prg1 gene which codes for a protein 55% similar to the human RING10, a catalytic subunit of the proteasome. This mutant arrests in a very similar $G_2$/M phase of the cell cycle. Deletion of *clb2*, a non-vital mitotic cyclin, in the *prg1* strain suppresses the phenotype at the restrictive temperature, suggesting that the mutation in these cells leads to the inability to degrade this cyclin [86]. The *nin1-1* mutant of *Saccharomyces cerevisiae,* which cannot perform the $G_1$/S or $G_2$/M transitions at the restrictive temperature, fails to activate CDC28 kinase and accumulates polyubiquitinated proteins [87]. NIN1 is a regulatory subunit of the 26S proteasome which may participate in degradation of an inhibitor of the CDC28–cyclin complex necessary for the $G_1$/S (for example p40) and $G_2$/M transitions.

In animals, cell cycle regulated activation of the proteasome has been reported in starfish oocyte maturation and in the ascidian cell cycle [88–90]. In starfish, a 650 kDa protease, the 20S proteasome according to the authors, is involved in oocyte maturation, i.e. exit from the first meiotic prophase [88]. In the ascidian cell cycle [89,90], the 26S proteasome is activated in prophase, in metaphase and at the metaphase/anaphase transition. This transition occurs after fertilization and is accompanied by an increase in intracellular calcium. When the eggs are first treated with a calcium-chelating agent, the proteasome activity is totally abolished, indicating that this activity is regulated by intracellular calcium mobilization. The maxima in activity of the 26S proteasome correspond to an increase in amount of 26S proteasome. Quantification experiments suggest that the activity of the 26S proteasome is regulated through interconversion between the 26S and the 20S proteasomes [90].

In addition to the changes in activity during the cell cycle, a modulated localization of the proteasomes exists in mammalian cells as well as in amphibians

[91–93]. In amphibian cells, the proteasomes are detected in the chromosome in early metaphase, but become abundant around each spindle pole and at the mitotic spindle in late metaphase [91]. In mammalian cells, proteasome localization, revealed by immunofluorescence microscopy, is found throughout the cell during interphase. In prophase, the staining is mostly restricted to the perichromosomal area. This staining increases in metaphase and early anaphase. In late anaphase, the proteasome seems to be localized close to the spindle fibres. At no stage of the cell cycle was co-localization of the proteasome with chromosomes observed [92,93]. However, their close proximity does not exclude indirect interactions.

# Ubiquitin and the plant cell cycle

Some components of the ubiquitin-dependent proteolytic pathway have been well documented in plants. In *Arabidopsis thaliana*, the ubiquitin gene family is composed of 14 members [94]. At least 15 different E2 enzymes [95–100] have been described. We described the first plant proteasome $\alpha$-type subunit and $\beta$-type subunit sequences [101,102]. Since then, 13 new sequences have been reported through the *Arabidopsis* systematic sequencing program. Although the high conservation of the different elements of the cell cycle between yeast, animal and plants [103] allows us to speculate that proteolysis may be involved in control of the plant cell cycle, there is no direct evidence for proteolytic events regulating it. However, some new results reinforce the idea of a link between ubiquitin and plant cell cycle control.

## Arguments for a role of ubiquitin in the cell cycle control
Two cell cycle related proteins, plant cyclins and histones, behave as putative substrates for the ubiquitin-dependent pathway
Almost nothing is known about the stability of plant cyclins during the cell cycle. However, some of them have been used successfully in complementation of yeast mutants or *Xenopus* extracts, suggesting they behave like their animal or yeast counterparts. Significantly, a putative 'destruction box' in all mitotic plant cyclins and 'PEST' sequences in the $G_1$ cyclins are conserved (see Fig. 3). In animals, both sequences have been shown to be involved in ubiquitin-dependent turnover of cyclins and the conservation of these elements is indicative of biological significance (see J.P. Renaudin et al., chapter 4).

Notably, ubiquitinated histones have been identified in *Chlamydomonas reinhardtii* [104], wheat [105] and *A. thaliana* [106]. The biological role of these conjugates is intriguing, since no such histone ubiquitination has been reported in yeast. Moreover, in yeast, mutation of the putative site of ubiquitination on H2A histone does not confer any phenotype [107]. Histones H2A and H2B are ubiquitinated in mammalian cells [108]. These histone conjugates are not intermediates in histone degradation, but rather fluctuate with respect to the cell cycle. In

**Fig. 3**            **'Destruction box' and 'PEST' sequence in plant cyclins**

| Species | Name (b.c) | Destruction box or 'PEST' motifs | Accession no. | Ref. |
|---|---|---|---|---|
| Antirrhinium majus | amcyl1/cyc1a-am | RRALGDIGN | X76122 | 127 |
|  | amcyl2/cyc1b-am | " " | X76123 | 127 |
| A. thaliana | cyc1At/cyc1a-at | RQVLGDIGN | X62279 | 128 |
|  | cycx1-ara/cyc1b-at | RRALGDIGN | L27223 | 129 |
|  | cycx3-ara/cyc1c-at | " " | L27224 | 129 |
|  | cyc2aAt/cyc3a-at | RRVLRVINQ + PEST | Z31400 | 130 |
|  | cyc2bAt/cyc3b-at | RRALGVINH + PEST | Z31401 | 130 |
|  | cyc3aAt/cyc2m-at | AKALGVSNS | Z31589 | 130 |
|  | cyc3bAt/cyc2n1-at | RAVLKDVSN | Z31402 | 130 |
|  | cycd1-ara/cyc4a-at | PEST | X83369 | 131 |
|  | cycd2-ara/cyc4b-at | PEST | X83370 | 131 |
|  | cycd3-ara/cyc4c-at | PEST | X83371 | 131 |
|  | J18 | none | n.r. | 103 |
| Brassica napus | bncyc1/cyc2m-bn | RAVLGDISN | L25405 | 132 |
|  | bncyc2/cyc2a-bn | RAQLGNITN + PEST | L25405 | 132 |
| Daucus carrota | C13-1/cyc2g-dc | RVVLGEISN | S49312 | 133 |
| Glycine max | S13-6/cyc1a1-gm | RKALGDIGN | S49316 | 133 |
| Medicago sativa | cycMs1/cyc3a1-ms | RRALGVINQ | X68740 | 134 |
|  | cycMs2/cyc3b2-ms | RRALGGINQ | X82040 | 135 |
|  | cycIIIMs/cyc3a2-ms | RRALGVINQ | X78504 | 136 |
|  | cycMs3/cyc2m-ms | RAILQDVTN | X85783 | 135 |
|  | cycMf/n.r. | " " | n.r. | 137 |
|  | cyc5/n.r. | " " | n.r. | 137 |
|  | cyc15-6/n.r. | " " | n.r. | 137 |
|  | cyc3B/n.r. | " " | n.r. | 137 |
|  | cyc15-3/n.r. | " " | n.r. | 137 |
|  | cyc9-1/n.r. | REILQDVTN | n.r. | 137 |
| N. tabacum | cyc1nt/cyc1a-nt | RRALGDIGN | Z37978 | 138 |
|  | tobcycC/cyc1b-nt | RKALGDIGN | D50737 | u.r. |
|  | tobcycA/cyc2a-nt | RPALTNISN | D50735 | u.r. |
|  | n.r./cyc2g-nt | RVVLGELPN | n.r. | p.c. |
|  | n.r./cyc2h-nt | RVVLGEIQN | n.r. | p.c. |
|  | NT19 related/n.r. | RVVLGEIRN | n.r. | p.c. |
|  | tobcycB/cyc2m-nt | RAVLKDMKN | D50736 | u.r. |
| Pisum sativum | cycPs1/n.r. | RAALHDIGN | n.r. | 103 |
|  | cycPs/n.r. | RAGLTDVTN | n.r. | 103 |
|  | cycPs3/n.r. | RAILHDVTN | n.r. | 103 |
| Zea mays | cyclaZm/cyc1a-zm | RAPLGDIGN | U10079 | 139 |
|  | cycIIZm/cyc2a-zm | RASVGSLGN | U10077 | 139 |
|  | cycIIIZm/cyc3a-zm | RRALGDIGN | U10076 | 139 |
|  | cyc1bZm/cyc1b-zm | " " | U10078 | 139 |

*The names are indicated by their original nomenclature given by the authors (b) and by the new nomenclature (c) proposed by J.P. Renaudin, Chapter 4. The Figure is adapted from Jacobs [103]. n.r., not reported; u.r., unpublished results; p.c., personal communication from N. Chaubet.*

synchronized cell cultures of *Physarum polycephalum* [109], the conjugates disappear in metaphase and reappear in anaphase. Histone ubiquitination may modify nucleosome structure to facilitate transcription or replication.

## Ubiquitin genes are highly expressed in dividing cells

Ubiquitin genes are organized in small multigene families that contain two types of basic structures, namely polyubiquitin genes, which code for tandem head-to-tail repeats, and ubiquitin fusion genes, which code for a ubiquitin monomer fused to a ribosomal protein, also called carboxyl extension protein (CEP).

Polyubiquitin genes. As far back as 1990 [110], we showed that polyubiquitin gene expression is cell cycle regulated in *Nicotiana sylvestris*. Indeed, in a cell-suspension culture, two size classes of polyubiquitin mRNA of 1.9 and 1.3 kb increased significantly by the seventh day of culture, i.e. when the division rate was the highest [110]. We further demonstrated that the 1.9 kb mRNA coding gene is stress regulated, whereas one of the 1.3 kb polyubiquitin mRNAs from the coding gene, *Ubi.U4*, was, in part, cell cycle regulated [111]. Indeed, expression of the *Ubi.U4* gene increased during a time course in a protoplast-derived culture. High expression of this gene occurred in meristematic tissues, as well as in the dividing cells of *Agrobacterium tumefaciens* infected stems [111]. In addition, the promoter of the gene is able to drive GUS (β-glucuronidase) reporter gene expression in meristematic tissue as can be seen in Fig.4 (B. Plesse, unpublished result). On the other hand, gel-shift experiments, using nuclear extracts from dividing cells (from cell culture, young leaves), show a specific pattern of retarded DNA–protein complex, distinct from that seen with nuclear extract from mature leaves (B. Plesse et al., unpublished work ). In *Arabidopsis*, too, polyubiquitin gene expression is related to cell division (Fig.5): in a cell culture, the different polyubiquitin genes are up-regulated [102].

More recent reports from other plant species have confirmed our results. In rice, expression of the GUS reporter gene under the control of the *Ubi-1* polyubiquitin promoter is increased 2-fold during the logarithmic growth phase of culture [112]. In two algae, *Volvox carteri* [113] and *Chlamydomonas reinhardtii* [114], polyubiquitin expression is also in part cell cycle dependent. Moreover, in *C. reinhardtii*, a temperature-sensitive cell cycle mutant that is blocked in the vegetative cell cycle has been isolated; this mutant cannot transcribe the 2,3 kb polyubiquitin mRNA [115].

Ubiquitin fusion genes (Ubi–CEP). The first report of a plant Ubi–CEP was made in 1986 for barley [116]. They have since been reported in *Nicotiana sylvestris* [110,111], *A. thaliana* [101–117], tomato [118], maize [119] and potato [120,121]. Their expression is also related to cell growth or cell division, in *N. sylvestris* [110], as well as in *A. thaliana* [111]. Indeed the 0.8 kb-encoded mRNA accumulated in (i) young leaves; (ii) cell cultures in logarithmic growth phase [102]; (iii) protoplast-derived cell cultures after one day of culture, i.e. preceding the first division; (iv) meristematic tissues [117]; (v) stem tumours induced by *A. tumefaciens*, where actively dividing cells are present and, finally, in (vi) leaf strips incubated in a culture medium suitable for division [102] (Fig.5). In this last case, inhibition of DNA replication with hydroxyurea does not disturb the Ubi–CEP mRNA accumulation, suggesting it is replication independent. However, since

**Fig. 4**          **Histochemical localization of ubiquitin transcription by GUS activity in transgenic *Nicotiana tabacum* plantlets harbouring a polyubiquitin Ubi.U4 promotor–GUS construction**

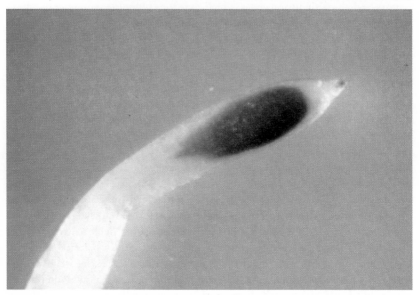

*Note the labelling in the root meristem.*

protein synthesis is high in these rapidly dividing cells, the expression of these *Ubi–CEP* genes might not be due to a requirement for the ubiquitin monomer in cell division, but for the ribosomal CEP.

## E2-type enzyme and proteasome subunits mRNA are highly expressed in dividing cells

We have also shown that three other mRNAs of the ubiquitin-dependent proteolytic pathway accumulate in *Arabidopsis* dividing cells.

ubc3 (cdc34) homologue. Two cDNAs coding for plant homologues of the *UBC3 (CDC34)* gene from budding yeast have been isolated from wheat [122] and *Arabidopsis* [99]. CDC34 is required for the $G_1$/S transition (see above). However, none of the plant counterparts to the CDC34 will necessarily have the same biological function since both lack the C-terminal extension domain, which in the yeast CDC34 gene product is necessary for its cell cycle function [78,79]. *UbcAt3*, the *Arabidopsis* homologue, is nevertheless regulated by the cell cycle [99] (Fig.5).

Proteasome α-type and β-type subunits. Both α- and β-subunits are highly similar in primary sequence to their vertebrate counterparts. In an actively growing *Arabidopsis* cell-suspension culture, the mRNAs of both subunits accumulate at the

**Fig. 5**         **Schematic representation of the ubiquitin pathway com-
                   ponents mRNA accumulation in leaf strips of A. thaliana
                   incubated in a culture medium adequate to trigger cell
                   division**

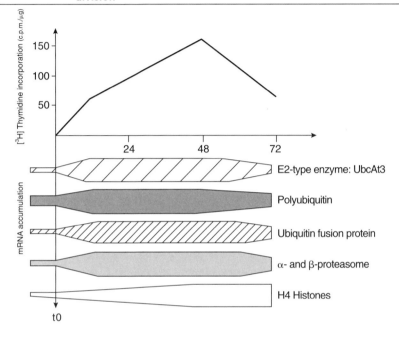

*The thickness of the drawings is based on Northern blot hybridizations. DNA synthesis was monitored in
parallel by [³H]thymidine incorporation.*

beginning of the logarithmic growth phase. In contrast, histone H4 mRNAs
accumulate later during midlog growth phase. An analogous situation is found in the
experimental system consisting of leaf-strips culture, where proteasome mRNAs
accumulate earlier than histone H4 mRNAs. Moreover, proteasome mRNA
accumulation, unlike histone H4 mRNA, is blocked by the addition of 100 mM
hydroxyurea (a concentration which is known to inhibit DNA synthesis). Their
expression is not strictly coupled to S phase, but rather to a very early phase, when
cells progress into $G_1$ phase. A similar accumulation of mRNA encoding different
proteasome subunits has been observed in malignant human cell cultures as well as in
tumours [123,124]. In addition, proteasome subunits are expressed at much higher
levels in immature mammalian cells than in mature differentiated cells [125,126].
Moreover, the expression of the human subunit C2 mRNAs occurred in parallel
with induction of cellular DNA synthesis in PHA (phytohaemagglutinin)-
stimulated cells [123].

We have recently used *in situ* hybridization to show that, in the *Arabidopsis* leaf-strips model, only the cells which are able to divide can accumulate proteasome mRNAs (D. Frauli and R. Bronner, unpublished work) in the same way that they accumulate histone H4 mRNAs.

## Conclusion

The structural conservation of the 'PEST' and 'destruction box' motifs in higher plant cyclins, and the significant accumulation of mRNAs encoding different components of the ubiquitin-dependent proteolytic machinery in the course of the cell cycle, argue strongly for a role of ubiquitin-dependent proteolysis in cell cycle control. However, it remains to be determined when and how this proteolytic pathway is involved in the plant cell cycle.

Since this selective proteolytic machinery is highly complex, and since multiple ubiquitination pathways may regulate the degradation of one single cyclin, there is a great deal of work to be done before the role of ubiquitin in the cell cycle is elucidated.

*We thank Gordon Simpson for critical reading of the manuscript and Jacqueline Marbach for help in manuscript preparation.*

## References

1.   Nigg, E.A. (1995) BioEssays 17, 471–480
2.   Nasmyth, K. (1993) Curr. Opin. Cell Biol. **5**, 166–179
3.   Pines, J. (1994) Nature (London) 371, 742–743
4.   King, R.W., Jackson, P.K. and Kirschner, M.W. (1994) Cell 79, 563–571
5.   Murray, A. (1995) Cell **81**, 149–152
6.   Barinaga, M. (1995) Science **269**, 631–632
7.   Hochstrasser, M. (1995) Curr. Opin. Cell Biol. **7**, 215–223
8.   Ciechanover, A. (1994) Cell 79, 13–21
9.   Scheffner, M., Nuber, U. and Huibregtse, J.M. (1995) Nature (London) 373, 81–83
10.  Evans, T., Rosenthal, E.T., Youngblom, J., Distel, D. and Hunt, T. (1983) Cell 33, 389–396
11.  Murray, A.W., Solomon, M.J. and Kirschner, M.W. (1989) Nature (London) 339, 280–286
12.  Glotzer, M., Murray, A.W. and Kirschner, M.W. (1991) Nature (London) 349, 132–138
13.  Hershko, A., Ganoth, D., Pehrson, J., Palazzo, R.E. and Cohen, L.H. (1991) J. Biol. Chem. **266**, 16376–16379
14.  Mahaffey, D.T., Yoo, Y. and Rechsteiner, M. (1995) FEBS Lett. **370**, 109–112
15.  Luca, F.C., Shibuya, E.K., Dohrmann, C.E. and Ruderman, J.V. (1991) EMBO J. **10**, 4311–4320
16.  Lorca, T., Devault, A., Colas, P., Van Loon, A., Fesquet, D., Lazaro, J.B. and Dorée, M. (1992) FEBS Lett. **306**, 90–93
17.  van der Velden, H.M.W. and Lohka, M.J. (1993) Mol. Cell. Biol. **13**, 1480–1488
18.  Stewart, E., Kobayashi, H., Harrison, D. and Hunt, T. (1994) EMBO J. **13**, 584–594
19.  van der Velden, H.M.W. and Lohka, M.J. (1994) Mol. Biol. Cell **5**, 713–724
20.  Holloway, S.L., Glotzer, M., King, R.W. and Murray, A.W. (1993) Cell 73, 1393–1402
21.  Minshull, J., Golsteyn, R., Hill, C.S. and Hunt, T. (1990) EMBO J. **9**, 2865–2875

22. Sudakin, V., Ganoth, D., Dahan, A., Heller, H., Hershko, J., Luca, F.C., Ruderman, J.V. and Hershko, A. (1995) Mol. Biol. Cell **6**, 185–197
23. Whitfield, W.G.F., Gonzalez, C., Maldonado-Codina, G. and Glover, D.M. (1990) EMBO J. **9**, 2563–2572
24. Hunt, T., Luca, F.C. and Ruderman, J.V. (1992) J. Cell Biol. **116**, 707–724
25. Minshull, J., Sun, H., Tonks, N.K. and Murray, A.W. (1994) Cell **79**, 475–486
26. Hershko, A., Ganoth, D., Sudakin, V., Dahan, A., Cohen, L.H., Luca, F.C., Ruderman, J.V. and Eytan, E. (1994) J. Biol. Chem. **269**, 4940–4946
27. King, R.W., Peters, J.-M., Tugendreich, S., Rolfe, M., Hieter, P. and Kirschner, M.W. (1995) Cell **81**, 279–288
28. Seufert, W. and Jentsch, S. (1990) EMBO J. **9**, 543–550
29. Seufert, W., Futcher, B. and Jentsch, S. (1995) Nature (London) **373**, 78–81
30. Tugendreich, S., Tomkiel, J., Earnshaw, W. and Hieter, P. (1995) Cell **81**, 261–268
31. Irniger, S., Piatti, S., Michaelis, C. and Nasmyth, K. (1995) Cell **81**, 269–278
32. Félix, M.-A., Labbé, J.-C., Dorée, M., Hunt, T. and Karsenti, E. (1990) Nature (London) **346**, 379–382
33. Lorca, T., Labbé, J.-C., Devault, A., Fesquet, D., Strausfeld, U., Nilsson, J., Nygren, P.-A., Uhlen, M., Cavadore, J.-C. and Dorée, M. (1992) J. Cell Sci. **102**, 55–62
34. Lorca, T., Fesquet, D., Zindy, F., Le Bouffant, F., Cerruti, M., Brechot, C., Devauchelle, G. and Dorée, M. (1991) Mol. Cell. Biol. **11**, 1171–1175
35. Surana, U., Amon, A., Dowzer, C., McGrew, J., Byers, B. and Nasmyth, K. (1993) EMBO J. **12**, 1969–1978
36. Gallant, P. and Nigg, E.A. (1992) J. Cell Biol. **117**, 213–224
37. Yen, T.J., Li, G., Schaar, B.T., Szilak, I. and Cleveland, D.W. (1992) Nature (London) **359**, 536–539
38. Brown, K.D., Coulson, R.M.R., Yen, T.J. and Cleveland, D.W. (1994) J. Cell Biol. **125**, 1303–1312
39. Yoon, H.J. and Carbon, J. (1995) Mol. Cell. Biol. **15**, 4835–4842
40. Amon, A., Irniger, S. and Nasmyth, K. (1994) Cell **77**, 1037–1050
41. Reed, S.I. (1991) Trends Genet. **7**, 95–99
42. Tyers, M., Tokiwa, G. and Futcher, B. (1993) EMBO J. **12**, 1955–1968
43. Richardson, H.E., Wittenberg, C., Cross, F. and Reed, S.I. (1989) Cell **59**, 1127–1133
44. Tyers, M., Tokiwa, G., Nash, R. and Futcher, B. (1992) EMBO J. **11**, 1773–1784
45. Wittenberg, C. and Reed, S.I. (1988) Cell **54**, 1061–1072
46. Yaglom, J., Linskens, M.H.K., Sadis, S., Rubin, D.M., Futcher, B. and Finley, D. (1995) Mol. Cell. Biol. **15**, 731–741
47. Goebl, M.G., Yochem, J., Jentsch, S., McGrath, J.P., Varshavsky, A. and Byers, B. (1988) Science **241**, 1331–1335
48. Deshaies, R.J., Chau, V. and Kirschner, M. (1995) EMBO J. **14**, 303–312
49. Rogers, S., Wells, R. and Rechsteiner, M. (1986) Science **234**, 364–369
50. Cross, F.R. and Blake, C.M. (1993) Mol. Cell. Biol. **13**, 3266–3271
51. Salama, S.R., Hendricks, K.B. and Thorner, J. (1994) Mol. Cell. Biol. **14**, 7953–7966
52. Peter, M. and Herskowitz, I. (1994) Cell **79**, 181–184
53. Chang, F. and Herkowitz, I. (1990) Cell **63**, 999–1011
54. McKinney, J.D., Chang, F., Heintz, N. and Cross, F.R. (1993) Genes Dev. **7**, 833–843
55. Schwob, E., Böhm, T., Mendenhall, M.D. and Nasmyth, K. (1994) Cell **79**, 233–244
56. Polyak, K., Lee, M.-H., Erdjument-Bromage, H., Koff, A., Roberts, J.M., Tempst, P. and Massagué, J. (1994) Cell **78**, 59–66
57. Pagano, M., Tam, S.W., Theodoras, A.M., Beer-Romero, P., Del Sal, G., Chau, V., Yew, R.P., Draetta, G.F. and Rolfe, M. (1995) Science **269**, 682–685
58. Ciechanover, A., DiGuiseppe, J.A., Bercovich, B., Orian, A., Richter, J.D., Schwartz, A.L. and Brodeur, G.M. (1991) Proc. Natl. Acad. Sci. U.S.A. **88**, 139–143
59. Nishizawa, M., Okazaki, K., Furuno, N., Watanabe, N. and Sagata, N. (1992) EMBO J. **11**, 2433–2446
60. Treier, M., Staszewski, L.M. and Bohmann, D. (1994) Cell **78**, 787–798
61. Shkedy, D., Gonen, H., Bercovich, B. and Ciechanover, A. (1994) FEBS Lett. **348**, 126–130

62.  Rolfe, M., Beer-Romero, P., Glass, S., Eckstein, J., Berdo, I., Theodoras, A., Pagano, M. and Draetta, G. (1995) Proc. Natl. Acad. Sci. U.S.A. **92**, 3264–3268

63.  Finley, D., Sadis, S., Monia, B.P., Boucher, P., Ecker, D.J., Crooke, S.T. and Chau, V. (1994) Mol. Cell. Biol. **14**, 5501–5509

64.  Finley, D., Ciechanover, A. and Varshavsky, A. (1984) Cell **37**, 43–55

65.  Ciechanover, A., Finley, D. and Varshavsky, A. (1984) Cell **37**, 57–66

66.  Zacksenhaus, E. and Sheinin, R. (1990) EMBO J. **9**, 2923–2929

67.  Kulka, R.G., Raboy, B., Schuster, R., Parag, H.A., Diamond, G., Ciechanover, A. and Marcus, M. (1988) J. Biol. Chem. **263**, 15726–15731

68.  Eki, T., Enomoto, T., Miyajima, A., Miyazawa, H., Murakami, Y., Hanaoka, F., Yamada, M. and Ui, M. (1990) J. Biol. Chem. **265**, 26–33

69.  Handley-Gearhart, P.M., Trausch-Azar, J.S., Ciechanover, A. and Schwartz, A.L. (1994) Biochem. J. **304**, 1015–1020

70.  Sudha, T., Tsuji, H., Sameshima, M., Matsuda, Y., Kaneda, S., Nagai, Y., Yamao, F. and Seno, T. (1995) Chromosome Res. **3**, 115–123

71.  Grenfell, S.J., Trausch-Azar, J.S., Handley-Gearhart, P.M., Ciechanover, A. and Schwartz, A.L. (1994) Biochem. J. **300**, 701–708

72.  Cook, J.C. and Chock, P.B. (1991) Proc. Natl. Acad. Sci. U.S.A. **88**, 11388–11392

73.  Cook, J.C. and Chock, P.B. (1995) Proc. Natl. Acad. Sci. U.S.A. **92**, 3454–3457

74.  Nagai, Y., Kaneda, S., Nomura, K., Yasuda, H., Seno, T. and Yamao, F. (1995) J. Cell. Sci. **108**, 2145–2152

75.  Ellison, K.S., Gwozd, T., Prendergast, J.A., Paterson, M.C. and Ellison, M.J. (1991) J. Biol. Chem. **266**, 24116–24120

76.  Al-Khodairy, F., Enoch, T., Hagan, I.M. and Carr, A.M. (1995) J. Cell. Sci. **108**, 475–486

77.  Plon, S.E., Leppig, K.A., Do, H.-N. and Groudine, M. (1993) Proc. Natl. Acad. Sci. U.S.A. **90**, 10484–10488

78.  Kolman, C.J., Toth, J. and Gonda, D.K. (1992) EMBO J. **11**, 3081–3090

79.  Silver, E.T., Gwozd, T.J., Ptak, C., Goebl, M. and Ellison, M.J. (1992) EMBO J. **11**, 3091–3098

80.  Papa, F.R. and Hochstrasser, M. (1993) Nature (London) **366**, 313–319

81.  Xiao, W., Fontanie, T. and Tang, M. (1994) Yeast **10**, 1497–1502

82.  Gordon, C., McGurk, G., Dillon, P., Rosen, C. and Hastie, N.D. (1993) Nature (London) **366**, 355–357

83.  Ghislain, M., Udvardy, A. and Mann, C. (1993) Nature (London) **366**, 358–362

84.  Bachmair, A. and Varshavsky, A. (1989) Cell **56**, 1019–1032

85.  Varshavsky, A. (1992) Cell **69**, 725–735

86.  Friedman, H. and Snyder, M. (1994) Proc. Natl. Acad. Sci. U.S.A. **91**, 2031–2035

87.  Kominami, K., DeMartino, G.N., Moomaw, C.R., Slaughter, C.A., Shimbara, N., Fujimuro, M., Yokosawa, H., Hisamatsu, H., Tanahashi, N., Shimizu, Y., Tanaka, K. and Toh-e, A. (1995) EMBO J. **14**, 3105–3115

88.  Sawada, M.T., Someno, T., Hoshi, M. and Sawada, H. (1992) Dev. Biol. **150**, 414–418

89.  Kawahara, H., Sawada, H. and Yokosawa, H. (1992) FEBS Lett. **310**, 119–122

90.  Kawahara, H. and Yokosawa, H. (1994) Dev. Biol. **166**, 623–633

91.  Kawahara, H. and Yokosawa, H. (1992) Dev. Biol. **151**, 27–33

92.  Amsterdam, A., Pitzer, F. and Baumeister, W. (1993) Proc. Natl. Acad. Sci. U.S.A. **90**, 99–103

93.  Palmer, A., Mason, G.G.F., Paramio, J.M., Knecht, E. and Rivett, A.J. (1994) Eur. J. Cell Biol. **64**, 163–175

94.  Callis, J., Carpenter, T., Sun, C.-W. and Vierstra, R.D. (1995) Genetics **139**, 921–939

95.  Sullivan, M.L. and Vierstra, R.D. (1991) J. Biol. Chem. **266**, 23878–23885

96.  Bartling, D., Rehling, P. and Weiler, E.W. (1993) Plant Mol. Biol. **23**, 387–396

97.  Girod, P.-A., Carpenter, T.B., van Nocker, S., Sullivan, M.L. and Vierstra, R.D. (1993) Plant J. **3**, 545–552

98.  Sullivan, M.L., Carpenter, T.B. and Vierstra, R.D. (1994) Plant. Mol. Biol. **24**, 651–661

99.  Genschik, P., Durr, A. and Fleck, J. (1994) Mol. Gen. Genet. **244**, 548–556

100. Watts, F.Z., Butt, N., Layfield, P., Machuka, J., Burke, J.F. and Moore, A.L. (1994) Plant Mol. Biol. **26**, 445–451

101. Genschik, P., Philipps, G., Gigot, C. and Fleck, J. (1992) FEBS Lett. **309**, 311–315

102. Genschik, P., Jamet, E., Philipps, G., Parmentier, Y., Gigot, C. and Fleck, J. (1994) Plant J. **6**, 537–546
103. Jacobs, T.W. (1995) Annu. Rev. Plant Physiol. Plant Mol. Biol. **46**, 317–339
104. Shimogawara, K. and Muto, S. (1992) Arch. Biochem. Biophys. **294**, 193–199
105. Ferguson, D.L., Guikema, J.A. and Paulsen, G.M. (1990) Plant Physiol. **92**, 740–746
106. Beers, E.P., Moreno, T.N. and Callis, J. (1992) J. Biol. Chem. **267**, 15432–15439
107. Swerdlow, P.S., Schuster, T. and Finley, D. (1990) Mol. Cell. Biol. **10**, 4905–4911
108. Bonner, W.M., Hatch, C.L. and Wu, R.S. (1988) in Ubiquitin (Rechsteiner, M. ed.), pp. 157–172, Plenum Press, New York
109. Mueller, R.D., Yasuda, H., Hatch, C.L., Bonner, W.M. and Bradbury, E.M. (1985) J. Biol. Chem. **260**, 5147–5153
110. Jamet, E., Durr, A., Parmentier, Y., Criqui, M.C. and Fleck, J. (1990) Cell Differ. Dev. **29**, 37–46
111. Genschik, P., Parmentier, Y., Durr, A., Marbach, J., Criqui, M.-C., Jamet, E. and Fleck, J. (1992) Plant Mol. Biol. **20**, 897–910
112. Takimoto, I., Christensen, A.H., Quail, P.H., Uchimiya, H. and Toki, S. (1994) Plant Mol. Biol. **26**, 1007–1012
113. Schiedlmeier, B. and Schmitt, R. (1994) Curr. Genet. **25**, 169–177
114. von Kampen, J. and Wettern, M. (1991) Eur. J. Cell Biol. **55**, 312–317
115. von Kampen, J., Nieländer, U. and Wettern, M. (1995) Planta **197**, 528–534
116. Gausing, K. and Barkardottir, R. (1986) Eur. J. Biochem. **158**, 57–62
117. Callis, J., Raasch, J.A. and Vierstra, R.D. (1990) J. Biol. Chem. **265**, 12486–12493
118. Hoffman, N.E., Ko, K., Milkowski, D. and Pichersky, E. (1991) Plant Mol. Biol. **17**, 1189–1201
119. Chen, K. and Rubenstein, I. (1991) Gene **107**, 205–212
120. Garbarino, J.E., Rockhold, D.R. and Belknap, W.R. (1992) Plant Mol. Biol. **20**, 235–244
121. Garbarino, J.E. and Belknap, W.R. (1994) Plant Mol. Biol. **24**, 119–127
122. Van Nocker, S. and Vierstra, R.D. (1991) Proc. Natl. Acad. Sci. U.S.A. **88**, 10297–10301
123. Kumatori, A., Tanaka, K., Inamura, N., Sone, S., Ogura, T., Matsumoto, T., Tachikawa, T., Shin, S. and Ichihara, A. (1990) Proc. Natl. Acad. Sci. U.S.A. **87**, 7071–7075
124. Kanayama, H., Tanaka, K., Aki, M., Kagawa, S., Miyaji, H., Satoh, M., Okada, F., Sato, S., Shimbara, N. and Ichihara, A. (1991) Cancer Res. **51**, 6677–6685
125. Shimbara, N., Orino, E., Sone, S., Ogura, T., Takashina, M., Shono, M., Tamura, T., Yasuda, H., Tanaka, K. and Ichihara, A. (1992) J. Biol. Chem. **267**, 18100–18109
126. Tanaka, K., Tamura, T., Yoshimura, T. and Ichihara, A. (1992) New Biol. **4**, 173–187
127. Fobert, P.R., Coen, E.S., Murphy, G.J.P. and Doonan, J.H. (1994) EMBO J. **13**, 616–624
128. Hemerly, A., Bergounioux, C., Van Montagu, M., Inzé, D. and Ferreira, P. (1992) Proc. Natl. Acad. Sci. U.S.A. **89**, 3295–3299
129. Day, I.S. and Reddy, A.S.N. (1994) Biochim. Biophys. Acta **1218**, 115–118
130. Ferreira, P., Hemerly, A., Engler-De Almeida, J., Bergounioux, C., Burssens, S., Van Montagu, M., Engler, G. and Inzé, D. (1994) Proc. Natl. Acad. Sci. U.S.A. **91**, 11313–11317
131. Soni, R., Carmichael, J.P., Shah, Z.H. and Murray, J.A.H. (1995) Plant Cell **7**, 85–103
132. Szarka, S., Fitch, M., Schaerer, S. and Moloney, M. (1995) Plant Mol. Biol. **27**, 263–212775
133. Hata, S., Kouchi, H., Suzuka, I. and Ishii, T. (1991) EMBO J. **10**, 2681–2688
134. Hirt, H., Mink, M., Pfosser, M., Bögre, L., Györgyey, J., Jonak, C., Gartner, A., Dudits, D. and Heberle-Bors, E. (1992) Plant Cell **4**, 1531–1538
135. Meskiene, I., Bögre, L., Dahl, M., Pirck, M., Ha, D.T.C., Swoboda, I., Heberle-Bors, E., Ammerer, G. and Hirt, H. (1995) Plant Cell **7**, 759–771
136. Savouré, A., Fehér, A., Kalò, P., Petrovics, G., Csanàdi, G., Szécsi, J., Kiss, G., Brown, S., Kondorosi, A. and Kondorosi, E. (1995) Plant. Mol. Biol. **27**, 1059–1070
137. Russinova, E., Slater, A., Atanassov, A.I. and Elliott, M.C. (1995) Cell. Mol. Biol. **41**, 703–714
138. Qin, L.X., Richard, L., Perennes, C., Gadal, P. and Bergounioux, C. (1995) Plant Physiol. **108**, 425–426
139. Renaudin, J.-P., Colasanti, J., Rime, H., Yuan, Z. and Sundaresan, V. (1994) Proc. Natl. Acad. Sci. U.S.A. **91**, 7375–7379

# 8

# Cell cycle dependent gene expression

**Masaki Ito**

Department of Biological Sciences, Graduate School of Science,
University of Tokyo, Hongo, Tokyo 113, Japan

## Cell cycle dependent control of gene expression

At specific times during the cell cycle, the regulatory systems involving promoters and transcription factors mediate a burst of gene expression. These regulatory systems should sense cell cycle positional information, and hence, the transcription factors responsible for periodic expression must be a target of the regulators that control the cell cycle progression. Thus, the studies of transcriptional regulation during the cell cycle can provide valuable information about the processes which control cell division. It is well known that most eukaryotic genes are expressed at roughly constant levels throughout the cell cycle, and only a small number of genes display periodicity in their rate of transcription [1,2]. The patterns of expression of some cell cycle regulated genes are well conserved in a broad range of taxonomically distant organisms from yeast to humans, and also in higher plants. For example, histone genes are expressed strictly in S phase in all organisms studied so far ([3]; see N. Chaubet and C. Gigot, chapter 13). Cyclins show different patterns of expression during the cell cycle depending on their type [4], but the same type of cyclin tends to exhibit the same pattern of expression in different organisms [5]. On the other hand, several genes being expressed periodically in some organisms are constantly expressed during the cell cycle in others. The less conserved genes include those involved in DNA replication. For example, expression of the gene for DNA polymerase $\alpha$ occurs only at $G_1/S$ boundary in budding yeast [6] and human cells [7], but is constant in fission yeast ([8]; see P. Sabelli et al., chapter 12).

Continuous over-expression of cell cycle regulated genes often causes lethal phenotype or abnormal cell cycles. For instance, many mitotic cyclin genes cause cell cycle arrest when their expression is driven by continuously strong promoters [9–11]. In the case of $G_1$ cyclins, over-expression leads to smaller cell size and shortens the $G_1$ phase resulting in premature entry into S phase in both yeast [12] and mammalian cells [13–16]. In a similar way, the increased dosage of the fission yeast gene *cdc25* (encoding a phosphatase that regulates function of the p34$^{cdc2}$ kinase) causes a corresponding decrease in the cell size at the beginning of mitosis, showing that the cells enter mitosis earlier than normal ([17,18]; see D. Francis, chapter 9). In another case, the unbalanced synthesis of histones, which are normally expressed in S phase of the cell cycle, causes chromosome loss [19]. So it seems that expression of the cell cycle regulated genes should not exceed certain levels. However, cells in which the cell cycle dependent genes are expressed constantly, but in proper levels, still progress through the cell cycle normally. Thus, the periodicity of expression of many essential cell cycle genes is not critical for cell cycle progression [2].

These observations have raised questions: why is this phenomenon of periodic cell cycle dependent expression so conserved in a wide range of organisms, and what is the role of this phenomenon in the cell cycle progression? There are only speculative answers. First, cell cycle dependent expression of certain genes can be important for the control of transitions in the cell cycle. In budding yeast, co-ordinate regulation of the DNA synthesis genes is mediated by a single transcription factor [20]. Overproduction of any individual DNA synthesis gene does not affect the timing of S-phase initiation, meaning that their products are not rate limiting. However, over-expression of all these genes together would accelerate the S phase. Thus, the co-ordinating regulatory factor may be rate limiting for the initiation of DNA synthesis. In this case, a single transcription factor, instead of many genes, can be a target for control of cell cycle transitions.

Second, cell cycle dependent expression of the genes can also play an important role in temporal co-ordination of cell cycle events. Given that a bifunctional protein controls a particular cell cycle event and, at the same time, activates the transcription factors, two different events can be directed in a temporally ordered manner, either by the protein itself, or by a product of target gene of the transcription factor. For example, the transcription factor called E2F/DP1 heterodimer plays a pivotal role in regulation of several cell cycle dependent genes, which include the genes involved in DNA replication [21]. A model is proposed that the activity of E2F/DP1 factor is regulated by cyclin/cyclin-dependent kinase (Cdk) complex. In $G_1$ phase, as the level of cyclin E/Cdk2 increases, it phosphorylates pRb (protein encoded by retinoblastoma susceptibility gene $RB$), causing the release of E2F/DP1 from pRb. Free E2F/DP1 becomes active for binding E2F sites in the promoters of several cell cycle genes and activates their transcription. Later in the S phase, cyclin A/Cdk2 complex appears and phosphorylates E2F/DP1 inactivating the heterodimer [22,23]. As a consequence, the transcription of genes regulated by E2F/DP1 is temporally controlled in the cell cycle, and the activation and inactivation of the transcription are co-ordinated with cell cycle transition driven by the cyclins A and E ([4]; see J.A.H. Murray et al., chapter 5).

The third possible role of cell cycle dependent gene expression is to ensure the maintenance of order in the cell cycle. As discussed earlier, periodic expression of any one gene is not critical for normal cell cycle progression. However, it is possible that there are several different levels of regulation that complement each other to ensure the normal order of events. So the constant expression of a particular gene which naturally is expressed periodically does not influence the cell cycle progression in general, but it is possible that if some other damage occurs simultaneously (for example a mutation in some genes), it can significantly influence or arrest cell cycle progression.

## Overview on plant cell cycle dependent genes

At present, little is known about the mechanisms of regulation of the cell cycle dependent genes in higher plants. Thus, it is necessary to begin with the identification of plant genes that are regulated periodically in the cell cycle, and it is very important to determine their timing of expression as precisely as possible. Plant cell cycle dependent genes identified to date encode histones, ribonucleotide reductase (RNR), a transcription factor that regulates histone H3 gene expression, mitogen-activated protein kinase (MAPK), and several different types of cyclins. In addition, some other isolated plant genes have yeast and mammalian homologues that are expressed in a cell cycle dependent manner, although the expression of these genes has not been examined in the plant cell cycle. In other cases, the periodicity of gene expression during the cell cycle is demonstrated only at the protein level or its enzymic activity. Such plant genes could be described as candidates for further studies of transcriptional regulation during the plant cell cycle.

The periodic transcription of plant genes can be analysed by direct and indirect methods. The indirect method developed by Fobert et al. [24] is to apply double-target *in situ* hybridization using a combination of two probes, one of them a gene marker specifically expressed at certain timings of the cell cycle (see J.H. Doonan, chapter 10). The direct method is based on analysis of RNA samples isolated from synchronous plant cell-suspension cultures. For this purpose, several established plant cell lines with high growth rate, including tobacco TBY-2 [25,26] and *Catharanthus roseus* (periwinkle) cells [27,28], are commonly used. The TBY-2 cell line is characterized by a remarkably high growth rate with average generation time of less than 16 h [26]. The most widely used method of synchronization of TBY-2 cells is reversible arrest of cell division by the DNA polymerase($\alpha$)inhibitor, aphidicolin [25]. When this inhibitor is removed from the culture, tobacco cells progress through the cell division cycle synchronously, and mitotic indices (MIs) up to 70–80 % have been recorded [25]. After aphidicolin treatment, the TBY-2 cells could be further synchronized by treatment with the antitubulin drug, propyzamide, to achieve a MI of 80–90% [29]. TBY-2 cells have several other features that make them very convenient for molecular studies of the plant cell cycle. For example, genetic transformation of TBY-2 cells can be achieved with high frequency by usage of *Agrobacterium tumefaciens* [30].

Another system applied to cell cycle studies, cultured *C. roseus* (periwinkle) cells, have slower growth rate than TBY-2 cells. However, periwinkle cell cultures can also be synchronized by two methods, double phosphate starvation [27] and auxin starvation [28]. The synchrony in *C. roseus* systems is comparable with TBY-2 cells [31].

## Histones

Histone genes constitute a complex multigene family the structure and organization of which have been investigated in a wide range of organisms [32,33]. Although most members of the gene family show S-phase specific expression, some do not (see N. Chaubet and C. Gigot and M. Iwabuchi et al., chapters 13 and 14, respectively). Practically, probing of RNA blots with either a histone gene coding region or a cDNA usually results in detection of the total amount of mRNAs from all members of the gene family. Thus, to understand the mechanisms of periodic expression of histone genes, the study should focus on an individual member of the gene family. Current studies of histone gene promoters make the situation more complex. One histone H4 gene promoter from *Arabidopsis* can direct both DNA-replication-dependent and -independent gene expression in transgenic tobacco [34]. Similarly, a wheat histone H3 promoter confers cell-division-dependent and -independent expression of a reporter gene in transgenic rice plants ([35]; see M. Iwabuchi et al., chapter 14). Thus, to study the cell cycle regulation of histone gene transcription it is necessary to characterize both the gene and its promoter in detail. From this point of view, the wheat histone H3 gene, TH012, is the best characterized. When rice cells have been stably transformed with promoter–GUS (glucuronidase) chimeric genes and synchronized by aphidicolin treatment, the promoter region of TH012 up to -185 could confer S-phase-specific expression of the GUS mRNA [36]. This region of the promoter contains four positive *cis*-acting elements which are responsible for the efficient transcription of the gene [37,38]. Among the *cis*-acting elements, the hexamer motif (ACGTCA) probably plays a major role in the S-phase-specific expression. Two sequence-specific DNA-binding proteins, HBP-1a and HBP-1b, which bind to the hexameric sequence [39,40] are also expressed specifically during S phase in both wheat suspension cultures and seedlings [41]. This co-ordinated expression of the histone gene and its possible transcription factors resembles the mechanisms of transcriptional regulation in yeasts and animal cells. However, HBP-1a and HBP-1b contain the bZIP domain [40], and have no structural similarity with yeast and mammalian factors that regulate cell cycle dependent expression. Thus, in plants, like in all other eukaryotes studied so far, histone gene expression is regulated by specific mechanisms (see M. Iwabuchi et al., chapter 14).

## Proliferating cell nuclear antigen

The proliferating cell nuclear antigen (PCNA) has been initially identified as a nuclear protein the expression of which is high in the S phase of the cell cycle [42–45]. The location of PCNA within the nucleus coincides with sites of ongoing DNA replication [46]. PCNA is known to be an auxiliary protein of DNA polymerase δ [43,47], and it is necessary for the replication of SV40 and BPV viral

DNA [48,49]. In *Saccharomyces cerevisiae*, the *POL*30 gene encoding PCNA is essential for cell growth, showing a requirement for PCNA in chromosomal DNA replication [50]. In addition, PCNA and DNA polymerase δ also play important roles in the repair of damaged DNA [51,52]. Current studies suggest a potential role of PCNA in response to cellular signals which regulate the progression of the cell cycle [53,54]. In normal non-tumorigenic human cells, PCNA is found in complexes with a variety of cyclins, CDK, and p21, an inhibitor of CDK [55,56]. In the experiments with SV40 DNA replicated *in vitro*, PCNA-dependent DNA replication is directly inhibited by p21. Furthermore, p21 blocks the ability of PCNA to activate DNA polymerase δ [53]. Thus, both CDK and PCNA are inhibited by the same protein p21, possibly enabling the co-ordination between DNA replication and cell cycle progression.

The expression of PCNA is regulated in a cell cycle dependent manner in both yeast [57] and mammalian cells [58]. Levels of PCNA mRNA increase specifically in the late $G_1$ phase of the cell cycle. The promoter region of *S. cerevisiae* PCNA contains the sequence called MCB (MluI cell cycle box) which can direct cell cycle regulated transcription of a reporter gene [59,60].

A DNA-binding protein complex called MBF (MCB-binding factor) binds to MCB elements [59,61]. MBF is regulated at the post-transcriptional level, and the activation of MBF requires CDC28 kinase activity [62]. The periodic expression of the PCNA gene is believed to be directed by MBF- and E2F-mediated mechanisms in *S. cerevisiae* and mammalian cells, respectively (see P. Sabelli et al., chapter 12).

The presence of PCNA in higher plants was first demonstrated by cloning the genomic fragment from rice [63]. Later, we isolated the periwinkle cDNA for PCNA by library screening with the rice homologue [64]. Putative protein structures of rice and periwinkle showed a high degree of similarity with PCNA proteins from other organisms [65,66]. In agreement with the structural conservation, the rice PCNA protein produced in bacteria can stimulate DNA synthesis catalysed by DNA polymerase δ from human cells, indicating that it is also functionally conserved [66]. We examined the patterns of expression of the PCNA gene during the cell cycle in synchronous cultures of periwinkle and TBY-2 cells. The maximum levels of expression were observed in late $G_1$/S phase of the cell cycle, showing that the cell cycle regulated pattern of PCNA expression is also conserved between plants and other eukaryotic cells [64]. In our further experiments, TBY-2 cells were transformed with the rice promoter–GUS chimeric gene constructs, and the promoter region up to -1500 was able to confer cell cycle dependent fluctuations of mRNA levels of the GUS reporter gene.

The promoter analysis of rice PCNA gene has been also performed by Kosugi et al. [67]. The PCNA promoter truncated to position -263 was shown to confer meristematic tissue-specific expression of the GUS gene in transgenic tobacco plants [67]. When mature leaf discs from the transgenic plants were incubated with phytohormones (auxin and cytokinin) to induce callus growth, a significant GUS activity was observed. Thus, the 263 bp promoter is responsible for the cell division associated expression of the rice PCNA. The cell division dependent pattern of PCNA transcription was also observed in our experiments with periwinkle seedlings, where the mRNA was shown to be abundant only in meristematic regions [64].

Kosugi et al. also have defined three *cis*-acting elements within the promoter region of the rice gene by DNaseI footprinting and gel-retardation analysis [68]. These *cis*-acting elements are critical for the transcriptional activation of the gene, but do not have sequence similarity to MCB, Swi4-Swi6 cell cycle box (SCB) or E2F sites. It is unclear if these elements have a function in meristem-specific or cell cycle dependent expression of the PCNA gene. Moreover, the elements directing the cell cycle dependent expression may differ from those responsible for meristem-specific expression.

## RNR

RNR is a key enzyme in the sequence of biochemical reactions leading to DNA synthesis, and its activity is strongly correlated with DNA replication [69,70]. This enzyme catalyses direct reduction of all four ribonucleotides to corresponding deoxyribonucleotides. RNR is a heterodimeric enzyme composed of a large subunit R1 and a smaller subunit R2, both of which are essential for activity [71]. In mammalian cells, the amount of R2 protein fluctuates during the cell cycle via S-phase-specific *de novo* synthesis [72], whereas R1 protein levels are constant [73]. However, mRNAs of the both genes are cell cycle regulated and are expressed in late $G_1$/S phase of the cell cycle [74]. In *Schizosaccharomyces pombe*, *cdc22*, the R1 counterpart, is periodically expressed in late $G_1$ [75]. *S. cerevisiae* counterparts, *RNR1 (R1)* and *RNR2 (R2)*, are both regulated in cell cycle dependent manner [76]. The *RNR1* mRNA level fluctuates more than 10-fold during the cell cycle, while the *RNR2* transcript shows only a modest 2-fold fluctuation. All of these genes, *cdc22* in *S. pombe*, and *RNR1* and *RNR2* in *S. cerevisiae*, contain MCB motifs in their promoters and are regulated by MBF-mediated mechanisms [57]. In mammalian cells, regulation of the gene for RNR is not so well studied. E2F sites are not reported to be present in the mammalian promoter. A possible mechanism for regulation of *R2* gene expression was suggested by Bjorkloud et al. [77]. The mouse *RNR* subunit *R2* promoter is inactive in $G_0$, but is activated early after growth stimulation, before the appearance of *R2* mRNA could be detected. In this early period, short premature *R2* transcripts are observed, which suggests the presence of a transcriptional block [77]. From these observations, a model was proposed that late $G_1$/S-phase-specific synthesis of mature full-length *R2* mRNA occurs as a result of a cell cycle dependent release from the transcriptional block as cells proceed into the S phase.

A cDNA clone encoding the RNR small subunit (R2) was isolated from *Arabidopsis* [78]. The encoded protein shows 64% sequence identity with the human and mouse R2. The gene for R2 is present as a single copy gene in

*Arabidopsis.* Cell cycle regulated expression of *RNR* was demonstrated in the synchronous cell cultures of TBY-2 cells [78]. The mRNA for *R2* increases in parallel to DNA synthesis in the same manner as in other eukaryotic cells. However, the promoter analysis of the *R2* gene remains to be done.

## DNA polymerase α

There are three DNA polymerases that are thought to be involved in DNA replication, DNA polymerase α, δ, ε [79]. Two other mammalian polymerases have been described, DNA polymerase γ, which is believed to be involved in mitochondrial DNA synthesis [80], and DNA polymerase β, which is considered to have a major role in DNA repair [81]. DNA polymerase α is essential for chromosomal DNA replication as demonstrated, for example, by analysis of a temperature-sensitive mouse mutant [82]. Furthermore, monoclonal antibodies against DNA polymerase α [83] inhibit DNA synthesis in permeabilized cells [84], or when microinjected into nuclei [85]. cDNAs for DNA polymerase α have been isolated from *S. cerevisiae, S. pombe,* human, *Drosophila* and mouse [86–90]. Alignment of amino acid sequences reveals seven highly conserved regions [87,91]. These regions are also conserved in DNA polymerases δ and ε. We isolated a rice cDNA for DNA polymerase α by the polymerase chain reaction (PCR) with degenerate oligonucleotides corresponding to the conserved regions. We have tested the cell cycle regulated expression of the plant gene for DNA polymerase α, because the genes encoding this protein show late $G_1$ specific expression in *S. cerevisiae* [6] and mammalian cells [7]. Total RNA fractions isolated from synchronous cultures of tobacco TBY-2 cells at several timepoints were analysed by hybridization with the rice cDNA clone containing the six conserved regions. In low stringent conditions, the rice cDNA hybridized to two mRNA species of different length. This additional mRNA possibly corresponds to either DNA polymerases δ or ε. The amounts of these two mRNAs changed in parallel and fluctuated during the cell cycle peaking at late $G_1$/S. In human cells, two other polymerases (δ and ε) are also expressed in a periodic manner in the cell cycle [7], with maximal expression in late $G_1$. Similarly, in *S. cerevisiae*, *POL1-3* genes which encode for DNA polymerases α, ε, δ, respectively, are induced in late $G_1$ by MBF-mediated mechanisms [57]. So it is likely that, in plants, other DNA polymerases (for example, δ or ε) will also show late-$G_1$-specific expression.

## Bifunctional dihydrofolate reductase and thymidylate synthase

Dihydrofolate reductase (DHFR) and thymidylate synthase (TS) have been studied from bacterial [92], avian [93], protozoal [94] and mammalian [95] sources. In most organisms including bacteria, yeast and vertebrates, DHFR is a monomeric enzyme, while TS is a homodimer made up of two identical subunits. Thus, the two enzymes are independent products of different genes in these organisms. In mammalian cells, the two enzymes can associate temporarily during the S phase as part of a large multi-enzyme complex [96,97]. In protozoa, this complex is stabilized through the formation of a covalently joined bifunctional enzyme [94]. Higher plants appear to have both the monofunctional DHFR and TS [98], as well as the bifunctional enzyme [99]. Recently, cDNA sequences of bifunctional DHFR–TS were isolated from *Arabidopsis*, carrot and soybean [100–102], although cDNAs for the monofunctional enzymes have not yet been isolated. The proteins encoded by these genes have two separate domains, a DHFR domain in the N-terminus and a TS domain in the C-terminus. Both show structural similarity to the corresponding monofunctional enzymes from other organisms. Expression of soybean DHFR–TS in bacteria confirmed the presence of both DHFR and TS activity in the single polypeptide.

Both enzymes are involved in the synthesis of DNA precursors, and the genes encoding for both enzymes are expressed at late $G_1$ in preparation for DNA synthesis in mammalian cells [103,104]. In *S. cerevisiae*, the gene for TS (*TMP1*) is cell cycle regulated [1], although the gene for DHFR (*DFR1*) lacks the MCB sequence and does not display periodic expression at the mRNA level [57]. In the case of the bifunctional enzyme, the presence of these two enzymic activities in a single polypeptide could make the reactions more efficient because the dihydro-folate produced by TS is the immediate substrate for DHFR. Hence, the rapid production of reduced folates for DNA synthesis can be achieved easily. These facts suggest that the gene for bifunctional DHFR-TS in plants might also be expressed just before DNA synthesis in the cell cycle. To confirm this idea, we isolated cDNA for this bifunctional enzyme from tobacco using *Arabidopsis* cDNA as a probe, and then examined the changes in mRNA levels during the cell cycle of cultured tobacco cells. In synchronous cultures, mRNA levels fluctuated, peaking in late $G_1$/S phase. We detected an additional mRNA in the Northern blot analysis, which was shorter in length than the major mRNA for bifunctional DHFR–TS. This minor mRNA of reduced length was expressed in parallel to the major mRNA which peaked in late $G_1$. We guess that the smaller mRNA might be monofunc-tional TS, since most of the isolated tobacco cDNA, utilized as a probe for Northern blot analysis, is composed of the TS domain. However, further analysis is required to clarify the situation.

## Cyclins

Recently, cyclin homologues were isolated from plants, including carrot and soybean [105,106], *Arabidopsis thaliana* [107,108], alfalfa [109–111], *Antirrhinum majus* [24], maize [112], etc. In many cases, plant cyclins show sequence similarity to both the mitotic cyclins A and B [24,112]. In animal cells, A-type cyclins are expressed earlier in the cell cycle and destroyed before B-type cyclins, so possible functions of plant cyclins and their classification to a certain type can be suggested from timing their appearance and disappearance during the cell cycle [106]. For example, using this approach, soybean mitotic cyclins, cyc3Gm and cyc5Gm, have been attributed to types A and B, respectively [106]. However, in some cases, expression patterns of cyclins are not consistent with the classification based on sequence similarity with mammalian cyclins. For example, soybean cyc1Gm (soybean homologue of carrot C13-1 [105]) is related to type A, but its expression is limited to a short period in S phase, and it could not be found in either $G_2$ or M phase [106]. Another cyclin, alfalfa cycMs3, although attributed to the group of mitotic cyclins, is present in all phases of the cell cycle approximately at the same levels [110]. Therefore, many authors propose rather complicated classification of plant mitotic cyclins into at least three groups ([111–113]; however, see J.-P. Renaudin et al., chapter 4). Another classification subdivides plant mitotic cyclins into four groups [106]. Recently, mitotic cyclins that share characteristics of both types A and B were found also in mammalian cells [114]. Some plant mitotic cyclins are suggested to be the closest relatives of these B3-type animal cyclins [111].

In our work, cDNAs of a carrot A-type cyclin, C13-1, and a soybean B-type cyclin, S13-6 (recent name cyc5Gm) [105], were used for screening a periwinkle cDNA library. The cDNA clones for two different mitotic cyclins named CYS and CYM were isolated [115]. To show that the products of *CYS* and *CYM* actually function as regulatory subunits for CDK, we have tested if the expression of these cDNAs complements the cyclin mutants of *S. pombe* and *S. cerevisiae*. Expression of the cDNAs in *S. pombe* at moderate levels did not complement the temperature-sensitive (ts) defect of a *cdc13* (the gene encoding mitotic cyclin B) mutant. Instead, the transformed cells carrying *CYM* or *CYS* grew more slowly than the control cells carrying the vector alone. The reduced rate of growth was also observed in wild-type cells expressing *CYM* or *CYS*. In addition, wild-type cells expressing periwinkle cyclins were elongated, indicating that progression through the cell cycle was delayed. The same results were observed when the *S. pombe* cyclin, *puc1*, was over-expressed in the *cdc13* mutant [10]. In these experiments, at the permissive temperature, cells were very elongated and arrested at $G_2$ phase. The heterologous cyclins expressed in *S. pombe* may out-compete the native B-type cyclin in associating with the Cdc2 kinase [10]. From this point of view, the periwinkle cyclins encoded by *CYS* and *CYM* are likely to possess binding ability to Cdc2 of *S. pombe*.

Constitutive over-expression of the full-length cDNAs of periwinkle cyclins in *S. cerevisiae* leads to lethal phenotypes. However, the truncated versions of the cDNAs encoding proteins with deleted N-terminal sequences had no effect on the growth of *S. cerevisiae*. To test if these truncated cyclins could functionally complement the function of $G_1$ cyclin genes (*CLN1–3*), we utilized a yeast mutant in which all three chromosomal *CLN* genes were deleted, but which carried a plasmid containing the *GAL10* promoter fused to the wild-type *CLN2*. The *GAL10* promoter directs high-level expression when cells are grown in galactose-containing medium, but it is repressed if the cells are incubated in the presence of glucose [116]. As a result, the mutant cells are viable if grown in galactose-based medium, but are arrested in $G_1$ upon transfer to glucose-based medium, owing to a lack of functional CLN gene products. Expression of truncated periwinkle cyclins could maintain viability of the transformed mutant in glucose-based medium, thus showing that the truncated versions of *CYS* and *CYM* could bind and activate CDC28 kinase. In a similar approach, Meskiene et al. [110] showed that the truncated version of *cycMs3* also can complement the yeast $G_1$–cyclin-deficient mutant (see H. Hirt et al., chapter 6). When the full-length alfalfa *cycMs3* gene was used to transform the $G_1$–cyclin-deficient yeast strain, it could functionally complement CLN cyclins, but the transformed cells grew more slowly [110].

To analyse structure–function relationships and the role of the N-terminus of CYM, we prepared a series of N-terminal deletions, and tested for their ability to complement CLN function. Large N-terminal deletion did not affect kinase binding and activation, unless the region of cyclin box was deleted. The CYM cyclin box begins with MRAIL, the consensus sequence, and the protein translated from the first methionine in the cyclin box could complement the CLN triple mutant. These observations are consistent with the report that the cyclin box in mammalian cyclins contains all of the information necessary for binding and activation of CDKs [117]. In addition, the finding that expression of full-length, but not N-terminal, deleted proteins resulted in the lethal phenotype suggests that the substrate specificity of CDKs may be determined by the N-terminal sequence of associated cyclins.

To speculate about the possible functions of an isolated cyclin, it is always necessary to analyse its pattern of expression during the cell cycle. In our work, the synchronous cultures of *C. roseus* cells and tobacco TBY-2 cells were used to characterize expression patterns of *CYS* and *CYM*. During the cell cycle, *CYM* was specifically expressed in $G_2$/M phase, whereas *CYS* mRNA appeared at the onset of DNA replication and then was present during S phase. Later, near the end of $G_2$, it disappeared just after the appearance of *CYM* mRNA. Thus, the expression of *CYS* occurred before the expression of *CYM* in the cell cycle, similarly to the expression of A-type and B-type cyclins in animals. *CYS* is homologous to soybean *cycGm1* and to carrot *C13-1*, whereas *CYM* is similar to the soybean *cycGm5* [105,106]. Using double-target *in situ* hybridization, Kouchi et al. [106] suggested that *cycGm1* is expressed during a short time in S phase, and *cycGm5* is expressed during $G_2$ and M phases [106]. These observations are consistent with our results obtained from using synchronous plant cell cultures. Recently, Setiady et al. [118] isolated tobacco mitotic cyclin genes and used synchronous cultures of TBY-2 cells to analyse their expression. Tobacco *Ntcyc29* is expressed in $G_2$/M phase, similarly to *CYM*. Other tobacco cyclin genes, *Ntcyc25* and *Ntcyc27*, which are classified as type A, are expressed at S/$G_2$ phase, earlier than *Ntcyc29* (see De Veylder et al., chapter 1). The pattern of *CYS* expression differs from these genes; it is expressed earlier than *Ntcyc25* and *Ntcyc27*. When TBY-2 cells are synchronized by aphidicolin, *CYS* mRNA is already present just after aphidicolin release, while the mRNAs for the two tobacco cyclin genes do not appear until later in the S phase. Thus, the expression of *CYS* seems to be different from that of typical A-type cyclins, suggesting a distinct function in the cell cycle.

Transcriptional regulation of plant cyclins, animal and yeast alike, seems to play an important role in the control of cell cycle progression. Our data suggest that *CYS* and *CYM* expression is transcriptionally regulated. Therefore, we analysed the activity of *CYM* and *CYS* promoters during the progression through the cell cycle in synchronous plant cell cultures. Chimeric genes composed of a cyclin promoter and GUS-reporter gene were introduced into tobacco TBY-2 cells. The transformants were cultured in suspension and synchronized by aphidicolin treatment. Levels of GUS mRNA were monitored by Northern blot analysis using the coding region of GUS as a probe. The GUS mRNA levels varied during the cell cycle in both the transformants carrying *CYS* promoter–GUS and *CYM* promoter–GUS constructs. *CYM* promoter conferred the $G_2$-specific expression of GUS mRNA, while *CYS* promoter activity correlated with the rate of DNA synthesis, with the highest levels of GUS mRNA occurring during the S phase.

## Mitogen activated protein kinases (MAPKs)

MAPKs comprise a family of serine/threonine kinases which are activated by a variety of mitogenic stimuli [119]. This class of kinases is characterized by their ability to phosphorylate microtubule-associated protein (MAP2) and myelin basic protein (MBP). MAPKs also phosphorylate and reactivate S6 kinase II [120] and transcription factors with roles in cell cycle activation [121]. Both $G_0$-arrested somatic cells and $G_2$-arrested oocytes use many of the same signalling mechanisms to break cell cycle arrest. $G_0$-arrested quiescent mammalian cells and $G_2$-arrested *Xenopus* oocytes contain MAPK mRNAs and inactive forms of the enzymes. In mammalian cells, MAPK activity is stimulated by mitogens, such as epidermal growth factor, nerve growth factor and insulin [119]. In oocytes, MAPK activity increases rapidly when they are stimulated to resume meiosis [122]. Maturation promoting factor, the CDK/cyclin complex, acts as an upstream activator of MAPK [123]. MAPKs have been identified from many organisms, including human and yeast, and show highly conserved structures [124]. Recently, cDNAs for MAPKs have been isolated from several plant species [125–127]. The plant genes for MAPKs are often expressed ubiquitously in all organs [125,127]. However, an alfalfa cDNA clone for MAPK, *MsK7*, showed cell cycle dependent expression in suspension-cultured alfalfa cells [126]. When cells were synchronized by aphidicolin treatment, *MsK7* mRNA was present at a very low level in $G_1$, and later at higher levels in S and $G_2$ phases of the cell cycle. This pattern of expression of *MsK7* in the plant cell cycle is contrary to that reported for *Xenopus* MAPK, where the level of mRNA is constant throughout the cell cycle [123]. In addition, *MsK7* gene does not show meristem-specific expression in alfalfa seedlings; instead it is expressed in the root and stem. These contradictory results require further studies to allow us to understand the physiological role of the cell cycle regulation of the *MsK7*.

## Topoisomerase II

Eukaryotic DNA topoisomerase II is a nuclear enzyme that catalyses the decatenation and the unknotting of topologically linked DNA circles and the relaxation of supercoiled DNA [128]. The enzyme is essential for the condensation of interphase chromatin into metaphase chromosomes [129–132] and for the separation of sister chromatids in anaphase [133]. In mammalian cells, protein levels of topoisomerase II are high in rapidly proliferated cells, and fluctuate during the cell cycle, peaking at $G_2$ phase [134]. The amount of topoisomerase II mRNA increases with the progression of the cell cycle and peaks in $G_2/M$ cells [135]. In *S. pombe* and *S. cerevisiae*, genes for topoisomerase II (TOP2) both have MCB motifs in their upstream sequences [1,57].

A cDNA clone for topoisomerase II has been isolated from *Arabidopsis* by PCR with two degenerate primers designed from conserved amino acid sequences found in the N-terminal region of human and *Drosophila* topoisomerase II [136]. Amplified fragments were used for the screening of a cDNA library to obtain a cDNA clone named AtTopII. The amino acid sequence of AtTopII shows about 53% identity with topoisomerases from other eukaryotic organisms. The use of antibodies against the AtTopII protein has demonstrated that the level of the protein is high in the tissues containing proliferating cells. The amount of AtTopII mRNA also correlates with cell division rate during the growth of seedlings. The gene for topoisomerase II is a candidate for the group of genes regulated in a cell cycle dependent manner in higher plants.

## Deoxyuridine triphosphatase

Deoxyuridine triphosphatase (dUTPase) catalyses the hydrolysis of dUTP to dUMP and PPi, and plays a critical role in DNA replication. The enzyme has a double function; first, it provides a source of dUMP for TS, and, secondly, it eliminates dUTP as a potential substrate for DNA polymerases and thereby prevents the incorporation of uracil into replicating DNA. In *S. cerevisiae*, the dUTPase (*DUT1*) gene is essential [137], so that a mutant with a disrupted *DUT1* gene is arrested in S phase of the cell cycle. This enzyme functions between RNR and TS in the dTTP biosynthetic pathway, and genes for both of these enzymes exhibit periodic expression during the cell cycle [1]. Therefore, in *S. cerevisiae*, the expression of the *DUT1* gene was expected to be late $G_1$ specific. However, in budding yeast, the *DUT1* gene exhibits a relatively constant level of transcription throughout the cell cycle [138]. In contrast, the dUTPase gene (*DUT1*) from *Candida albicans*, when introduced into *S. cerevisiae*, displays fluctuations of the mRNA levels with a peak at late $G_1/S$ phase [139]. In mammalian cells, accumulation of dUTPase is dependent on growth stimulation; *de novo* synthesis of dUTPase occurs in late $G_1$ phase, and the rate of synthesis decreases substantially as cells progress through S and into $G_2/M$ phase [140]. The gene for dUTPase has been identified also in higher plants [141]. The product of the gene was found as a soluble protein which is present specifically in the meristem of the *anantha* mutant of tomato. Both an enzymic test and the conservation of sequence motifs verified that the isolated cDNA represents a tomato dUTPase. Immunogold localization and *in situ* hybridization experiments showed that proteins and transcripts of dUTPase are present at high levels in apical meristematic cells of vegetative and floral meristems, suggesting that expression of the gene is associated with cell proliferation. A similar pattern of dUTPase activity has been observed in onion (*Allium cepa*) roots [142]. dUTPase activity progressively decreases from the proliferating apical root segment toward the mature root, and it is very low in terminally differentiated cells. In onion, dUTPase activity is cell cycle dependent: it increases at the $G_1/S$ boundary; remains high during the S phase, and decreases rapidly in $G_2$ and M phases during the cell cycle in the system of synchronized root meristems. We examined the changes in the levels of dUTPase mRNA in the synchronous cultures of TBY-2 cells. As expected, the levels of dUTPase mRNA showed clear S-phase-dependent expression, just like other DNA synthesis genes, such as those encoding PCNA and DNA polymerase $\alpha$.

## Genes isolated by differential screening

In addition to the isolation of plant homologues of yeast and animal cell cycle dependent genes, another useful approach is to identify plant genes of interest using their periodical expression in plant synchronous cell cultures [143,144]. With this strategy, it might be possible to isolate novel plant cell cycle dependent genes, or genes which have not been recognized as cell cycle dependent in yeast or mammalian cells. Two-dimensional gel electrophoresis demonstrated the existence of several *in vitro* translated polypeptides, which displayed fluctuation in their levels during the cell cycle, in synchronous cultures of periwinkle cells [143]. To identify genes encoding such polypeptides, we performed differential hybridization to screen cDNA libraries prepared from synchronous cultures of periwinkle cells at various times during the cell cycle [144,145]. As a result of screening, we isolated several cDNA clones, which were named *cyc* genes.

### cyc19

The deduced amino acid sequence of *cyc19* encodes for heat shock protein 90 (HSP90). The expression of *cyc19* was high in apical meristems and root tips, where active cell division occurred. The level of *cyc19* mRNA correlated with the rate of DNA synthesis in suspension cultures of *C. roseus* cells synchronized by two different methods, double-phosphate starvation and auxin starvation. A similar pattern of cell cycle dependent expression has been previously reported for the HSP90 gene in cultured chicken cells [146]. In vertebrates, two isoforms of HSP90 are encoded by two closely related genes: *HSP90a* and *HSP90b*. In chicken cell cultures synchronized by nocodazole or aphidicolin, *HSP90a* mRNA accumulation is specific to $G_1/S$ transition. Recently, an unexpected role of HSP90 in cell cycle control has been revealed in *S. pombe*. The $G_2/M$ transition in fission yeast is regulated by the Cdc2 kinase complexed with a B-type cyclin (encoded by *cdc13* gene) [147]. This complex is inactive until the onset of mitosis owing to phosphorylation in the ATP-binding domain of the Cdc2 kinase [148]. At the $G_2/M$ transition, dephosphorylation by the Cdc25 phosphatase activates the cyclin B–Cdc2 complex [149]. In opposition to Cdc25, Wee1 kinase phosphorylates Cdc2 and inhibits its activity, thus functioning as a negative regulator of mitosis [150,151]. A member of the HSP90 family encoded by a gene named *swo1* (suppressor of *wee1* over-expression) is identified as a positive regulator of Wee1 protein kinase [152]. It is suggested that formation of the active Wee1 kinase requires interaction with the HSP90, and that this protein might control the cell cycle through regulation of Wee1 activity. Although no cell cycle dependent changes in Wee1 activity have been observed in *S. pombe*, the kinase activity of a human Wee1 homologue (WEE1Hu) varies during the cell cycle with the maximum in S and $G_2$ phases [153]. A plant homologue of the Wee1 kinase is not yet cloned. However, our data show that the plant HSP90, Cyc19, is expressed in a pattern that resembles the cell cycle dependent changes in Wee1 activity in animal cells. Thus, we speculate that, in plants, Wee1 kinase activity might be controlled by the S-phase-specific expression of HSP90, and that this mechanism might regulate Cdk activity.

### cyc15 and cyc17

Deduced amino acid sequences for both *cyc15* and *cyc17* contain pentapeptide repeats Ser-Pro-4 which are common for cell wall proteins called extensins. Extensins compose the major class of hydroxyproline-rich glycoproteins (HRGPs) [154]. The expression of both *cyc15* and *cyc17* fluctuated during the cell cycle in synchronous cultures of periwinkle cells. The mRNA levels increased during the S phase and then decreased rapidly in $G_2/M$. In asynchronous cultures, however, the expression of these genes did not correlate with cell proliferation, i.e. the levels of mRNA were high in the stationary phase of growth when cell proliferation ceased. These apparently conflicting patterns of extensin gene expression can be explained by the dual function of the protein. Extensins are considered to have a structural role in the cell wall as part of the glycoprotein network complementing the cellulose mesh [155,156]. When extensins are deposited in the cell wall, they are desolubilized by isodityrosine cross-links and by the covalent cross-linkage to some cell wall carbohydrates. Increased deposition of extensins and increased extensin cross-linkage should strengthen the walls [157,158]. Thus, the incorporation of extensins in the cell wall may be associated with the cessation of extension growth [156,159]. Such active incorporation of extensin is expected to occur in the stationary phase of asynchronous cultures. In addition, extensins are required also for the proper assembly of cell wall components which are necessary for continued mitosis [160]. Treatment of regenerating tobacco protoplasts with a selective inhibitor of prolyl hydroxylase, which hydroxylates the Pro residues of

Ser-Pro-4 units to produce hydroxyproline, severely inhibits protoplast development. Cells develop an abnormal morphology as a consequence of abnormal cell wall structure, and both cytokinesis and nuclear division of these cells are blocked. The inhibitor also blocks growth of cultured soybean cells and induces the disappearance of the major salt-extractable HRGPs [161]. These observations indicate that extensins are required for normal assembly of other cell wall polymers, and that specific HRGPs are required for nuclear division and cytokinesis [160]. Thus, the cell cycle dependent expression of *cyc15* and *cyc17* may contribute to the normal progression through the cell cycle by providing the necessary cell wall structure.

## cyc07

A novel plant cell cycle dependent gene, *cyc07*, encodes a basic protein of 35 kDa [145]. Antibodies raised against the fusion protein expressed in bacteria showed that the Cyc07 protein is preferentially present in actively proliferating cells of periwinkle and tobacco suspension-cultured cells [31,115]. The protein is localized in interphase nuclei. However, in mitotic cells, it is dispersed into the cytoplasm rather than associated with chromosomes. *cyc07* may encode a DNA-binding protein, since it binds to a DNA cellulose column and is released only by 300 mM NaCl. The possible DNA-binding ability was also supported by the observation that *cyc07* protein was released from the isolated nuclei by digestion of DNA with DNaseI [115].

Southern blot analysis of yeast genomic DNA demonstrated that *S. cerevisiae* had two copies of the homologous genes (*PLC*1 and *PLC*2), and both genes were isolated from a yeast genomic library [162]. Alignment of the protein sequences showed that periwinkle Cyc07 and yeast homologues shared 65% similar amino acids. To determine if *PLC1* and *PLC2* are essential for cell division, we have analysed the phenotype of yeast cells caused by disruption of *PLC1, PLC2* or both. Although cells carrying the disruption in the either one of the *PLC* genes were viable, they showed a slower rate of growth than the wild type [162]. Simultaneous disruption of both *PLC1* and *PLC2* resulted in a lethal phenotype. Thus, these two genes are functionally redundant, and their function is essential for cell proliferation of *S. cerevisiae*. Since the reduced gene copy number decreased the rate of cell proliferation, it can be speculated that the level of expression of the *PLC* gene family can control cell cycle progression. The reduced growth rate caused by disruption of *PLC2* was reversed by the expression of periwinkle *cyc07* cDNA in yeast [115]. Thus, the *cyc07* gene and yeast *PLC* gene family have a conserved function in the cell cycle, and *cyc07* might also be essential for cell cycle progression in higher plants.

The expression of *cyc07* was correlated with active cell proliferation [145]. This correlation was also observed in the expression of *PLC1* and *PLC2* in *S. cerevisiae*. In synchronous cultures of periwinkle cells, the *cyc07* mRNA specifically accumulated in S phase of the cell cycle [145], and *de novo* synthesis of

Cyc07 protein was also cell cycle dependent [115]. In contrast, the amounts of Cyc07 protein were relatively constant throughout the cell cycle. However, protein synthesized *de novo* may be functionally distinct from the protein generated in the preceding cell cycle, as in the case of *cdc2* in *S. pombe* [163]. This hypothesis is also supported by our observation that yeast spores which lack both *PLC1* and *PLC2* genes could not complete the first round of the cell cycle, even though such spores possibly contain the Cyc07 protein inherited from diploid cytoplasm.

To study the regulation of cell division associated expression of the *cyc07* gene, we constructed the promoter fusion with the GUS reporter gene. The chimeric gene was introduced into *Arabidopsis* and the distribution of GUS activity in transgenic plants was examined [164]. In transgenic *Arabidopsis*, the spatial pattern of expression directed by the promoter was closely correlated with meristematic activity. The *cyc07* promoter sequence truncated to position -589 was sufficient to confer the meristem-specific expression. However, the truncated promoter did not direct cell cycle dependent expression; the level of GUS mRNA was constant throughout the cell cycle in synchronous cultures of transgenic tobacco cells. Cell cycle dependent changes in the level of GUS mRNA were observed only when the 2.0 kb full-length *cyc07* promoter was used to direct the GUS expression. Thus, the promoter region between -589 and -1 contains sufficient information for directing meristem-specific expression, whereas some elements that direct periodic expression in the cell cycle are located in the region between -2015 and -589. Apparently, in the case of the *cyc07* promoter, the mechanisms for the cell cycle dependent and meristem-specific expression differ from each other.

## Concluding remarks

The mechanisms of cell cycle dependent gene expression are largely different between lower eukaryotes and mammals, in contrast to the conserved cell cycle control machinery involving CDK and cyclin complexes. In mammalian cells, late $G_1$ genes are regulated by E2F-mediated mechanisms, whereas in yeasts, the transcription of these genes is controlled by MBF or SBF (SCB-binding factor; [165,166]). E2F has no structural resemblance to yeast transcription factors such as MBF or SBF. Moreover, the sequence motifs recognized by mammalian and yeast factors diverged during evolution. However, in related groups of organisms the regulation is well conserved; for example, MCB elements in the promoters of late $G_1$ genes are similar in many fungal species. Therefore, it is probable that both conserved mechanisms and diverged mechanisms are operating for the regulation of cell cycle dependent gene transcription in plant cells.

Our studies on the cell cycle dependent genes in higher plants are just at the beginning. We have identified several plant cell cycle dependent genes. Among them are those encoding PCNA, DNA polymerase α, bifunctional DHFR–TS,

**Table 1    Cell cycle dependent genes in higher plants**

| Gene product | Gene name (source) | Timing of expression | Ref. |
|---|---|---|---|
| Histone H1, H2A, H2B, H3, H4 | Many genes | S | [36, 41] |
| Putative transcription factor | HBP1a, HBP1b (wheat) | S | [41] |
| Ribonucleotide reductase (small subunit) | TAI251 (Arabidopsis) | S | [78] |
| MAP kinase | MsK7 (alfalfa) | S/G2 | [126] |
| Function unknown | cyc07 (periwinkle) | S | [145] |
| HSP90 | cyc19 (periwinkle) | S | This study |
| Extensin | cyc15, 17 (periwinkle) | S | This study |
| dUTPase | p18 (tomato) | S | This study |
| Topoisomerase II | AtTopII (Arabidopsis) | ? | [136] |
| PCNA | cyc16 (periwinkle) | Late $G_1$/S | [64], this study |
| DNA polymerase α | (tobacco) | Late $G_1$/S | This study |
| Bifunctional DHFR-TS | (tobacco) | S | This study |
| Mitotic cyclins | cycIIIMs (alfalfa) | $G_2$/M | [111] |
|  | cyclin1, cyclin2 (Antirrhinum) | $G_2$/M | [24] |
|  | cycMs1, cycMs2 (alfalfa) | $G_2$/M | [109] |
|  | cyc5Gm (soybean) | $G_2$/M | [106] |
|  | cyc3Gm (soybean) | Late S/$G_2$ | [106] |
|  | cyc1Gm (soybean) | S | [106] |
|  | Ntcyc25, Ntcyc27 (tobacco) | Late S/$G_2$/M | [118] |
|  | Ntcyc29 (tobacco) | $G_2$/M | [118] |
|  | CYS (periwinkle) | Early S/G2 | This study |
|  | CYM (periwinkle) | $G_2$/M | This study |
| G1 cyclin | cyclin δ3 (Arabidopsis) | early S | [168] |
|  | cycMs4 (alfalfa) | $G_1$/S | [169] |

dUTPase, two different types of cyclins (*CYM* and *CYS*), Cyc07, HSP90 and extensin. In synchronous plant cell cultures, the genes involved in DNA synthesis (PCNA, bifunctional DHFR–TS, DNA polymerase $\alpha$) are expressed just before or coincidentally with, DNA synthesis. RNR, which is also involved in DNA synthesis, by producing DNA precursors, is reported to be expressed co-ordinately with histone genes. In our experiments, A-type cyclin *CYS* is also expressed mainly in S phase. The cell cycle dependent genes so far identified in plants are listed in Table 1. We have demonstrated that the promoters of the gene for PCNA, CYS and CYM are able to direct periodic transcription during the cell cycle. In the rice PCNA promoter, several MCB-like sequences are present. However, according to the report of Kosugi et al. [67], the *cis*-acting elements identified by DNaseI footprinting do not contain MCB-like motifs. The promoter sequences of *cyc07* from periwinkle, HSP90 from rice, maize and *Arabidopsis* do not contain consensus motifs such as MCB, SCB and E2F-binding sites. In contrast, the CYS promoter contains several repeated motifs, one of which, CTCGAAA, is similar to SCB (CACGAAA). The promoter region of an *Arabidopsis* cyclin, which is expressed in late $G_1$ [167], also contains two almost perfect SCB motifs. At present, we do not know if these elements in plant promoters function in the same way as in yeast, and if they are responsible for cell cycle dependent expression. We need to identify a larger number of cell cycle dependent genes from plants, and determine if promoters of such genes can confer periodic expression to a reporter gene. Accumulation of data about promoter sequences of plant cell cycle dependent genes, and comparison with the sequence motifs of yeasts and mammals, would answer the question about the existence of both conserved and specific mechanisms in plants. It is also necessary to identify the elements which are sufficient for cell cycle dependent expression. Identification of transcription factors that interact with plant promoters would provide a clearer view of the transcriptional control of plant cell cycle dependent genes and enable us to study the regulation of transcription factors during the cell cycle or upon stimulation of proliferation. In mammalian cells, the transcription factor E2F is regulated by physical interactions with pRb which has a central role in the control of the cell cycle. The *RB* gene seems to be absent in lower eukaryotes, such as yeasts. In plants, if the specific mechanisms do exist, they should involve special factors that also may be absent in animal or yeast cells. Information about transcriptional regulation of plant genes during the cell cycle will help us to understand the molecular basis of specific features of cell cycle control in plants.

*The study of plant cyclins was done together with T. Ohno, M.-C. Criqui, M. Sakabe, S. Hata, H. Kouchi, J. Hashimoto, A. Watanabe, A. Komamine and H. Fukuda. The DNA polymerase α project was carried out together with M. Yokoi, M. Izumi, F. Hanaoka, and H. Nakai. The work on HSP90 and extensin was done with H. Kodama, N. Ohnishi, and A. Komamine.*

# References

1. Andrews, B.J. and Herskowitz, I. (1990) J. Biol. Chem. **265**, 14057–14060
2. McKinney, J.D. and Heintz, N. (1991) Trends Biochem. Sci. **16**, 430–435
3. Nigg, E.A. (1993) Trends Cell Biol. **3**, 296–301
4. Lew, D.J. and Reed, S.I. (1992) Trends Cell Biol. **2**, 77–81
5. Norbury, C. and Nurse, P. (1992) Annu. Rev. Biochem. **61**, 441–470
6. Johnston, L.H., White, J.H.M., Johnson, A.L., Lucchini, G. and Plevani, P. (1987) Nucleic Acids Res. **15**, 5017–5030
7. Zeng, X.-R., Hao, H., Jiang, Y. and Lee, M.Y.W.T. (1994) J. Biol. Chem. **269**, 24027–24033
8. Park, H., Francesconi, S. and Wang, T.S.-F. (1993) Mol. Biol. Cell. **4**, 145–157
9. Lew, D.J., Dulic, V. and Reed, S.I. (1991) Cell **66**, 1197–1206
10. Forsburg, S.L. and Nurse, P. (1991) Nature (London) **351**, 245–248
11. Bueno, A., Richardson, H., Reed, S.I. and Russell, P. (1991) Cell **66**, 149–159
12. Nash, R., Tokiwa, G., Anand, S., Erickson, K. and Futcher, A.B. (1988) EMBO J. **7**, 4335–4346
13. Ando, K., Ajchenbaum-Cymbalista, F. and Griffin, J.D. (1993) Proc. Natl. Acad. Sci. U.S.A. **90**, 9571–9575
14. Musgrove, E.A., Lee, C.S.L., Buckley, M.F. and Sutherland, R.L. (1994) Proc. Natl. Acad. Sci. U.S.A. **91**, 8022–8026
15. Quelle, D.E., Ashmun, R.A., Shurtleff, S.A., Kato, J., Bar–Sagi, D., Roussel, M.F. and Sherr, C.J. (1993) Genes Dev. **7**, 1559–1571
16. Resnitzky, D., Gossen, M., Bujard, H. and Reed, S.I. (1994) Mol. Cell. Biol. **14**, 1669–1679
17. Russell, P. and Nurse, P. (1986) Cell **45**, 145–153
18. Moreno, S., Nurse, P. and Russell, P. (1990) Nature (London) **34**, 549–552
19. Meeks-Wagner, D. and Hartwell, L.H. (1986) Cell **44**, 43–52
20. Koch, C. and Nasmyth, K. (1994) Curr. Opin. Cell Biol. **6**, 451–459
21. La Thangue, N.B. (1994) Curr. Opin. Cell Biol. **6**, 443–450
22. Dynlacht, B.D., Flores, O., Lees, J.A. and Harlow, E. (1994) Genes Dev. **8**, 1772–1786
23. Wittenberg, C., Sugimoto, K. and Reed, S.I. (1990) Cell **62**, 225–237
24. Fobert, P., Coen, E.S., Murphy, G.J.P. and Doonan, J.H. (1994) EMBO J. **13**, 616–624
25. Nagata, T., Okada, K. and Takebe, I. (1982) Plant Cell Rep. **1**, 250–252
26. Nagata, T., Nemoto, Y. and Hasegawa, S. (1992) Int. Rev. Cytol. **132**, 1–30
27. Amino, S., Fujimura, T. and Komamine, A. (1983) Physiol. Plant. **59**, 393–396
28. Nishida, T., Ohnishi, N., Kodama, H. and Komamine, A. (1992) Plant Cell Tissue Org. Cult. **28**, 37–43
29. Kakimoto, T. and Shibaoka, H. (1988) Protoplasma Suppl. **2**, 95–103
30. An, G. (1985) Plant Physiol. **79**, 568–570
31. Ito, M. and Komamine, A. (1993) J. Plant Res. (Special Issue) **3**, 17–28
32. Chaubet, N., Chaboute, M.E., Philipps, G. and Gigot, C. (1987) Dev. Genet. **8**, 461–473
33. Hentschel, C.C. and Brinstiel, M.L. (1981) Cell **25**, 301–313
34. Lepetit, M., Ehling, M., Chaubet, N. and Gigot, C. (1992) Mol. Gen. Genet. **231**, 276–285
35. Terada, R., Nakayama, T., Iwabuchi, M. and Shimamoto, K. (1993) Plant J. **3**, 241–252
36. Ohtsubo, N., Nakayama, T., Terada, R., Shimamoto, K. and Iwabuchi, M. (1993) Plant Mol. Biol. **23**, 553–565
37. Nakayama, T., Ohtsubo, N., Mikami, K., Kawata, T., Tabata, T., Kanazawa, H. and Iwabuchi, M. (1989) Plant Cell Physiol. **30**, 825–832
38. Nakayama, T., Sakamoto, A., Yang, P., Minami, M., Fujimoto, Y., Ito, T. and Iwabuchi, M. (1992) FEBS Lett. **300**, 167–170
39. Minami, M., Tabata, T., Kawata, T., Nakayama, T. and Iwabuchi, M. (1987) FEBS Lett. **223**, 273–278
40. Tabata, T., Nakayama, T., Mikami, K. and Iwabuchi, M. (1991) EMBO J. **10**, 1459–1467
41. Minami, M., Huh, G.H., Yang, P. and Iwabuchi, M. (1993) Plant Mol. Biol. **23**, 429–434
42. Almendral, J.M., Huebsh, D., Blundell, P.A., MacDonald-Bravo, H. and Bravo, R. (1987) Proc. Natl. Acad. Sci. U.S.A. **84**, 1575–1579
43. Bravo, R., Frank, R., Blundell, P.A. and MacDonald-Bravo, H. (1987) Nature (London) **326**, 515–517
44. Bravo, R. and MacDonald-Bravo, H. (1985) EMBO J. **4**, 655–661

45.  Celis, J.E. and Madsen, P. (1986) FEBS Lett. **209**, 277–283
46.  Madsen, P. and Celis, J.E. (1985) FEBS Lett. **193**, 5–11
47.  Prelich, G., Tan, C.K., Kostura, M., Mathews, M.B., So, G.A., Downey, K.M. and Stillman, B. (1987) Nature (London) **326**, 517–520
48.  Muller, F., Seo, Y.S. and Hurwitz, J. (1994) J. Biol. Chem. **269**, 17086–17094
49.  Prelich, G., Kostura, M., Marshak, D.R., Mathews, M.B. and Stillman, B. (1987) Nature (London) **326**, 471–475
50.  Bauer, G.A. and Burgers, P.M.J. (1990) Nucleic Acids Res. **18**, 261–265
51.  Shivji, M.K.K., Kenny, M.K. and Wood, R.D. (1992) Cell **69**, 367–374
52.  Zeng, X.R., Jiang, Y., Zhang, S.J., Hao, H. and Lee, M.Y. (1994) J. Biol. Chem. **269**, 13784–13791
53.  Waga, S., Hannon, G.J., Beach, D. and Stillman, B. (1994) Nature (London) **369**, 574–578
54.  Chen, J., Jackson, P.K., Kirschner, M.W. and Dutta, A. (1995) Nature (London) **374**, 386–388
55.  Xiong, Y., Zhang, H. and Beach, D. (1992) Cell **71**, 505–514
56.  Zhang, H., Xiong, Y. and Beach, D. (1993) Mol. Biol. Cell **4**, 897–906
57.  McIntosh, E.M. (1993) Curr. Genet. **24**, 185–192
58.  Muller, R. (1995) Trends Genet. **11**, 173–178
59.  Lownes, N.F., Johnson, A.L. and Johnson, L. (1991) Nature (London) **350**, 247–250
60.  McIntoch, E., Atkinson, T., Storms, R. and Smith, M. (1991) Mol. Cell. Biol. **11**, 329–337
61.  Drick, L., Moll, T., Auer, H. and Nasmyth, K. (1992) Nature (London) **357**, 508–513
62.  Marini, N.J. and Reed, S.I. (1992) Genes Dev. **6**, 557–567
63.  Suzuka, I., Daidoji, H., Matsuoka, M., Kadowaki, K., Takasaki, Y., Nakane, P.K. and Moriuchi, T. (1989) Proc. Natl. Acad. Sci. U.S.A. **86**, 3189–3193
64.  Kodama, H., Ito, M., Ohnishi, N., Suzuka, I. and Komamine, A. (1991) Eur. J. Biochem. **197**, 495–503
65.  Suzuka, I., Hata, S., Matsuoka, M., Kosugi, S. and Hashimoto, J. (1991) Eur. J. Biochem. **195**, 571–575
66.  Matsumoto, T., Hata, S., Suzuka, I. and Hashimoto, J. (1994) Eur. J. Biochem. **223**, 179–187
67.  Kosugi, S., Suzuka, I., Ohashi, Y., Murakami, T. and Arai, Y. (1991) Nucleic Acids Res. **19**, 1571–1576
68.  Kosugi, S., Suzuka, I. and Ohashi, Y. (1995) Plant J. **7**, 877–886
69.  Thelander, L. and Reichard, P. (1979) Annu. Rev. Biochem. **48**, 133–158
70.  Reichard, P. (1988) Annu. Rev. Biochem. **57**, 349–374
71.  Elledge, S.J., Zhou, Z. and Allen, J.B. (1992) Trends Biochem. Sci. **17**, 119–123
72.  Eriksson, S., Graslund, A., Skog, S., Thelander, L. and Tribukait, B. (1984) J. Biol. Chem. **259**, 11695–11700
73.  Mann, G.J., Musgrove, E.A., Fox, R.M. and Thelander, L. (1988) Cancer Res. **48**, 5151–5156
74.  Bjorklund, S., Skog, S., Tribukait, B. and Thelander, L. (1990) Biochemistry **29**, 5452–5458
75.  Lowndes, N.F., McInerny, C.J., Johnson, A.L., Fantes, P.A. and Johnston, L.H. (1992) Nature (London) **355**, 449–453
76.  Elledge, S.J. and Davis, R.W. (1990) Genes Dev. **4**, 740–751
77.  Bjorklund, S., Skogman, E. and Thelander, L. (1992) EMBO J. **11**, 4953–4959
78.  Philipps, G., Clement, B. and Gigot, C. (1995) FEBS Lett. **358**, 67–70
79.  Linn, S. (1991) Cell **66**, 185–187
80.  Burgers, P.M., Bambara, R.A., Campbell, J.L., Chang, L.M.S., Downey, K.M., Hubscher, U., Lee, M.Y.W.T., Lin, S.M., So, A.G. and Spadari, S. (1990) Eur. J. Biochem. **191**, 617–618
81.  Wilson, S.H. (1990) in The Eukaryotic Nucleus: Molecular Biochemistry and Molecular Assemblies (Strauss, P.R. and Wilson, S.H., eds.), vol. 1, pp. 199–234, Telford Press, Caldwell
82.  Takada-Takayama, R., Hanaoka, F., Yamada, M. and Ui, M. (1991) J. Biol. Chem. **266**, 15716–15718
83.  Tanaka, S., Hu, S.-Z., Wang, T.S.-F. and Korn, D. (1982) J. Biol. Chem. **257**, 8386–8390
84.  Miller, M.R., Ulrich, R.G., Wang, T.S.-F. and Korn, D. (1985) J. Biol. Chem. **260**, 134–138
85.  Kaczmarek, L., Miller, M.R., Hammond, R.A. and Mercer, W.E. (1986) J. Biol. Chem. **261**, 10802–10807
86.  Lcchini, G., Francesconi, S., Foiani, M., Badaracco, G. and Plevani, P. (1992) EMBO J. **6**, 737–742

87. Wong, S.W., Wahl, A.F., Yuan, P.-M., Arao, N., Pearson, B.F., Arai, K., Korn, D., Hunkapiller, M.W. and Wang, T.S.-F. (1988) EMBO J. **7**, 37–47

88. Damagnez, V., Tillit, J., De Recondo, A.-M. and Baldacci, G. (1991) Mol. Gen. Genet. **226**, 182–189

89. Hirose, F., Yamaguchi, M., Nishida, Y., Masutani, M., Miyazawa, H., Hanaoka, F. and Matsukage, A. (1991) Nucleic Acids Res. **19**, 4991–4998

90. Miyazawa, H., Izumi, M., Tada, S., Takada, R., Masutani, M., Ui, M. and Hanaoka, F. (1993) J. Biol. Chem. **268**, 8111–8122

91. Spicer, E.K., Ruth, J., Fung, C., Reha–Krantz, L.J., Karam, J.D. and Konigsberg, W.H. (1988) J. Biol. Chem. **263**, 7478–7486

92. Fierk, C.A., Johnson, K.A. and Benkovic, S.J. (1987) Biochemistry **26**, 4085–4092

93. Jarabak, J. and Bachur, N.R. (1971) Arch. Biochim. Biophys. **142**, 417–425

94. Ivanetich, K.M. and Santi, D.V. (1990) FASEB J. **4**, 1591–1597

95. Prendergast, N.J., Delcamp, T.J., Smith, P.L. and Freisheim, J.H. (1988) Biochemistry **27**, 3664–3671

96. Allen, J.R., Lasser, G.W., Goldman, D.A., Booth, J.W. and Mathews, C.K.(1983) J. Biol. Chem. **258**, 5746–5753

97. Noguchi, H., Reddy, G.P. and Pardee, A.B. (1983) Cell **32**, 443–451

98. Toth, I., Lazar, G. and Goodman, H.M. (1987) EMBO J. **6**, 1853–1858

99. Cella, R., Nielsen, E. and Parisi, B. (1988) Plant Mol. Biol. **10**, 331–338

100. Lazar, G., Zhang, H. and Goodman, H.M. (1993) Plant J. **3**, 657–668

101. Luo, M., Piffanelli, P., Rastelli, L. and Cella, R. (1993) Plant Mol. Biol. **22**, 427–435

102. Wang, M., Ratnam, S. and Freisheim, J.H. (1995) Biochim. Biophys. Acta **1261**, 325–336

103. Hofbauer, R. and Denhardt, D.T. (1991) Crit. Rev. Eukaryot. Gene Exp. **1**, 247–300

104. Feder, J.N., Guidos, C.J., Kusler, B., Carswell, C., Lewis, D. and Schimke, R.T. (1990) J. Cell. Biol. **111**, 2693–2701

105. Hata, S., Kouchi, H., Suzuka, I. and Ishii, T. (1991) EMBO J. **10**, 2681–2688

106. Kouchi, H., Sekine, M. and Hata, S. (1995) Plant Cell **7**, 1143–1155

107. Hemerly, A., Bergounioux, C., Van Montagu, M., Inze, D. and Ferreira, P. (1992) Proc. Natl. Acad. Sci. U.S.A. **89**, 3295–3299.

108. Soni, R., Carmichael, J.P., Shah, Z.H. and Murray, J.A.H. (1995) Plant Cell **7**, 85–103

109. Hirt, H., Mink, M., Pfosser, M., Bogre, L., Gyorgyey, J., Jonak, C., Gartner, A., Dudits, D. and Heberle-Bors, E. (1992) Plant Cell **4**, 1531–1538

110. Meskiene, I., Bogre, L., Dahl, M., Pirck, M., Ha, D.T.C., Swoboda, I., Heberle-Bors, E., Ammerer, G. and Hirt, H. (1995) Plant Cell **7**, 759–771

111. Savoure, A., Feher, A., Kalo, P., Petrovics, G., Csanadi, G., Szecsi, J.,Kiss, G., Brown, S., Kondorosi, A. and Kondorosi, E. (1995) Plant Mol. Biol. **27**, 1059–1070

112. Renaudin, J.-P., Colasanti, J., Rime, H., Yuan, Z., and Sundaresan,V. (1994) Proc. Natl. Acad. Sci. U.S.A. **91**, 7375–7379

113. Ferreira, P.C.G., Hemerly, A.S., de Almeida Engler, J., Bergounioux, C., Burssens, S., Van Montagu, M., Engler, G. and Inze, D. (1994) Proc. Natl. Acad. Sci. U.S.A. **91**, 11313–11317

114. Gallant, P. and Nigg, E.A. (1994) EMBO J. **13**, 595–605

115. Kodama, H., Ito, M. and Komamine, A. (1994) Plant Cell Physiol. **35**, 529–537

116. Johnston, M. and Davis, R.W. (1984) Mol. Cell. Biol. **4**, 1440–1448

117. Kobayashi, H., Stewart, E., Poon, R., Adamczewski, J.P., Gannon, J. and Hunt, T. (1992) Mol. Biol. Cell **3**, 1279–1294

118. Setiady, Y.Y., Sekine, M., Hariguchi, N., Yamamoto, T., Kouchi, H. and Shinmyo, A. (1995) Plant J. **8**, 949–9579

119. Ruderman, J.V. (1993) Curr. Opin. Cell Biol. **5**, 207–213

120. Sturgill, T.W., Ray, L.B., Erikson, E. and Maller, J.W. (1988) Nature (London) **334**, 715–718

121. Pulverer, B.J., Kyriakis, J.M., Avruch, J., Nikolakaki, E. and Woodgett, J.R. (1991) Nature (London) **353**, 670–674

122. Ferrell, J.E., Wu, M., Gerhart, J.C. and Martin, S.G. (1991) Mol. Cell. Biol. **11**, 1965–1971

123. Gotoh, Y., Moriyama, K., Matsuda, S., Okumura, E., Kishimoto, T., Kawasaki, H., Suzuki, K., Yahara, I., Sakai, H. and Nishida, E. (1991) EMBO J. **10**, 2661–2668

124. Pelach, S.L.and Shanghera, J.S. (1992) Trends Biochem. Sci. **17**, 233–238

125. Wilson, C., Eller, N., Gartner, A., Vicente, O. and Heberle-Bors, E. (1993) Plant Mol. Biol. **23**, 543–551
126. Jonak, C., Pay, A., Borge, L., Hirt, H. and Heberle-Bors, E. (1993) Plant J. **3**, 611–617
127. Stafstrom, J.P., Altschuler, M. and Anderson, D.H. (1993) Plant Mol. Biol. **22**, 83–90
128. Wang, J.C. (1985) Annu. Rev. Biochem. **54**, 665–697
129. Wood, E.R. and Earnshaw, W.C. (1990) J. Cell Biol. **111**, 2839–2850
130. Adachi, Y., Luke, M. and Laemmli, U.K. (1991) Cell **64**, 137–148
131. Uemura, T., Ohkura, H., Adachi, Y., Morino, K., Shiozaki, K. and Yanagida, M. (1987) Cell **50**, 917–925
132. Hirano, T. and Mitchison, T.J.J. (1993) J. Cell Biol. **120**, 601–612
133. Shamu, C.E. and Murray, A.W. (1992) J. Cell Biol. **117**, 921–934
134. Kimura, K., Saijo, M., Ui, M. and Enomoto, T. (1994) J. Biol. Chem. **269**, 1173–1176
135. Ramachandran, C., Mead, D., Wellham, L.L., Sauerteig, A. and Krishan, A. (1995) Biochem. Pharmacol. **49**, 545–552
136. Xie, S. and Lam, E. (1994) Nucleic Acids Res. **22**, 5729–5736
137. Godsden, M.H., Mcintosh, E.M., Game, G.C., Wilson, P.J. and Haynes, R.H. (1993) EMBO J. **12**, 4425–4431
138. McIntosh, E.M., Gadsden, M.H. and Haynes, R.H. (1986) Mol. Gen. Genet. **204**, 363–366
139. McIntosh, E.M., Looser, J., Haynes, R.H. and Pearlman, R.E. (1994) Curr. Genet. **26**, 415–421
140. Strahler, J.R., Zhu, X.-X., Hora, N, Karen Wang, Y., Andrews, P.C., Roseman, N.A., Neel, J.V., Turka, L. and Hanash, S.M. (1993) Proc. Natl. Acad. Sci. U.S.A. **90**, 4991–4995
141. Pri-Hadash, A., Hareven, D. and Lifschitz, E. (1992) Plant Cell **4**, 149–159
142. Pardo, E.G. and Gutierrez, C. (1990) Exp. Cell Res. **186**, 90–98
143. Kodama, H., Kawakami, N., Watanabe, A. and Komamine, A. (1989) Plant Physiol. **89**, 910–917
144. Kodama, H., Ito, M., Hattori, T., Nakanura, K. and Komamine, A. (1991) Plant Physiol. **95**, 406–411
145. Ito, M., Kodama, H. and Komamine, A. (1991) Plant J. **1**, 141–148
146. Jerome, V., Vourc'h, C., Baulieu, E.-E. and Catelli, M.-G. (1993) Exp. Cell Res. **205**, 44–51
147. Moreno, S., Hayles, J. and Nurse, P. (1989) Cell **58**, 361–372
148. Gould, K.L. and Nurse, P. (1989) Nature (London) **342**, 39–45
149. Xu, M., Sheppard, K.-A., Peng, C.-Y., Yee, A.S. and Piwnica-Worms, H. (1994) Mol. Cell. Biol. **14**, 8420–8431
150. Parker, L.L., Atherton, F.S., Lee, M.S., Ogg, S., Falk, J.L., Swenson, K.I. and Piwnica-Worms, H. (1991) EMBO J. **10**, 1255–1263
151. Parker, L.L., Atherton-Fessler, S. and Piwnica-Worms, H. (1992) Proc. Natl. Acad. Sci. U.S.A. **89**, 2917–2921
152. Aligue, R., Akhavan-Niak, H. and Russell, P. (1994) EMBO J. **13**, 6099–6106
153. Watanabe, N., Broome, M. and Hunter, T. (1995) EMBO J. **14**, 1878–1891
154. Kieliszewski, M. and Lamport, D.T.A. (1994) Plant J. **5**, 157–172
155. Cassab, G.L. and Varner, J.E. (1988) Annu. Rev. Plant Physiol. Plant Mol. Biol. **39**, 321–353
156. Wilson, L.G. and Fry, J.C. (1986) Plant Cell Environ. **9**, 238–260
157. Showalter, A.M. (1993) Plant Cell **5**, 9–23.
158. Keller, B. (1993) Plant Physiol. **101**, 1127–1130
159. Varner, J.E. and Lin, L.-S. (1989) Cell **56**, 231–239
160. Cooper, J.B., Heuser, J. and Varner, J.E. (1994) Plant Physiol. **104**, 747–752
161. Schmidt, A., Datta, K. and Marcus, A. (1991) Plant Physiol. **96**, 656–659
162. Ito, M., Yasui, A. and Komamine, A. (1992) FEBS Lett. **301**, 29–33
163. Broek, D., Bartlett, R., Crawford, K. and Nurse, P. (1991) Nature (London) **349**, 388–393
164. Ito, M., Sato, T., Fukuda, H. and Komamine, A. (1994) Plant Mol. Biol. **24**, 863–878
165. Andrews, B.J. (1992) Nature (London) **355**, 393–394
166. Breeden, L. and Nasmyth, K. (1987) Cell **48**, 389–397
167. Ferreira, P., Hemerly, A., Van Montague, M. and Inze, D. (1994) Plant Mol. Biol. **26**, 1289–1303

# 9

# Cell size and organ development in higher plants

**Dennis Francis**

School of Pure and Applied Biology, University of Wales,
Cardiff CF1 3TL, U.K.

## Introduction

Polarizing forces establish the root and shoot meristems soon after the initial divisions of the zygote. The molecular identity of factors which induce polarity is unknown, but $Ca^{2+}$ gradients are implicated strongly in the response [1,2]. The earliest morphological evidence of polarity is seen at the heart-shaped (dicots) or torpedo stages (monocots) of embryogenesis, where the root and shoot meristems grow away from each other. Hence, the meristems are true embryonic growth and organizing centres which persist for the life of the plant. Descendants of the meristem give rise to the various organs and tissues of the plant through the processes of cell division, cell expansion and cell differentiation. Arguably, the cell cycle is solely a mechanism for partitioning cytoplasm and generating more cells, an activity that predominates in meristems. It would follow that pattern formation arises after the differentiation of cells as the meristem grows away from them. In other words, the cell cycle is a mechanism that simply provides more cells which then expand, lead to growth and, through cytodifferentiation, result in the formation of different tissues. This would be in keeping with the organismal view of development, i.e. that it is the organism that makes cells [3,4] and not, as the cell theory of development claims [5], cells that make the organism. I have always regarded attempts to discriminate between cell division and cell expansion as wasteful because both contribute to growth and development of a normal healthy plant—with neither cell division nor cell expansion there is no growth. This view is somewhat at odds with Barlow's view that cells are an 'optional accompaniment of growth' [6]. I find this a difficult concept to grasp because each growing or developing organ comprises cells. If the organ gets bigger, it does so either because more cells are produced (e.g. a newly forming leaf primordium) or because cells elongate and expand (e.g. a leaf nearing maturity) (see J. Traas and P. Laufs, chapter 16). Clearly, the initial prompt for organ initiation could be a change in surface tension followed, soon after, by the partitioning of protoplasm. Here, I am more in agreement with Barlow [6] that mitosis 'may provide the conditions that permit morphogenesis'. The often-cited case, that claims to eliminate cell division from development, is that of gamma-irradiated wheat seedlings which made a leaf in the absence of cell division (the latter is bludgeoned by the treatment) [7]. In this case, the shoot meristem exhibited a primordial bulge, but, judging from the photomicrograph, the primordium was abnormal. I suspect that, in the absence of cell division, some sort of internal osmotic adjustment forced the apex to bulge, but the primordium failed to develop normally because cell division was blocked.

In a very recent paper, the cell cycle was apparently uncoupled from plant development by mutagenizing one of the central genes of the cell cycle, $cdc2$, and by studying the development of transgenic $Arabidopsis$ and tobacco plants which expressed the mutant gene ([8]; see also, L. De Veylder et al., chapter 1). A major conclusion was that alterations in the pattern of cell division in meristematic regions of the transgenic plants bear no relationship to the timing of developmental programmes compared with the wild type [8]. Clearly, it would be interesting to measure rates of cell division on these transgenic plants.

As illustrated elsewhere in this volume, the cell cycle has two principal control points, one in late $G_1$, and the other in late $G_2$, respectively [9]. Central to these major checkpoints is the 34 kDa protein kinase encoded by $cdc2$ [10]. However, the multitude of checkpoints that function during interphase [11], reported on elsewhere in this volume, indicate that the cell is nurtured through interphase so that it reaches an optimal size for division [11]. How this is achieved seems to differ in unicellular eukaryotes compared with multicellular eukaryotes. In budding yeast, a major nursery exists in late $G_1$/START [12]. Beyond START, and beyond a critical minimum size, the cell replicates its nuclear DNA, traverses $G_2$ and enters mitosis [13]. A similar requirement is met for fission yeast [14], but another molecular checkpoint in late $G_2$ governs the cell's entry into mitosis. In particular, the up- and down-regulation of two genes, $wee1$ and $cdc25$, which effectively govern cell size at mitosis [10], are critical [14] and will be discussed in depth later. In multicellular eukaryotes, cells can arrest in both $G_1$ ($G_0$) and $G_2$ of the cell cycle (e.g. epidermal cells in mouse [15], root meristems of $Vicia faba$ [16] and $Pisum sativum$ [17]). Cells that arrest in $G_2$ presumably do so by defaulting from the cell growth cycle. This can be easily understood for callus cells $in vitro$, where cessation of division for one cell will often be of no consequence to neighbouring cells. However, in plant meristems, cells that arrest in $G_2$ may well continue to grow by symplastic growth to keep pace with dividing cells in adjacent files. In other words, if they did not grow, small cells would be observed alongside large cells, or some cells would be separated from others, perhaps torn away, as a result of the growth of the entire organ. The end-result would be disjointed growth and air-spaces would form at random between cells; this is not observed in meristems. Indeed, when observing longitudinal sections through the meristem, one should never fail to be impressed by the organized files of the cells that exist. This is particularly so in the root. Although less obvious in the shoot, cells can be delineated into different zones, particularly during vegetative growth (see J.

Doonan, chapter 10). These observations are made possible because cells of different size occupy different domains within the meristem. To what extent is cell size important for the proper functioning of the meristem? This paper will address the question by assessing meristem function from the cellular through to the molecular levels, and I will try to identify the regulatory features of the cell cycle which could be regarded as pivotal, not only in controlling the cell cycle, but also in prompting developmental programmes. In this respect, I shall examine meristem function in vegetative and floral shoots, in leaf meristems and in roots of higher plants. The hypothesis under scrutiny is whether a sizer control of cell division can be considered as a component of endogenous developmental programmes. The aim will be to establish whether there are unique characteristics inherent in cell size and the cell cycle that would contribute to a model which would help to explain cell cycle function in meristems in relation to the development of associated organs.

## The shoot meristem

The shoot apex, taken to include the apical dome, the youngest primordia and associated tissues, typically exhibits iterative growth—the apical dome enlarging to a critical size, initiating leaf primordia, falling back to a minimum size before growing again to a critical size and so on [18]. The entire shoot apex is an exponential growth system so that, at the start of the plastochron, the dome increases in size exponentially, but so too does the newly formed primordium and sub-apical region (the phytomer). Each successive phytomer increases in size exponentially as the whole system expands to include a very large population of cells over a 2–3 cm distance (e.g. calculated to be as high as 2–3 million cells [19]). Cell division is central to the maintenance of this system. During vegetative growth, different populations of cells from different regions can be recognized in median sections of stained apices because the cells exhibit differential staining. For example, in the region above the youngest primordium (the apical dome), cells at the summit of the apex often appear to be the largest, and stain less intensely than adjacent cells. Those at the top of the dome are the cells of the central zone (CZ) which overlie smaller cells on the flanks [the peripheral zone (PZ)]. Below the CZ and adjacent to the PZ is the pith rib meristem (PRM) (Fig. 1). Sometimes called the vacuolating meristem, the cells in this zone are typically of intermediate size compared with those of the central and peripheral zones.

Such broad descriptions are plentiful in the literature, but to what extent is a size gradient a regular feature of the shoot apex during vegetative growth? Does the size gradient alter during the developmental switch to floral growth? In particular, a well-characterized feature of the transition to floral growth is faster cell cycles in each of the zones of the apex, together with an eventual loss of the zonation characteristic of the vegetative meristem. How does cell size change during the floral transition? Is there any significance to changes in cell size in relation to faster cell cycles in the prefloral apex? To tackle these questions, data have been collated from a range of species both with respect to cell size (Table 1) and

**Fig. 1.** **Schematic reconstructions of photomicrographs of median longitudinal sections of vegetative and prefloral apical domes [20]**

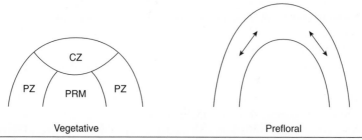

Vegetative                                                      Prefloral

On the basis of histological staining, the vegetative apical dome (left-hand side) is divided into the central zone (CZ), the peripheral zone (PZ) and the pith rib meristem (PRM). The prefloral meristems in [20] show that this zonation fades, the dome enlarges and cells in the epidermal and sub-epidermal regions become more isodiametric. This suggests a greater frequency of anticlinal division in this region, so that the cells in this region can be depicted (right-hand side) as a tunica-like region parallel to the epidermis (see text for further details). Typical vegetative apical domes are about 50–200 μm in diameter, whereas prefloral domes are much larger (e.g. 3000 μm in Chrysanthemum [84]).

to cell cycle times (Table 2). For the former, the review of Nougarède [20] was invaluable because it contains a series of photomicrographs of median longitudinal sections of both vegetative and prefloral shoot meristems. The data in Table 1 result from tracings of cell areas on enlargements of the photomicrographs and are consistent in showing a cell size gradient for the vegetative meristems.

**Table 1** **Mean cell areas (arbitrary units±S.E.) in the CZ, PZ and PRM of vegetative apical domes, and the central area (C) and peripheral area (P) of prefloral apical domes of seven angiosperms recorded from tracings of enlarged photocopies of photomicrographs in Nougarède (1967)**

| Vegetative | | | | Prefloral | |
|---|---|---|---|---|---|
| | CZ | PZ | PRM | C | P |
| Chrysanthemum segetum | 32 ± 2 | 17 ± 2 | 33 ± 5 | 19 ± 10 | 16 ± 1 |
| Teucrium scorodonia | 39 ± 7 | 21 ± 2 | 34 ± 2 | 1 ± 4 | 30 ± 4 |
| Trapaeolum majus | 27 ± 3 | 30 ± 1 | | | |
| Coleus blumei | 77 ± 4 | 40 ± 4 | 54 ± 5 | | |
| Lupinus albus | 52 ± 5 | 25 ± 1 | 43 ± 6 | | |
| Perilla nankinensis | 32 ± 2 | 38 ± 3 | | | |
| Jessicua grandiflora | 68 ± 6 | 70 ± 4 | | | |

All photocopies were to the same magnification (n = 10).

**Table 2**     Duration of the mean cell-doubling time (h) in the **CZ, PZ** and **PRM** of vegetative and prefloral apical domes

| Vegetative | Prefloral | | | | | |
|---|---|---|---|---|---|---|
| | **CZ** | **PZ** | **PRM** | **C** | **P** | **Ref.** |
| *Chrysanthemum segetum* | 139 | 48 | 70 | 54 | 48 | [84] |
| *Coleus blumei* | 237 | 157 | | | | [85] |
| *Datura stramonium* | 76 | 36 | | | | [86] |
| *Lolium temulentum* wt | 76 | 34 | 73 | | | [39] |
| s-t-g* | 82 | 41 | 177 | | | |
| *Rudbeckia bicolor* | >40 | 30 | | | | [87] |
| *Pisum sativum* | 69 | 28–30 | | | | [27] |
| *Sinapis alba* | 288 | 157 | | 108 | 47 | [88] |
| *Silene coeli-rosa* | 20 | 20 | | 10 | 10 | [89,90] |
| *Trifolium repens* | 108 | 69–87 | 136 | | | [91] |

*s-t-g, slow-to-green mutant of the C3 class, L.temulentum Ba3081.

In each case, mean cell area is significantly larger in the CZ compared with the PZ ($P <0.05$). Moreover, cell size in the PRM is also significantly larger than that in the PZ, but not significantly different to that in the CZ. Overall, there could be different sizer controls in different zones of the vegetative apex—the net result would be cells of different size in each zone as discussed above. Alternatively, the data in Table 1 could indicate that all cells grow to the same size before dividing and that subsequently cells at the summit expand more than those on the flanks. The former seems more likely given the exponential nature of growth at the apex (see above) and, for example, that the size measurements for *Chrysanthemum* (Table 1) were solely from mitotic cells [20].

The data collated from vegetative meristems on cell-doubling times (Table 2) and on cell size (Table 1) conform to a simple relationship: CZ—larger cell size— longer cell cycle; PZ—smaller cell size—shorter cell cycle.

If all cells in the apical dome grow at the same rate, then those in the PZ reach a critical size for division sooner than those in the CZ and, hence, exhibit shorter cell cycle times. The maintenance of constant cell growth rates throughout the dome would help to explain the continuous and rhythmic, increases and decreases in the size of the apical dome from one plastochron to the next, while different threshold sizes for division would result in the differences in sizes of cells in the different zones. Such a simple relationship is more difficult to deduce from the cell cycle and cell size data for the PRM. More often than not, the cells are more similar in size to those in the CZ, but the cell cycle is not always as long as that in corresponding cells in the CZ. To complicate matters further, cells of the PRM, particularly those towards the basal margins of this zone, are beginning to expand as they give rise to cells of the sub-apical region.

Upon the arrival of the floral stimulus, events which commit the apex to make a flower are defined as part of floral evocation [21]. Faster cell cycles (see Table 2) and higher rates of respiration are features of evocation in a range of unrelated species and these changes are all geared to make the apex increase in size to accommodate the floral primordia [22]. In the shoot meristem of prefloral plants, and consistent with the fading of zonation described by Nougarède [20], cell size is more uniform across the apex from the summit to the flanks (Table 1). However, the photomicrographs of the prefloral meristems depicted in Nougarède [20] also seem to reveal a distinct change in the orientation of cells in the apex (Fig. 1) The pattern of cell sizes in the prefloral meristem conforms to the equivalent of a wide tunica. This loss of zonation must involve a change in the plane of cell division. The only logical way that this could come about would be by an increase in the frequency of anticlinal divisions in the peripheral layers. Clearly, changes in the plane of cell division involving reorientation of the preprophase bands are of critical importance (see J.H. Doonan, chapter 10 and V. Sundaresan and J. Colasanti, chapter 3). A more uniform cell size in a prefloral apex compared with a cell size gradient in the vegetative apex suggests that a new threshold for size at division is established at the time of flowering. It is difficult to imagine that this change *per se* has any strategic role in flowering. However, coupled with a change in the plane of cell division for those cells at the periphery of the apex, it may predispose the apex to the change in morphogenesis to follow, which involves a rapid initiation of primordia, generally smaller at initiation, compared with vegetative primordia [23]. Another interpretation is that the change in morphogenesis, manifest in the initiation of smaller primordia, leads to a change in form which is then filled by cell division. This would confirm Barlow's theory [6] that new cells, arising from cell division, fill in the volume created by growth of the organ. However, without cell division, and indeed faster rates of cell division (see Table 2), the apex would not enlarge to accommodate the new primordia. On this basis, the cell and organismal theories of development are not mutually exclusive.

The change in the pattern of cell sizes from a 'summit-to-flanks' gradient to a more uniform cell size across both regions, may be part of a mechanism that predisposes the apex to initiate primordia that contribute to the floral structure. Admittedly, this conclusion is based on measurements of only a limited number of species (Table 1) and not in all of these does the terminal apex initiate the first floral parts. Moreover, the change in the plane of cell division is but only a small component of the overall change at the apex which leads to positioning of floral primordia. Green [24] described the onset of floral morphogenesis as the formation of a new reinforcement pattern of surface microtubules common to entire groups of epidermal cells of the flower meristem. It may be that the more uniform cell size which leads to a more prominent 'tunica' is in fact a reflection of a change in orientation of microtubules in epidermal cells. In other words, changes in surface microstructure may be accompanied by similar changes in the layers immediately below the surface. As noted by Lyndon [18], such a change in reinforcement hoops on the surface results in a series of concentric rings of microtubules in which incipient sepals and petals alternate. The nature of the molecules that bring about these changes in cellular architecture is unknown.

Little is known about the molecular basis of floral evocation, but meristem identity genes such as *flo* of *Arabidopsis* are clearly central to this phase of flowering ([25], see J.H. Doonan, chapter 10). Perhaps the *flo* gene product, or gene products downstream of *flo*, are instrumental in bringing about major changes in planes of cell division which lead to the change in the pattern of cells in the apex. Changes in cell cycle regulation which lead to a new threshold size for division again could be an important part of the prefloral stage. Clearly, the data in Table 2 indicate that the cell cycle shortens in the prefloral compared with the vegetative meristem. In this case, the smaller cells in the central region of the prefloral meristems are dividing at a faster rate than the larger cells in the central zone of the vegetative meristem. However, as will be evident elsewhere in this chapter, there are no general rules about the size of cells and the rate at which they divide. Indeed in unicellular organisms such as *Chlamydomonas*, larger cells tend to divide faster than smaller ones [26]. The cell cycle may have a role in determining cell type, particularly in relation to the rate of cell division in the different types of floral organ [18].

## Leaf meristems and leaf form

As with all other aspects of morphogenesis, leaf development is the product of cell division, changes in the plane of cell division, changes in surface extensibility, and increases in the rate of cell expansion as the leaf takes on its final shape and appearance. Are there changes in cell size, as a consequence of altered cell cycles, which play a part in, or are a feature of, leaf morphogenesis?

Leaf primordia arise from the sides of the apical dome. The primordium originates from a cluster of cells on the flanks of the meristem in the peripheral zone (see above). In *Pisum sativum*, the region where the next primordium will form is characterized by an increase in the frequency of cell division [27]. In longitudinal section, incipient leaf primordial cells were identified as a cluster of periclinal cell divisions which increased in density from about two-thirds of the way through a plastochron [27]. It could be assumed that this increase in the rate of cell division coupled with the change in the plane of cell division are driving forces for leaf initiation. The problem with this generalization is that not all species exhibit faster rates of growth at the sites of leaf initiation. For example, Lyndon [18] cites the cases of the shoot apices of the pteridophytes, *Isoetes* and *Polypodium*, where the rate of growth is similar in both the axial region and at the sites of primordium initiation [28,29]. Conversely, in the angiosperm, *Silene coeli-rosa*, localized cell division may be more important at the sites of leaf initiation [19]. In both *Pisum* and *Trifolium*, the initial outgrowth of the primordium seems also to depend on increased rates of cell division internal to the primordium [27,30]. We know very little about the nature of signalling molecules that bring about the increase in the rate of cell division or those that cause changes in the plane of cell division in the shoot meristem, although auxin has always been strongly implicated. Not only is this because of auxin's known effect of stimulating cell division *in vitro* ([18]; see A. Fehér et al., chapter 11), but also because localized applications of auxin on the sides of the apex can influence the siting of new primordia [31]. The general, but as yet unproven, idea is that auxin increases in concentration, causes a primordium to be initiated and then prevents other primordia forming within its vicinity, rather like an inhibitory diffusion field [32]. Inhibitors of auxin transport such as *N*-1-naphthylphthalamic acid (NPA) cause primordia to become large at initiation, while α-4-chlorophenoxy isobutyric acid (CPIB; an auxin inhibitor) causes the primordium to become smaller [33]. As mentioned above, cell size is smaller in the peripheral zone and this is probably the only major change in cell size threshold which precedes the initiation of leaf primordia.

Are changes in the rate and plane of cell division primary factors which lead to leaf initiation? One view is that changes in the rate and plane of cell division are enabling rather than causal factors for leaf initiation [34]. This conclusion takes into account the lack of a requirement for a localized increase in growth rate during leaf initiation in some species. According to Green [35], primordia form when changes to the surface microstructure permit a release of restriction growth resulting in a bulge on the side of the apex. Careful analysis of changes in surface microstructure of the shoot apex of *Graptopetallum* through a plastochron indicated changes in hoop reinforcements of microfibrils, which facilitate the initiation of leaf primordia [35]. Clearly, much still needs to be learned about the nature of the biophysical control, and the morphogens which interact and determine the sites at which primordia are initiated. However, the case for a cell sizer regulating leaf initiation is not strong.

Once a primordium is initiated, leaf development follows very divergent pathways in dicots compared with monocots. In the former, cell division is initially uniform over the entire organ, but, gradually, with time, becomes restricted to islands of mitotic activity. Essentially, once the leaf lengthens to about 2–4 mm, cell division stops and cell expansion starts, to give final leaf size and shape [36]. During the same phase of development of the monocot leaf, cell division becomes restricted to a basal meristem. For example, the grass leaf represents a gradient of cell age and has become something of a model system for investigating the cessation of cell division, and the cessation of cell expansion in relation to the development of the photosynthetic machinery [37]. In these studies, developmental mutants have been particularly valuable. The *slow-to-green* mutant of the C3 grass, *Lolium temulentum* Ba3081, arose from a single nuclear gene mutation as a result of somaclonal varation in an embryogenic callus. The fully fertile mutant exhibits a delayed and incomplete greening which occurs from the leaf tip downwards. During growth, leaf length, absolute growth rate, and number of cells per unit length at the leaf base, are greatly reduced compared with the wild type, but relative growth rate, duration of leaf growth and length of the cell division zone are little affected [38]. Indeed, the length of the cell division zone at the leaf base is remarkably similar in leaves at known stages of development [38a]. However, typically, the final length of the leaf in the mutant is only a quarter, to a third, of that in the wild type [38]. We wondered whether these changes were a function of an altered cell number, altered cell size or both. In the basal meristem, mean cell area was 25% larger in the mutant compared with the wild type, but there were about 50% fewer cells in this meristem of the mutant. This suggests that a larger size for cell division is required in the mutant, but that a fewer number of these larger cells give rise to the leaf [39]. The tentative conclusion is that a sizer control for division is related to final size of the leaf. However, more substantial account of our findings is in preparation, more measurements are required on a wide range of angiosperms to test the hypothesis that a cell sizer has a strong influence on final leaf form.

## Root meristem

The root meristem could be regarded as a simpler structure than the shoot meristem on the basis that it does not give rise to other organs. This is a delusion, because in reality the root meristem is complex, and is heterogeneous both with respect to cell cycle duration [40], and to the spatial distribution of cells in different tissue domains within the root [41]. Moreover, superimposed on this complexity is a population of slowly dividing cells at the summit of the meristem named the quiescent centre by Clowes [42]. This population of slowly cycling cells is something of a mystery because, arguably, these cells are simply a function of the geometric alignment of neighbouring cells and are held quiescent by surrounding cells [43]. On the other hand, the cells of the quiescent centre exhibit various features which argue strongly for them being founder cells [44].

As mentioned earlier, an angiosperm root apex is notable for the high degree of organization of cells into different tissue domains. This is most easily observed in median longitudinal sections of the angiosperm root. Upon examining cell sizes in the different tissue domains, cells are largest in the stele and smallest in the cortex. For example, in diploid cells of the root meristem of *Zea mays*, the mean sizes for xylem parenchyma, metaxylem, endodermis and pericycle were 760, 4990, 1110 and 3780 $\mu m^3$, respectively [45]. In *Pisum sativum*, longitudinal sections of the root tip showed that the central vascular system comprised narrow elongated cells, while the dermal tissue comprised cubical cells. Transverse sections emphasized the narrowness of the vascular cells, while the centrally located procambium started to vacuolate [41]. Notably, the plane of cell division in the stelar region was predominantly longitudinal so that cell length greatly exceeded transverse width [46]. The longer and thinner cells within the stele help explain the characteristic parabolic shape of the root meristem, as viewed in longitudinal section [47]. Barlow [48] collated data on the number of cell divisions that cells undergo in different regions of root meristems. The data were consistent with the idea that cells in a particular lineage, traceable from the proximal margin of the quiescent centre, undergo a finite number of cell divisions. For example, in *Vicia faba*, apical initials undergo about seven divisions and, hence, derivatives of the apical initials have a determinate proliferative life-span. Hence, it would follow that elongation of the longer cells before their division will result in a lengthening of the meristem in that region so that tissue-specific sizer controls for division result in the meristem acquiring its familiar shape. Whether the increased size of the cells in the stelar region has anything to do with the subsequent differentiation of these cells is questionable because it may well be that stelar cells start to elongate before cortical and epidermal cells and, hence, sizer controls become irrelevant. However, in *Allium cepa*, the frequency of cell division was highest at the proximal margins of the stelar region of the meristem (between 1300 and 1500 μm from the root tip), while the greatest concentration of cortical cell divisions was between 800 and 1000 μm from the tip [49]. In other words, stelar cells towards the proximal edge of the meristem, which would be long and thin, were dividing, while cortical cells at a comparable position had stopped.

Nuclear size probably remains fairly constant irrespective of tissue domain. The fact that cells in meristems typically exhibit nuclear DNA contents of between 2C and 4C makes this a reasonable assumption, even though xylem cells may go polyploid upon exit from the meristem [45]. Hence, because stelar cells in the meristem are long and thin, it means that the ratio of cytoplasmic to nuclear volume increases markedly. Not withstanding the dangers of a circular argument, this change alters the cytoplasmic concentration of regulatory molecules that may predispose the stelar cells to enter a programme of differentiation, which for xylem cells will include a sequence of proteolysis and programmed cell death. The regulatory molecules which commit such cells into their canalized pathway of development are unknown, but plant growth regulators [PGR(s)] are the obvious candidates. In *Lactuca sativa*, pith explants exhibited xylem differentiation only after a 2–3 day sub-culture on a medium containing indol-3-ylacetic acid (IAA) and zeatin [50]. In other words, 2–3 days of induction by the PGRs was necessary for the explants to acquire competence for xylem diferentiation. Cell division occurred on all media and may be required for the acquisition of competence for differentiation [19]. In my view, the establishment of a particular level of each PGR at a particular location in the root was best explained [48] by a gradient of a cytokinin from the quiescent centre (source) to dividing cells (sink), and an auxin gradient (which is known to be basipetal in plants [51,52]), moving from mature or maturing cells (source) towards the quiescent centre [48]. On the one hand, the concentration of cytokinins maybe supraoptimal for cell division in the quiescent centre and, on the other, optimal for cell division outside of the quiescent centre [48]. Moreover, at the margins of the meristem, the concentration gradient of both substances would cause cells to stop dividing and start differentiating. In the context of xylem differentiation from the root apex, I suggest the following : (i) descendant cells from the apical initials elongate to a particular threshold for division and through a temporal control are committed to undergo a number of cell divisions; and (ii) at the proximal edge of the meristem, and under the control of a particular cell sizer for division, they are long and thin, and are then induced to differentiate by a particular balance of auxins and cytokinins.

The extent to which the elongation of cells to a finite length (size) for division is incidental or crucial to this presumed sequence of events is unknown, but the cells in the stele are probably elongating at similar rates to those in other tissues, but divide at a larger threshold size. If so, then stelar cells may have somewhat longer cell cycles than those in the cortex. Davidson [53] concluded that longer cell cycles could provide the time required for programming of differentiation. A test of the relative importance of differential cell size and differential rates of cell division in different domains of the root would be to perturb cell size at division and then discover whether smaller cells (or larger cells) are as competent to differentiate as wild-type cells, when placed on inductive media.

To consider root growth and development solely by assessing the apical meristem is to ignore the overwhelming amount of growth which is accounted for by branching of the primary into secondaries into tertiaries, and so on. Secondary roots are initiated from the pericycle, a seemingly innocuous outermost layer of the vascular stele. The cells of the pericycle exhibit meristematic potential and, in my view, are undifferentiated, while neighbouring cells within the stele are programmed to differentiate as xylem, phloem and associated tissues. This means that there is a vast difference in morphogenetic potential across only about two cell files. Hence, lineages of pericycle cells which stretch away from the meristem have proliferative potential, exhibit high levels of Cdc2 kinase activity (see L. De Veylder et al., chapter 1) and, in effect, have an indeterminate lifespan. The cells of the pericycle are smaller than those of the metaxylem, but can be larger than those in the endodermis [45]. However, in roots of carrot (*Daucus carrota*), in regions in which lateral roots are about to be initiated, the pericycle comprises cells which are about a third to half the size of adjacent endodermal cells, as observed in longitudinal section [54]. Moreover, when the pericycle initiates secondary roots, a group of mother cells partition. Hence, at this stage, the pericycle cells become still smaller. In the carrot, these newly divided cells are about one-fifth of the length of the endodermal cells [54]. Transverse sections of primary roots of *Zea mays* also showed smaller pericycle cells compared with adjacent xylem and endodermal cells [55]. Clearly, pericycle cells are small compared with their neighbours.

For cells of the pericycle, something stops them from elongating to the same extent as neighbouring cells. The same something, or something else, stops them from dividing until they are displaced a certain distance from the root apex. In *Vicia faba*, primordia only begin to be initiated at about 1 cm from the root tip [56], while the equivalent distance in roots of *P. sativum* is 2–3 cm [57]. It seems likely that some sort of negative regulator prevents the pericycle cells from growing to a threshold size for cell division and that signalling molecules affect the balance of gene products that regulate cell size for division. Hence, the intricate balance mediated by likely plant homologues to *cdc25* and *wee1*, which regulate cell size at division (at least they do in fission yeast [10]), would be skewed towards the imposition of a negative regulation of cell size. Quite likely, an auxin delivered via the xylem to presumptive primordial sites activates the normal sizer control for division and this is followed rapidly by the formative divisions in the pericycle that lead to the formation of a lateral root primordium. This explanation fails to explain characteristic patterns of lateral roots comprising clearly separated clusters of lateral roots. In other words, why don't all pericycle cells respond to activator molecules? Possibly, once a primordium is initiated, it prevents others forming in its immediate vicinity—not unlike the field of inhibition that is presumed to surround a newly initiated leaf primordium. So, the conclusion here is that those cells held quiescent remain under the influence of negative regulators and hence fail to reach a critical size for division.

# Cell sizer models

## Unicellular algae

Lest the reader gains the false impression that the idea of a sizer control for cell division has emerged *de novo* from this paper, it is now timely to recall extensive work that has addressed sizer controls for cell division in a range of unrelated organisms. Because of the complexities of multicellular meristems, P.C.L. John pioneered the use of the unicellular algae, *Chlamydomonas* and *Chlorella*. Notably, John et al. [58] stressed the importance of the attainment of a minimum size for division and that growth beyond this minimum size commits the cell to division. In these algal cells, synchronized by light/dark cycles, and grown on a mineral salts medium, a late $G_1$ control was discovered which was equivalent to that in yeasts [59]. By separating large and small cells of *Chlamydomonas* (by centrifugation), a higher division number was recorded in the larger cells and was related to their larger size [26]. When the cells were grown at either 20 or 30 °C, or when they were transferred between the two temperatures, the growth rate was 2-fold higher at the higher compared with the lower temperature; the difference was equally great between cells transferred from 20 to 30 and those transferred from 30 to 20 °C. However, the commitment to divide, which was measured following transfer of the cells to darkness, was the same regardless of temperature treatment. This was consistent with the involvement of an endogenous timer determining cell cycle duration [26]. The larger cells exhibited a longer $G_1$-phase resulting in large mother cells, which were then committed to a second round of DNA replication and division [60], but the largest went through further commitments and further cell divisions [26]. The timer for commitment proved to be more complex and involves at least two circadian-based mechanisms analogous to hour-glass timers [61]. The attainment of commitment is dependent on cell growth and, according to John et al.

[58], a timed extension to $G_1$ phase occurs which results in larger cells, but is counterbalanced by these larger cells exhibiting successive commitments resulting in multiple fission at the end of the cycle. In other words, an endogenous timer dictates the duration of $G_1$ that is required for a cell to be committed to the next division. The cell grows to a minimum size which then signals the start of DNA replication, $G_2$ and mitosis. In this system, the latter phases are remarkably constant. Hence, these studies have shown the importance of a minimum size for division and the importance of endogenous processes that regulate the time taken to reach the commitment phase.

For dividing cells in plant meristems, cells double their volume once every cell cycle. Clearly, in proliferative cells, cell growth and cell proliferation are usually co-ordinated for this to occur. While recognizing the importance of the attainment of a minimum size for progression through the cell cycle, many differentiated cells grow to enormous lengths (e.g. metaxylem cells in plants and neurons in animals). If the attainment of a minimum size is important for cell division, why do some cells grow beyond the threshold, but do not divide? This is explainable in terms of a molecular mechanism. The minimum size requirement of late $G_1$ must be regulated by a molecular checkpoint governed by a cyclin-dependent protein kinase (Cdk)–cyclin heterodimer of the type outlined by Scherr ([62]; see also D. Dudits et al., chapter 2; J.P. Renaudin et al., chapter 4). A cell which is programmed to differentiate grows and passes through the checkpoint because it could be deficient in that particular Cdk or that particular cyclin or, conceivably, deficient in the Cdk-activating kinase that is expressed in proliferative cells. In other words, although the cell reaches the minimum size requirement for division, key substrates for the next stage of the cell cycle are either absent or are not phosphorylated. Perhaps the signals for the cessation of cell division operate in a feed-back loop which represses one or more of the CDK or cyclin genes which are normally expressed in late $G_1$.

## Yeasts

As mentioned at numerous points in this volume, most is known about the regulation of the cell cycle in late $G_2$ in fission yeast. Numerous reviews have described the network of genes and gene products that function in late $G_2$ and which lead to the entry of a cell into mitosis. Indeed, so well characterized are these genes and gene products that elegant explanations for the control of the cell cycle now exist in standard undergraduate textbooks [63]. In the context of this paper, I wish to focus on those genes which have been shown to affect cell size at mitosis and, incidentally, the understanding of which was a major springboard for the wealth of information that has now been gathered for the eukaryote cell cycle.

Size mutants of fission yeast (*Schizosaccharomyces pombe*) were isolated which undergo mitosis at about half the size of the wild-type cells. Christened by Paul Nurse as 'wee' mutants, they divide prematurely at a reduced cell size [64,65]. The genes, *wee1* and *cdc25*, have crucial roles in regulating cell size at division [10]. The *wee1* gene product, a 107 kDa protein kinase, acts as a negative regulator, and the *cdc25* gene product, an 80 kDa tyrosine phosphatase [66], is a positive regulator of Cdc2 kinase. The Wee1 kinase phosphorylates whilst the Cdc25 phosphatase dephosphorylates Tyr-15 on the ATP-binding domain of p34[cdc2] [67]. In humans, Thr-14 is also an important amino acid residue for these phosphorylation/dephosphorylation mechanisms [68,69]. Apparently, in higher plants only the Tyr-15 is suspected as the active site [8], although a dephosphorylation/phosphorylation mechanism has yet to be demonstrated in higher plants (see L. De Veylder et al., chapter 1).

In fission yeast, five tandem copies of a DNA fragment containing *cdc25* were integrated at the *cdc25* chromosomal locus. In these transformants, cell length was reduced to 76% of that of the wild-type strain [70]. Conversely, mutants isolated with multiple integrated copies of *wee1* were 24% longer than the wild type [71]. The third member of this now-familiar network was another dosage-dependent *new inducer of mitosis*, *nim1*. Fusing the *nim1* to the strong *adh* promoter, and then integrating a plasmid with this construct into the *nim1* chromosomal locus, resulted in about a 150% reduction in cell length compared with the wild type [72].

Subsequent biochemical analysis showed that the Cdc25 phosphatase operates on an independent biochemical pathway to that of the Wee1 and Nim1 kinases [66]. However, the exact mechanism by which cell size is regulated is unknown, although recent evidence suggests that genes of this late $G_2$ network exert some sort of feed-back control on earlier events of the cell cycle. For p34[cdc2] to exhibit protein kinase activity, it must bind to cyclin B encoded in fission yeast by *cdc13* [73]. This heterodimer is more commonly referred to as maturation promotion factor (MPF), named as such because of the original discovery of a protein extract from *Xenopus* which when injected into immature oocytes resulted in their premature meiotic maturation [74]. Independent biochemical and genetic analyses landed on the same spot, indicating that the heterodimer encoded by *cdc2* and *cdc13* in fission yeast [73] was remarkably similar to MPF [75]. In cell-free extracts of *Xenopus*, increased tyrosine phosphorylation of p34[cdc2] occurred when nuclear DNA was under-replicated [76]. Clearly, in this phosphorylated state, p34[cdc2] does not exhibit protein kinase activity and mitosis is prevented. Treatment with okadaic acid or caffeine (uncouplers of mitosis from S phase) resulted in higher levels of unphosphorylated p34[cdc2]. The data are consistent with a model whereby p34[cdc2], as part of MPF, exerts some sort of feed-back control, perhaps upregulating the activity of the Wee1 kinase which then acts as a negative regulator of p34[cdc2] and, hence, preventing mitosis until DNA replication is complete [76]. The entire cascade of protein kinases and phosphatase that may be acting in the phosphorylation cascade between S phase and the late $G_2$ network has yet to be identified. The genes *dis2* and *sds21* are the candidate genes for encoding phosphatases involved on this cascade; in fission yeast, *dis2* mutants have blocked chromosome disjunction [77]. The Nim1 kinase is suspected strongly as the protein kinase that imposes a negative regulation on the Wee1-like protein kinase that inactivates MPF [76]. What these data illustrate is the nature of the molecular dependency of mitosis on a normal S phase, and also point towards the type of phosphorylation cascade that may link the commitment phase of the cell cycle, in late $G_1$, to the onset of mitosis. What they also illustrate is the complex biochemistry of the mechanisms which operate to ensure that a proliferating cell enters mitosis at an optimum size.

## Perturbation of cell size at mitosis in higher plants

If, as I have suggested, cell size at division is an important component of plant development, then the simplest strategy would be to perturb cell size and study the development of the transgenic plants. The ideal situation would be to up- or down-regulate plant homologues to the fission yeast genes, *cdc25*, *wee1* and *nim1*. Despite the identification of a cDNA clone in maize with partial sequence homology to *cdc25* (see P. Sabelli et al., chapter 12), neither a full-length clone of this gene, nor *wee1* nor *nim1*, have been identified in higher plants. Faced with this problem, we opted to manipulate cell size by transforming plant cells with the fission yeast cell sizer genes; *cdc25* proved to be more user-friendly than either *wee1* or *nim1* (M. Bell, unpublished work) and was integrated into tobacco leaf disks behind the 35S gene promoter. Regenerated transgenic plantlets were obtained after a double kanamycin selection. Of the nine primary transformants, eight exhibited develop-mental abnormalities, including pocketed leaves, and aberrant flowers which formed alongside normal ones. The aberrant flowers comprised a normal whorl of sepals, but not petals, while the inner whorls were fused. However, the majority of flowers were fertile despite the presence of the aberrant ones [78]. In fact, the transgenic plants flowered precociously and formed more flowers than the wild type. Clearly, such changes could have been due to somaclonal variation or could have resulted from the Ti transformation. However, Southern analyses confirmed that all primary tranformants exhibited *cdc25* stably integrated into the tobacco genome. Careful choice of primers which were specific to either end of a 400 bp fragment of the *cdc25* sequence were used for a reverse transcription–polymerase chain reaction (RT–PCR)-based expression analysis. Of the nine primary transformants, one was used for cell size analyses, and RT–PCR was applied to the remaining eight, of which seven displayed the developmental abnormalities described above, but one was normal in its morphology and timing of develop-mental programmes. *cdc25* transcripts were detected in 'the seven', but not in the normal-looking plant. In other words, these observations showed a clear link between *cdc25* expression, and the perturbations of development described above. In the root meristem of secondary roots, cell size was also smaller both at birth and at division compared with the wild type [78].

The alterations in development in the transgenic plants could be partly attributed to a reduced cell size. For example, in one of the few studies of rates of cell division in flowering apices, oscillations between longer and shorter cell cycles were recorded as each whorl of the flower was initiated [36]. If the rhythmic appearance of a set of floral organs was dependent on these oscillating cell cycles, then perturbation of cell size may have resulted in the formation of abberant flowers. Why only some flowers were malformed is less easy to explain particularly in view of the deployment of the 35S gene promoter.

We sought a method that would enable us to perturb cell size in a more controlled manner and, more recently, we have integrated *cdc25* into a tetracycline (TET)-inducible cassette [79], and used this to transform tobacco (R. McKibbin, N.G. Halford and D. Francis, unpublished work). Transgenic tobacco roots were cultured for 28 days in darkness in the presence or absence of tetracycline. We found a greater frequency of lateral root primordia per cm of primary root in the + TET compared with the − TET treatment. Moreover, the primordia in the + TET treatment were smaller than those in the − TET and exhibited smaller cells at division. Using RT–PCR, a 5–10-fold increased in the level of *cdc25* transcripts was detected in the + TET compared with the − TET treatment. The data are consistent in showing that premature cell division, and reduced cell size at division, affected the pattern of lateral root development in these cultured roots.

A recent paper [8] demonstrated that tobacco plants over-expressing a mutated form of the plant Cdc2 (referred to as *cdc2a*) did not alter the timing of development in transgenic tobacco and *Arabidopsis.* Two forms of mutation were employed. First, the *cdc2* gene was either dominant–negative mutated so that the gene product exhibited Asn instead of Thr-147 and Ala instead of Thr-223 (residues in mammalian p34$^{cdc2}$ known to be important for the correct positioning of ATP). Secondly, the gene was mutated at Thr-14 and Tyr-15. Two-week-old tobacco seedlings with low levels of mutated Cdc2a exhibited shorter roots with a reduced number of lateral roots. Moreover, cell size was larger in all tissues of the mutants compared with the untransformed plants. Thus, an altered cell size was linked to altered morphogenesis. As far as they could determine, the frequency of cells in division was the same in mutants and wild type, but with moderate levels of Cdc2aN147.A223 expression, a reduced frequency of cell division was found [8]. They did not observe any changes to the timing of developmental events. For example, the onset of flowering was the same in the mutant and wild-type plants.

Taken together [8,78], these observations support the hypothesis that a cell sizer control can influence developmental programmes.

## Conclusions

The model (Fig. 2) recognizes the importance of a requirement for a minimum cell size which commits the cell to the next division, and emphasizes that a second network of genes, marshalled by *cdc2*, ensures that the cell is of the correct size for cell division. Hence, cell size is the morphological marker of an optimum level of those proteins that will activate and ensure completion of a normal mitosis. Through the Cdc2 kinase (MPF), there is at least one feed-back control that checks on cell growth and ensures that DNA replication is complete. The deployment of

**Fig. 2**      **A model which features a cell sizer control during a proliferative cell cycle, and during the commitment to different differentiation pathways—A, B or C (see text for further details)**

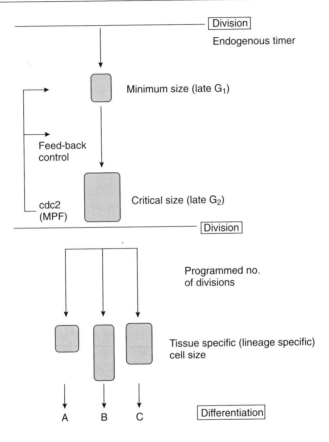

an endogenous timer will dictate the acquisition of particular sizes that the cell should grow for the next checkpoint of the cell cycle. The model dictates that in plant meristems, tissue-specific sizer controls are operating at both late $G_1$ and late $G_2$ checkpoints. The end-result will be cells of different size and, hence, different titratable amounts of gene products which programme the cell for a particular pathway of differentiation. Hence a cell size requirement is a component of a developmental programme. However, cell size does not drive the programme but nor, in my view, is it incidental. As a consequence of changes in the regulation of the cell cycle, cell size alters to meet the demands of different tissue domains.

Clearly, premature cell division resulting in a reduced cell size can disrupt normal development (see above). However, such a heterologous approach does not lead to simple explanations about the regulation of the endogenous cell cycle or about how changes in cell growth rates influence development. One of the *Drosophila* homologues to *cdc25*, *string*, features in a developmental model which has pattern formation genes regulating *string* transcription which is independent of cell cycle progression [80–82]. In other words, *string* expression, and both the timing and the pattern of cell division, are under a complex developmental control [82]. Jacobs [83] noted the impact that manipulating cell size could have in relation to developmental studies but, rightly, emphasized the need to understand the biochemical regulation of cell size (and the cell cycle) in plants. This has to be a major target to be met to test further the intricate relationships between cell-size-mediated processes in relation to the development of the higher plant. The various observations set out in this paper argue for a cell sizer control as a component of developmental programmes which emanate from plant meristems. In a developmental context, the extent to which signal transduction chains impinge on the plant cell cycle with consequent effects on cell size represents an exciting challenge.

## References

1.   Jaffe, L.F. and Nuccitelli, R. (1974) J. Cell Biol. **63**, 614–628
2.   Racusen, R.J. and Schiavone, F.M. (1990) Cell Differ. Dev. **30**, 159–169
3.   Kaplan, D.R. and Hagemann, W. (1991) BioScience **41**, 693–703
4.   Hagemann, W. (1992) Int. J. Plant Sci. **153**, 538–548
5.   Evered, D. and Marsh, J.  (1989) The Cellular Basis of Morphogenesis. J.Wiley, Chichester, U.K.
6.   Barlow, P.W. (1995) in Shape and Form in Plants and Fungi, (Ingram, D.S. and Hudson, A., eds.), pp. 169–193, Academic Press, London
7.   Foard, D.E. (1971) Can. J. Bot. **49**, 1601–1603
8.   Hemerly, A., Engler, J. de A., Bergounioux, C., Van Montagu, M., Engler, G., Inzé, D. and Ferreira, P. (1985) EMBO J. **14**, 3925–3936
9.   Van't Hof, J. (1973) in Advances in Radiation Research, Biology and Medicine (Duplan, J.F. and Chapiro, A., eds.), pp. 881–894, Gordon and Breach Sci. Pub., New York, London, Paris
10.  Nurse, P. (1990) Nature (London) **344**, 503–508
11.  Murray, A.W. (1992) Nature (London) **359**, 599–604
12.  Nasmyth, K. (1993) Curr. Opin. Cell Biol. **5**, 166–179
13.  Fosbury, S.L. and Nurse, P. (1991) Annu. Rev. Cell Biol. **7**, 227–256
14.  Norbury, C. and Nurse, P. (1992) Annu. Rev. Biochem. **61**, 441–470
15.  Gelfant, S. (1962) Exp. Cell Res. **26**, 395–400
16.  Van't Hof, J., Hoppin, D.P. and Yagi, S. (1973) Am. J. Bot. **60**, 889–895
17.  Ivanov, V.B. (1986) Tsitologiya **28**, 295–302
18.  Lyndon, R.F. (1994) New Phytol. **128**, 1–18
19.  Lyndon, R.F. (1976) in Cell Division in Higher Plants (Yeoman, M.M., ed.), pp. 285–314, Academic Press, London
20.  Nougarède, A. (1967) Int. Rev. Cytol. **21**, 203–351
21.  Evans, L.T. (1969) in The Induction of Flowering, Some Case Histories (Evans, L.T., ed.), pp. 457–480, MacMillan, Melbourne
22.  Bernier, G. (1988) Annu. Rev. Plant Physiol. Plant Mol. Biol. **39**, 175–219
23.  Battey, N.H. and Lyndon, R.F. (1988) Ann. Bot. **61**, 9–16
24.  Green, P.B. (1988) Planta **175**, 153–169
25.  Coen, E.S. (1991) Annu. Rev. Plant Physiol. Plant Mol. Biol. **42**, 241–279

26. Donnan, L. and John, P.C.L. (1983) Nature (London) **304**, 630–633
27. Lyndon, R.F. (1970) Ann. Bot. **34**, 1–17
28. Michaux-Ferrière, N. (1980) Can. J. Bot. **58**, 2506–2512
29. Michaux-Ferrière, N. (1981) Can. J. Bot. **59**, 1811–1816
30. Denne, M.P. (1966) N.Z. J. Bot. **4**, 300–314
31. Snow, M. and Snow, R. (1933) Philos. Trans. R. Soc. **222**, 353–400
32. Schwabe, W.W. (1979) in Differentiation and the Control of Development in Plants – Potential for Chemical Modifications (George, E.C., ed.), pp. 75–86. British Plant Growth Regulation Group, Monograph No. 3, Wessex Press, Wantage, UK
33. Meicenheimer, R.D. (1981) Am. J. Bot. **68**, 1139–1154
34. Lyndon, R.F. and Cunninghame, M.E. (1986) in Plasticity in Plants, Symposia of the Society for Experimental Biology (Jennings, D.H. and Trewavas, A.J., eds.), vol. 40, pp. 233–255, Company of Biologists, Cambridge
35. Green, P.B. (1994) J. Exp. Bot. **45** (Special issue), 1775–1788
36. Dale, J.E. (1988) Annu. Rev. Plant Physiol. Plant Mol. Biol. **39**, 267–295
37. Ougham, H.J. and Francis, D. (1992) in Crop Photosynthesis: Spatial and Temporal Determinants (Baker, N.R. and Thomas, H., eds.), pp. 313–336, Elsevier Science Publishers, Amsterdam, London, New York
38. Ougham, H.J., Thomas, M.M., Thomas, B.J., Roberts, P.C., Mutinda, C., Hayward, M.D. and Dalton, S.J. (1992) New Phytol. **122**, 261–272
38a Moses, L., Ougham, H.J. and Francis, D. (1997) New Phytol. **135**, 51–57
39. Moses, L. (1995) Ph.D. Thesis, Univ. Wales, Cardiff
40. Webster, P.L. and MacLeod, R.D. (1980) Env. Exp. Bot. **20**, 335–358
41. Steeves, T.L. and Sussex, I.M. (1987) Patterns in Plant Development, 2nd edn, Cambridge University Press, Cambridge
42. Clowes, F.A.L. (1954) New Phytol. **53**, 108–116
43. Barlow, P.W. (1995) in Plant Molecular Biology NATO-ASI Symposium (Coruzzi, G. and Puigdomeaech, P., eds.), pp. 17–30, Springer Verlag, Berlin
44. Barlow, P.W. (1996) in Stem Cells (Potten, C.S., ed.), pp. 29–57, Academic Press, London
45. Baluska, F. and Kubica, S. (1992) J. Exp. Bot. **43**, 991–996
46. Steeves, T.L. and Sussex, I.M. (1972) in Patterns in Plant Development, 1st edn., Cambridge University Press, Cambridge
47. Luxová, M. and Murin, A. (1973) Biol. Plant. **15**, 37–43
48. Barlow, P.W. (1976) J. Theor. Biol. **57**, 433–451
49. Jensen, W.A. and Kalvajian, L.G. (1958) Am. J. Bot. **45**, 365–372
50. Tucker, W.Q.J., Warren Wilson, J. and Gresshof, P.F. (1986) Ann. Bot. **57**, 675–679
51. Jacobs, W.P. (1952) Am. J. Bot. **39**, 301–309
52. Jacobs, W.P. and Morrow, I.B. (1957) Am. J. Bot. **44**, 823–842
53. Davidson, D. (1991) in Plant Physiology, vol. X, Growth and Development (Steward, F.C., eds.), pp. 341–436, Academic Press, New York
54. Esau, K. (1940) Hilgardia **13**, 175–226
55. Bell, J.K. and McCully, M. (1970) Protoplasma **70**, 179–205
56. MacLeod, R.D. (1976) Am. Bot. **40**, 551–562
57. Popham, R.A. (1955) Am. J. Bot. **42**, 529–540
58. John, P.C.L., Zhang, K. and Dong, C. (1993) in Molecular and Cell Biology of the Plant Cell Cycle (Ormrod, J.C. and Francis, D., eds.), pp. 9–34, Kluwer Academic Publishers, Dordrecht
59. John, P.C.L. (1987) in Algal Development (Molecular and Cellular Aspects) (Wiessner, W., Robinson, D.G. and Starr, R.C., eds.), pp. 9–16, Springer-Verlag, Berlin
60. Spudich, J.L. and Sagar R. (1980) J. Cell Biol. **85** 136–145
61. McAteer, M., Donnan, L. and John, P.C.L. (1985) New Phytol. **99**, 41–56
62. Scherr, C. (1993) Cell **73**, 1059–1065
63. Alberts, B., Bray, D., Lewis, J., Raff, M., Roberts, K. and Watson, J.D. (1994) Molecular Biology of the Cell, 3rd edn, Garland Publishers Inc., New York, London
64. Nurse, P. (1975) Nature (London) **256**, 547–551
65. Fantes, P.A. and Nurse, P. (1981) in The Cell Cycle (John, P.C.L., ed.), pp. 11–33, Cambridge University Press, Cambridge
66. Kumagai, A. and Dunphy, W.G. (1992) Cell **70**, 139–151

67. Hayles, J. and Nurse, P. (1993) in Molecular and Cell Biology of the Plant Cell Cycle (Ormrod, J.C. and Francis, D., eds.), pp. 1–8, Kluwer Academic Publishers, Dordrecht
68. Krek, W. and Nigg, E.A. (1991) EMBO J. **10**, 305–316
69. Norbury, C., Blow, J. and Nurse, P. (1991) EMBO J. **10**, 3321–3329
70. Russell, P. and Nurse, P. (1986) Cell **45**, 145–153
71. Russell, P. and Nurse, P. (1987) Cell **49**, 559–567
72. Russell, P. and Nurse, P. (1987) Cell **49**, 569–576
73. Booher, R. and Beach, D. (1988) EMBO J. **8**, 2321–2327
74. Wasserman, W. and Masui, Y. (1975) Exp. Cell Res. **91**, 381–388
75. Labbe, J.-C., Capony, J.-P., Caput, D., Caradore, J.-C., Derancourt, J., Kaghad, M., Lelias, J.-M., Picard, A. and Doree, M. (1989) EMBO J. **8**, 3053–3058
76. Smythe, C. and Newport, J.M. (1992) Cell **68**, 787–798
77. Ohkurea, H., Adachi, Y., Kinoshita, N., Niwa, O., Toda, T. and Tanagida, M. (1988) EMBO J. **7**, 1465–1473
78. Bell, M.H., Halford, N.G., Ormrod, J.C. and Francis, D. (1993) Plant Mol. Biol. **23**, 445–451
79. Gatz, C., Kaiser, A. and Wendenburg, R. (1991) Mol. Gen. Genet. **227**, 229–237
80. Edgar, B.A., Sprenger, F., Duronio, R.J., Leopold, P. and O'Farrel, P.H. (1994) Genes Dev. **8**, 440–452
81. Edgar, B.A., Lehman, D.A. and O'Farrell, P.H. (1994) Development **120**, 3131–3143
82. Reed, B.H. (1995) BioEssays **17**, 553–556
83. Jacobs, T. (1995) Annu. Rev. Plant Mol. Biol. Plant Physiol. **46**, 317–339
84. Nougarède, A. and Rembur, J. (1977) Z. Pfl. Physiol. **85**, 283–295
85. Saint-Côme, R. (1969) C.R. Seances Acad. Sci. (Ser. D) **268**, 508–511
86. Corson, G.E. (1969) Am. J. Bot. **56**, 1127–1134
87. Jacqmard, A. (1967) C.R. Seances Acad. Sci. (Ser. D) **264**, 1282–1285
88. Bodson, M. (1975) Am. Bot. **39**, 547–554
89. Miller, M.B. and Lyndon, R.F. (1975) Planta **126**, 37–43
90. Miller, M.B. and Lyndon, R.F. (1976) J. Exp. Bot. **27**, 1142–1153
91. Denne, M.P. (1966) N.Z. J. Bot. **4**, 300–314

# 10

# Cell division during floral morphogenesis in *Antirrhinum majus*

John H. Doonan

John Innes Centre, Colney Lane, Norwich NR4 7UH, U.K.

## Introduction

During the development of multicellular organisms, cell proliferation is restricted both spatially and temporally. In plants, most cell proliferation occurs in specialized regions called meristems, but the controls which modulate proliferation are not well understood. The purpose of this chapter is to review cell proliferation in the context of apical meristem development of the snapdragon, *Antirrhinum majus*. This plant has proven to be a useful model system in which to study factors that are important for plant development. Floral development in *Antirrhinum*, as in several other species, has been genetically dissected, providing a large collection of mutants which affect meristem behaviour [1]. Several of these genes have now been characterized at the molecular and cellular levels, providing insights into the regulatory mechanisms underlying meristem growth. The floral spike, or inflorescence, of *Antirrhinum* contains many floral meristems at different stages of development in a well-ordered spiral allowing the observations to be made at many stages of development. The availability of morphologically abnormal mutants, combined with the defined and characterized development of a convenient-sized inflorescence, make *Antirrhinum* an attractive system in which to study cell cycle control within a developmentally relevant context. The floral meristem provides a useful system for elucidating some of the constraints on cell proliferation, because a number of mutants partly or completely restore indeterminacy on what is a determinate meristem. Understanding how these genes act on cell proliferation may provide some insight into how cell division is regulated *in planta*.

## Apical meristems: an overview

Plant development depends on the behaviour of meristems. Factors that control the division, growth and shape of meristems determine the form of plant structures.

Plants contain a number of meristematic regions, but the behaviour of two terminally positioned meristems, the root and shoot (or apical) meristems, are critically important for plant development. The developmental potential and cellular organization of the root and apical meristems are quite different. The root meristem has a very ordered cellular structure, with a very defined developmental programme [2]. Although the meristem does not produce lateral organs, it generates distinct files of cells which run the length of the root. In contrast, shoot apical meristem organization is somewhat less ordered. The apical meristem produces on its periphery lateral organs whose identities and relative positions change during development (Fig. 1). This indicates that the structure and function of the apical meristem also changes as development progresses, leading to different patterns of cell proliferation. However, some general features of the apical meristem organi-

**Fig. I**     **Apical meristem development in *Antirrhinum* as shown by tracings of medial longitudinal sections through different meristems**

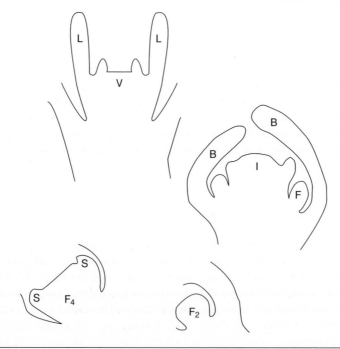

*The vegetative meristem (V) produces opposing leaves (L); the inflorescence meristem (I) produces bracts (B) in a spiral fashion, while floral meristems (F) are produced in the axils of the bracts. The shape of the floral meristem undergoes considerable alteration as it matures: stage 2 ($F_2$) are loaf shaped, while at stage 4 ($F_4$), the meristem has become broader and sepal initials (S) can be observed on the flanks.*

zation remain constant during development. All apical meristems are organized into three distinct cell layers: L1, L2 and L3. The outermost layer, L1, is composed of a single layer of cells which divide anticlinally and give rise to the epidermis of the plant. L2 also comprises a single layer of cells and gives rise to sub-epidermal mesophyll. The L3 layer is comprised of several layers of cells which divide in all planes and give rise to most of the internal and vascular tissues. The apical meristem can also be subdivided into three main regions [3]: the central zone, the peripheral zone and the midrib. In this model, the central zone is formed by initial cells that divide and are involved in the maintenance of the meristem. The central zone has been called the *'meristeme d'attente'* or quiescent centre because of apparently slow division rates. The peripheral zone, located to the side and beneath the central zone, is the region where organ initiation takes place. Finally, the rib zone, the boundary between the meristem and the rest of the plant, gives rise to the interior tissues of the stem (see D. Francis, Chapter 9).

Other features of the apical meristem do change during development, sometimes radically. First, as a meristem changes from vegetative to an inflorescence and finally to a floral state, or direct to a terminal flower the potential for growth decreases. The vegetative meristem is indeterminate, whereas the inflorescence or floral meristem is usually somewhat more limited. The floral meristem has a very defined developmental programme producing a genetically determined number of organs before growth ceases. Secondly, the identity of organs produced by the three different meristems is distinctive for each meristem. Leaves are produced by the vegetative meristem, bracts by the inflorescence meristem and various floral organs by the floral meristem. Both phyllotaxy and internode length, which determine the spacing of organs, also differ in the three meristems. Since many of these structural differences between the three types of meristems reflect changes in cell proliferation patterns, understanding how cell division is co-ordinated with development may provide some insight into the cellular basis of plant development.

## Floral development in *Antirrhinum*

After a period of vegetative growth, or upon switching plants from short days to long days, the *Antirrhinum* shoot apex converts to making an inflorescence. The flowers are borne in a spiral up the stem in a racemose inflorescence. Whereas the vegetative meristem produces pairs of leaves at each node in decussate phyllotaxy, the inflorescence meristem produces small leaf-like structures (bracts) in a spiral arrangement. The first visible bract primordium is referred to as being at node 0, and the position of successively older bracts are numbered sequentially. Floral meristems develop in the axils of bracts and produce zygomorphic flowers consisting of four concentric whorls of organs [4]. The first and outermost whorl, the calyx, consists of five sepals. The corolla (whorl 2) has five petals, fused for part of their length to form a tube that has five lobes. Whorl 3 consists of four mature stamens and one stamen initial which fails to complete development. This aborted stamen is positioned nearest the main stem of the inflorescence, opposite the bract. Two fused carpels occupy whorl 4. Estimates of growth rate at 25°C suggest that one node is generated on average every 10 h [5].

Floral meristem development has been divided into six stages based on the physical appearance of the meristem. Under standard conditions these stages occupy characteristic nodal positions on the floral spike [6]. Stage 0 is defined as small mounds of cells which eventually form the bract primordia. These are observed on the periphery of the inflorescence meristem and are the sole visible features. Stage 1 floral meristems comprise an eye-shaped group of cells, formed between the bract primordia and the main stem at nodes 4–7. By stage 2 (nodes 7–10), the floral meristem becomes raised up like a loaf of bread, and, by stage 3 (nodes 10–12), the pentagonal symmetry of the flower is apparent. At stage 4 (nodes 12–14), also known as the floritypic stage, five sepal primordia (whorl 1) emerge on the flanks of the meristem. The central region of the meristem, now relatively flat, will give rise to the petals, stamens and carpels (whorls 2, 3, and 4, respectively). Petal and stamen primordia are visible as small mounds by stage 5 (nodes 14–18). Several floral meristems are seen in each longitudinal section, with the youngest stages of development near the top of the inflorescence, and progressively older stages below this. Because of the spiral arrangement of flowers and their subtending bracts, not all stages of development are seen in each section. Consequently, the node numbers are determined by serially reconstructing apices as described in Bradley et al. [7]. There are typically 45–60 nodes between node 0 and the first fully opened flower.

The switch from a vegetative state to flowering is under both environmental and genetic control. *Antirrhinum* requires long days in order to flower, so growth under short days results in a persistent vegetative state. This is useful experimentally since it permits synchronization of development. Many genes affect floral development. Several homoeotic mutants that alter the course of flower development have been described [1] and their corresponding genes cloned. The *floricaula* (*flo*) gene controls the switch from an indeterminate inflorescence meristem to a floral meristem [8]. Four other genes, *deficiens* (*def*), *globosa* (*glo*), *plena* (*ple*) and *sepaloidea* (*sep*) control the identity of floral organs within specific whorls: mutations in these genes change both the time and location at which particular organs develop. For example, mutations in *ple* result in the development of petals instead of stamens in whorl 3 and variable petaloid/carpeloid/sepaloid tissues in whorl 4. Whorls 1 and 2 are not affected. An interesting side-effect of several of the homoeotic mutants is that determinacy of the meristem is also changed: mutations in *ple*, *flo* and *sep* all induce the appearance of extra organs. *flo* mutations lead to leafy shoots instead of flowers, while *ple* and *sep* lead to the proliferation of petaloid and/or sepaloid whorls. Perhaps all three of these genes are required either to limit the otherwise continuous production of organ primordia on the meristem, or to limit the continued growth of the meristem itself. Ultimately, these genes appear to act to restrict the continued growth of the meristem and are therefore directly, or more likely indirectly, inhibitors of cell proliferation.

## Estimating rates of cell division during floral morphogenesis

Cell division rates in the L1 layer have been estimated by measuring the width of pigmented sectors formed in flowers when the transposon, *tam3*, excises from the *pal-2* gene. By exploiting a naturally temperature-sensitive transposon in plants, where floral development was synchronized by exposure to long days, Vincent et al. [5] have produced sectors at different, but defined, stages of development. By comparing the final size of sectors formed at two different times, estimates of the number of cell divisions occurring between those two times could be made. Sectors induced when a meristem was at stage 0 were eight times as wide as those induced at stage 5, suggesting that about three generations of cell division had occurred in this dimension. Since the L1 layer divides in two dimensions, with periclinal divisions (in the third dimension) very rare, Vincent et al. concluded that six generations had occurred during this interval and cell number had increased by a factor of $2^6$ ($= 64$) over about 18 nodes. Given the rate of new node formation, this suggests that the average rate of cell division in the L1 layer is one division every 30 h. However, cell division rates are probably not constant during floral development; while analysing the rate of sector formation, Vincent et al. noted significant differences in the numbers of sectors formed at different stages of floral development. Since excision of *Tam3* is thought to occur during DNA replication, rapidly dividing cells should be in this phase more often, providing a greater opportunity for excision, and it was argued that the rate of sector formation should therefore reflect the rate of cell division in the L1 layer. Based on this assumption, the rate of sector formation was calculated, for different stages of flower development, and relative rates of cell division deduced. Before stage 0, sector formation was below average and increased to average during stage 0, at about the time when transcription of *flo*, a meristem identity gene, first could be detected. This suggests that cell divisions contributing specifically to the meristem first occur at about the same time as this meristem is initiated and acquires its identity. Rates of division then fell to a very low level during stage 1 and scanning electron microscope studies confirmed that most of the epidermal growth at this stage was due to cell enlargement rather than cell division [6]. Rates then increased during stage 2 to a maximum, in stage 3, of an estimated one division every 15 h at the time when sepal primordia are being initiated. At this time, organ identity genes are switched on and cell identities become fixed. Comparison between wild-type and *flo* mutant meristems show that this rapid burst of division does not occur in the mutant meristems [6], suggesting that cell division rates are regulated, directly or indirectly, by developmental control genes. Cell division rates decline to a low level by stage 5 and then return to the average level. Other approaches, used to measure cell division rates in species such as *Silene* and *Sinapis* [9,10] also suggest that cell division rates vary during floral development. Work on *Antirrhinum* confirms these observations and, furthermore, provides evidence that these variations are due to the activity of the meristem identity gene, *flo*.

## Localization of cell division during development

Genes that control meristem structure may also influence where cells divide. The cell cycle presents one possible target of the developmental genes, but no clear molecular mechanisms have been proposed by which floral homoeotic genes might influence genes controlling the cell cycle. One attractive idea, for which there is not yet any evidence, is that cell cycle genes might be transcriptionally regulated by homoeotic genes. Several of the genes involved in floral development are predicted to encode proteins possessing characteristics typical of transcription factors [4,11]. The products of at least two, *def* and *glo*, form a heterodimer capable of binding specific DNA sequences *in vitro* [12]. RNA *in situ* hybridization experiments indicate that these genes are expressed in young meristems in regions where the organs they affect will eventually arise. Homoeotic genes may, therefore, act alone or in combination in different regions of the meristem to modulate the expression of target genes and hence the developmental fate of that region. Many target genes may be expressed generally within the meristem, and even outside it, but modulated to different degrees in the different regions. Candidate genes include those controlling cell shape, division and elongation. Cell division cycle genes are particularly interesting candidates, since these genes, in many species including plants, are subject to regulation at the transcriptional level. Such genes may also provide useful tools with which to characterize cell cycle activity within meristems.

　　While there are many genes involved in progression through the cell cycle, perhaps the key regulators of cell cycle events are the cyclin-dependent protein kinases (CDK). These kinases are involved in critical stages of cell progression, such as the decision to enter the division cycle and the commitment to start mitosis (see L. De Veylder et al., chapter 1). The first of this family to be characterized was the p34$^{cdc2}$ from fission yeast [13]. Activity of p34$^{cdc2}$ depends on it being associated with a cyclin (reviewed in [14,15]). This complex is subject to multiple forms of regulation: reversible phosphorylation carried out by several kinases and phosphatases, and a number of accessory proteins whose number and type may vary. Additional complexity is added by the existence of multiple kinase and cyclin proteins in most organisms, not all of which function in cell cycle regulation.

　　Although activity of the p34$^{cdc2}$ kinase ultimately depends on post-translational modifications co-ordinated via cell cycle checkpoints (reviewed in [16, 17]), both *cdc2*-like and cyclin genes have been shown to be transcriptionally regulated during the cell cycle [18–23]. This suggests that there is the potential to regulate the cell cycle at the transcriptional level, as has been suggested for control of the $G_1/S$ transition in yeasts [24–26] and control of $G_2/M$ by the *nimA* protein kinase in *Aspergillus nidulans* [27]. As a first step towards understanding how cell cycle progression is regulated, a number of CDK and cyclin genes have been isolated and their patterns of expression characterized.

## Phase-specific gene expression

Transcripts of at least five cell cycle related genes, including two CDK and two cyclin genes, accumulated at distinct and sometimes different phases of the cell cycle in *Antirrhinum* meristems [28], inferring that transcriptional regulation of cell cycle genes may play an important role in plant cell cycle progression (Fig. 2). Transcripts of a further cyclin-A-like gene were expressed throughout the cycle, but accumulated to somewhat higher levels at certain stages, probably S phase (V. Gaudin, unpublished work). These probes provide a useful insight into the regulation of the cell cycle and the spatial distribution of dividing cells within apical meristems. RNA *in situ* hybridization experiments indicated that these genes were expressed in a strikingly different pattern from floral homoeotic genes which labelled most, if not all, cells within particular sub-regions of floral meristems [4,7]. Instead, isolated cells within the meristem and other proliferating regions expressed these cell cycle related genes, producing a spotty pattern. The simplest explanation for the spotty hybridization pattern is that the transcripts were abundant for a limited period of the cell cycle. Isolated cells were labelled because cell divisions are poorly synchronized in meristems and hence neighbouring cells are unlikely to be in the same phase of the cycle. That these genes were transcribed in specific phases of the cell cycle was supported by the correlation between expression of three genes, cyclin B1, cyclin B2 and *cdc2d*, and the occurrence of mitosis. As judged by DAPI staining, less than 2% of meristematic cells were in mitosis, yet these three genes were expressed in over 90% of mitotic cells and only 5–10% interphase cells. The cyclin B transcripts decreased as the cells entered anaphase and were almost completely absent from telophase cells. This mirrors the dynamics of cyclin protein accumulation and destruction reported in *Drosophila* [29,30]. The time of expression of the cyclin B genes correlates well with the established timing of cyclin function [14] and suggests that these genes may have an homologous function in *Antirrhinum*. This seems likely since cyclin B is known to be rate limiting or at least required for entry in M phase in many species.

    The histone H4 gene also produced a similar pattern where the transcript was confined to isolated cells or clusters of cells. In this case, however, almost no mitotic cells were labelled compared with over 30% of interphase cells. Studies in a number of organisms have indicated that the histone H4 gene is expressed primarily in S phase [31,32]. Double labelling of the same tissue with a histone probe and a

**Fig. 2**          **Expression of phase-specific genes within inflorescences of**
                        ***Antirrhinum***

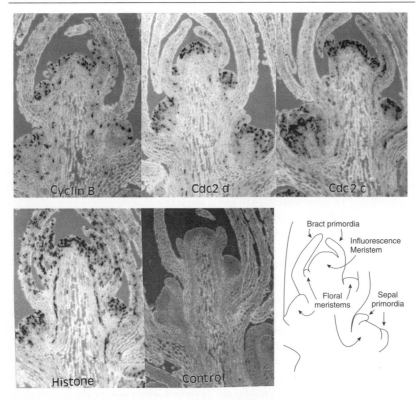

*Longitudinal sections of the inflorescence were probed with antisense cyclin B, cdc2c, cdc2d and histone, or sense histone (control) dig-labelled probes. The spotty pattern of transcription indicates that these transcripts are present for a limited period during the cell cycle.*

cyclin probe demonstrated that these genes label different populations of cells, supporting the idea that they are expressed at discrete stages of the cell cycle.

The known *cdc2* gene family from *Antirrhinum* currently comprises four members; *cdc2a* and *b* encode proteins which contain well-conserved PSTAIRE motifs, while those of *cdc2c* and *d* genes have poorly conserved motifs. The expression patterns of the two groups are dramatically different: *cdc2a* and *cdc2b* were expressed at a low level throughout the cell cycle, and at even lower levels in non-cycling cells, while *cdc2c* and *d* were expressed during a limited phase or phases. Both *cdc2c* and *cdc2d* were expressed at the $G_2/M$ transition as judged by DAPI staining and decreased during anaphase very much like cyclin B. However, *cdc2c* is expressed additionally during S phase: using *cdc2c* and histone probes on the same material showed that expression of the two genes overlapped extensively, but not completely. All cells expressing *cdc2d* also expressed *cdc2c*, confirming that *cdc2c* was also expressed at the $G_2/M$ boundary. Either *cdc2c* is expressed from S phase until mitosis, or *cdc2c* is expressed in early S phase and again at the $G_2/M$ transition. The *cdc2* gene of *Schizosaccharomyces pombe* encodes the p34[cdc2] protein kinase required twice in the cell cycle: first for progression into S phase and secondly for entry into mitosis [33]. In animal cells, p34[cdc2] appears to be required solely for entry into mitosis [15,17] and another kinase, Cdk2, acts at $G_1/S$ [34]. The functions of the plant *cdc2* genes are unknown, but when expressed in fission yeast all have the capacity to interfere with size control. Those with a conserved PSTAIR are more toxic than *cdc2c* and *cdc2d* [35].

Genes whose transcripts accumulate at specific phases of the cycle provide useful markers *in situ* for the identification of cells at various phases of the cell cycle in plant meristems. For example, cyclin B transcripts accumulate during $G_2/M$ and so provide a useful cytological marker for this phase of the cycle. Histone genes are useful to mark S-phase cells because they have been shown to be preferentially expressed during S phase in many systems [31,32]. The availability of probes for specific phases of the cell cycle provides a useful tool to study the spatial distribution of dividing cells. There is a clear advantage over other techniques, such as DAPI staining, in that a higher proportion of cells are labelled and one is no longer restricted to detecting cells in a single, and perhaps short, phase of the cell cycle. Preferential labelling of the same regions by all the cell cycle probes we have used indicates that they provide a convenient and reproducible method to identify regions of high cell cycle activity. All the *Antirrhinum* cell cycle genes, which are expressed in a phase-specific manner, examined to date seem to be expressed in all tissue types analysed (root, vegetative, inflorescence and floral meristems), so it is likely that they provide reliable indicators of cell cycle events throughout development.

These probes have provided useful general information of cell cycle activity during floral development. First we confirmed the observations that the highest rates of division in plants occur in meristems and young primordia [35a,36], since these regions contain the highest proportion of labelled cells (see D. Francis, Chapter 9). The lack of synchrony between neighbouring cells was also striking. To examine this in more detail, only the two outer layers, L1 and L2, were considered. These two layers are thought to behave as distinct cell lineages, because most cell divisions are anticlinal (new cell walls are at right angles to the surface; [37]). The probability that two neighbours are at the same stage of the cell cycle can be calculated assuming either no synchrony or full synchrony and compared with actual results. This indicated that there is a significant, but rather low, level of synchrony in these regions and that, following mitosis, daughter cells become rapidly unsynchronized. In other regions, such as microsporocytes and tapetal cells of developing anthers, there were broad domains of expression of cell cycle related genes, indicating a high level of local synchrony (Fig. 3; [38,39]).

**Fig. 3          Expression of phase-specific genes in anthers of *Antirrhinum***

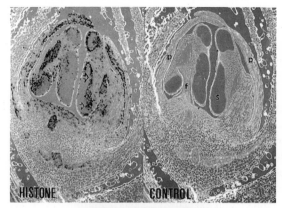

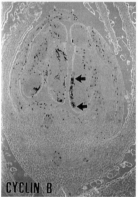

*Sections through anthers, probed with dig-labelled antisense cyclin B, and histone or sense histone (control) probes, show that some regions have a spotty pattern of transcription, while others display a more uniform expression (see region between the closed arrows for cyclin and between the open arrows for histone), indicating that cells at the edge of the anther sac(s) are locally synchronized, while cells in other regions, such as the filament (f) and petal (p), are asynchronous.*

Phase-specific probes also provide detailed spatial information of cell division patterns in meristems at different stages of development. A high rate of division, as inferred from the presence of cyclin B transcript, was observed at the apex of the inflorescence within the top 6–7 layers of small, isodiametric cells; and in the larger cells, 1–2 layers immediately beneath (see D. Francis, Chapter 9). The large cells in the remainder of the main stem axis divided much less frequently. During floral development the spatial distribution of cell divisions changes as the meristem changes shape. In the young meristem, stages 1 and 2, cells expressing cyclin transcripts are found throughout the depth of the structure, presumably allowing the loaf-shaped meristem to swell. As meristems enter stage 3, cells expressing cyclin tended to be found closer to the surface and the periphery of the meristem. This coincided with the development of the flat dish of the floritypic stage. Thereafter, transcripts were found mainly in the organ primordia, the centre of the meristem containing fewer dividing cells.

## Genes expressed in proliferating cells

Several other cell cycle related genes were expressed in a phase-independent manner. These included two *cdc2*-like genes, *cdc2a* and *cdc2b*, which contain a well-conserved PSTAIR motif (P. Fobert et al. unpublished work), a *nimA* protein kinase-like transcript [40], a protein phosphatase in *Arabidopsis* [41] and two D-type cyclin genes (V. Gaudin et al. unpublished work). The *cdc2a* and *cdc2b* genes were weakly expressed in all cells in the meristem, but were also expressed outside the meristem and were present in all cells at a very low level. The expression levels of a *cdc2* homologue in *Arabidopsis* have been linked to the competence of cells to divide [42] and a pattern of expression reciprocal to that of the homoeotic genes *agamous* and *apetala3* has been reported [43]. Apparent differences between the species may reflect the possibility that the *cdc2* gene family is rather more complex and orthologues have not been compared. The *nimA*-like gene and the protein phosphatase were also expressed in a similar manner to the *cdc2a/b* genes, but at a higher level in dividing cells. From their expression patterns, all these genes seem to be expressed in a proliferation-dependent manner, but the increased expression in dividing cells may simply reflect the increased metabolic activity associated with cell proliferation. It will be difficult to discriminate between these two possibilities until it is possible to isolate mutations in such genes.

A new class of plant cyclins, originally termed delta because they were closely related to the mammalian D cyclins, have recently been isolated from *Arabidopsis* by their ability to rescue yeast lacking cyclin(s) of budding yeast (CLN) cyclin function [44]. These cyclins have been renamed D1, D2 and D3 (J.P. Renaudin et al., Chapter 4) to make the nomenclature consistent with that of animal cyclins. Two structurally similar genes, cyclin D3a and cyclin D3b, have been isolated from *Antirrhinum* on the basis of homology to the *Arabidopsis* genes (V. Gaudin et al., Chapter 4). The proteins encoded by these genes contain a cyclin box, but no mitotic destruction motif. Instead they contain extensive PEST sequences, indicating that they may be subject to rapid turnover. Additionally, they contain a retinoblastoma-binding motif near the N-terminus, suggesting that they may be involved in re-entry into the cell cycle. Expression of both D cyclins was restricted to dividing cells, as judged by staining for histone H4 transcript (Fig. 4), but was not phase specific. Despite being closely related at the structural level, the two genes were expressed in markedly different patterns. The cyclin D3b gene seems to be

**Fig. 4**         Expression of D-cyclin genes, compared with expression of histone H4, in consecutive longitudinal sections through the inflorescence of *Antirrhinum*

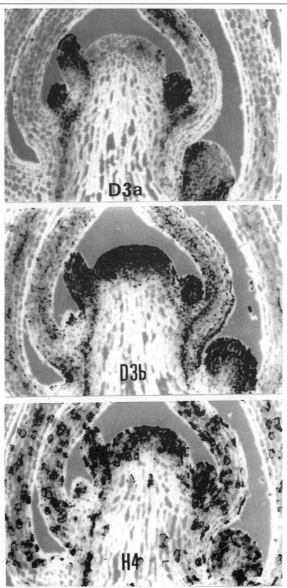

*Cyclin D3a (D3a) is expressed only in peripheral regions of the meristem, whereas cyclin D3b (D3b) is expressed in all proliferating regions as shown by histone probes (H4).*

expressed in all proliferating cells in the shoot. Expression was highest in the meristem, and young primordia, where division rates are highest. The other cyclin gene, D3a, was only expressed in a subset of proliferating cells, usually those in the periphery of meristems, in young primordia and primordial meristems and in vascular tissues. This gene represents perhaps the first molecular marker for those cells which have been partitioned out of the central region of the meristem into lateral organs and lends some support to the notion, proposed for the zonal model of the meristem, that cells in different zones are indeed physiologically distinct (see D. Francis, Chapter 9). The other intriguing observation during floral meristem development was that expression of cyclin D3a is remarkably similar to that of the floral identity gene, *flo*. Both *flo* and cyclin D3a were expressed throughout the young floral meristem (stage 1), but, during stage 2, transcript levels in the central region of the meristem diminished leaving a ring around the periphery [45]. The similarity between these two expression patterns suggests that, during this stage of development, these two genes could be under similar regulation. It is also possible that *flo* may act, directly or indirectly, to modulate cyclin D3a expression.

The spatial and temporal association between cyclin D3b and cell proliferation continues throughout development and provides a useful marker for studying how cell proliferation changes during morphogenesis. For instance, during late petal formation, regions of the petal, which bend to produce the complex face of the *Antirrhinum* flower, had increased levels of cell proliferation and increased levels of cyclin D3b transcript. Conversely, the aborted stamen, which ceases development before the other four stamens, showed a reduced level of cell proliferation and reduced cyclin D3b transcript. That this was not owing to general transcript destruction is shown by the expression of floral homoeotic genes, such as *def*, in all five stamens. Levels of expression did not vary between neighbouring cells within a particular region, suggesting that the level of expression is constant throughout the cell cycle. However, expression levels did vary from region to region, sharply declining in regions between organs. At about the stage when sepal primordia first form (stage 4), and lineage restriction boundaries are being set up between the sepals and internal organs (4), cyclin D3b expression decreases in a band of cells (about 3–4 cells wide) lying internal to the whorl 1. A similar reduction in expression levels was observed as each whorl of organs is formed. This gene therefore represents a useful indicator of cell proliferation within regions, even small ones such as boundaries between organs. Whether cyclin D3b has a direct role in cell proliferation, and what that role is, remains to be determined. The related *Arabidopsis* cyclin D3 is expressed in proliferating cells and in non-proliferating ones which have been exposed to cytokinin [44]. Another related gene, cyclin D2, is expressed in a similar manner, but in response to sucrose. These data suggest that different cyclin D genes may be expressed in response to different extracellular signals associated with re-entry into the cell cycle. (see J.A.H. Murray et al., Chapter 5)

## Conclusions

During *Antirrhinum* development, both the rate and position of cell proliferation within the apical meristem varies. These variations may be regulated by the action of developmental controls on the cell cycle machinery. Expression patterns for several putative cell cycle genes suggest that there are at least three different types of transcriptional regulation (Fig. 5). First there are those genes whose transcription is restricted to a particular phase of the cell cycle. These include histone, an important component of chromatin, expressed in S phase; B-type cyclins expressed in G2/M, and a vital regulatory component of the G2/M kinase, and two cyclin-dependent protein-kinase-like genes, *cdc2c* (S phase and $G_2$/M) and *cdc2d* ($G_2$/M). Secondly, there are those transcripts, while present in most cells, which appear to be more highly expressed in meristematic zones. These include two *cdc2*-like genes, *cdc2a* and *cdc2b*, and a NimA-like protein kinase. Transcriptional regulation may not play an important role in regulating the function of these gene products. The idea that *cdc2a* and *cdc2b* are not rate limiting for the functions they are involved in remains to be demonstrated, but seems a likely possibility, since the abundance of several cyclin transcripts is stringently regulated at the transcript level. Such cyclins may be rate limiting and therefore are good candidates for modulation by developmental control genes. A final class, comprising the D cyclins, present only in proliferating cells, but not in a phase-dependent manner, may be regulated in a tissue-dependent manner. The expression patterns, and known functions of related proteins in other systems, suggest that D-cyclins may allow cross-talk between development and the cell cycle. These studies only implicate, by association, several of these genes as being involved in cell proliferation; the next challenge is to use reverse genetics to examine the function of these genes during plant development and confirm their proposed functions.

**Fig. 5**     **Timing of the expression of cell cycle related genes during the cell cycle in *Antirrhinum* apical meristems**

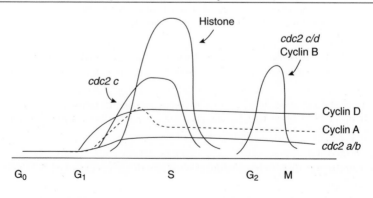

# References

1.  Carpenter, R. and Coen, E.S. (1993) Plant Cell **5**, 1175–1181
2.  Dolan, L., Janmaat, K., Willemsen, V., Linstead, P., Poethig, S. and Roberts, K. (1993) Development (Cambridge, U.K.) **119**, 71–84
3.  Esau, K. (ed) (1960) Anatomy of seed plants, John Wiley & Sons, Chichester
4.  Coen, E.S., Romero, J.M., Doyle, S., Elliott, R., Murphy, G. and Carpenter, R. (1990) Cell **63**, 1311–1322
5.  Vincent, C.A., Carpenter, R. and Coen, E.S. (1995) Curr. Biol. **5**, 1449–1458
6.  Carpenter, R., Copsey, L., Vincent, C.A., Doyle, S., Magrath, R. and Coen, E.S. (1995) Plant Cell **7**, 2001–2011
7.  Bradley, D., Carpenter, R., Sommer, H., Hartley, N. and Coen, E. (1993) Cell **72**, 85–95
8.  Coen, E.S., Romero, J.M., Doyle, S., Elliott, R., Murphy, G. and Carpenter, R. (1990) Cell **63**, 1311–1322
9.  Gonthier, R., Jacqmard, A. and Benier, G. (1987) Planta **170**, 55–59
10. Miller, M.B. and Lyndon, R.F. (1975) Planta **126**, 37–43
11. Coen, E.S. and Meyerowitz, E.M. (1991) Nature (London) **353**, 31–37
12. Schwarz-Sommer, Z., Hue, I., Huijser, P., Flor, P.J., Hansen, R., Tetens, F., Lönning, W.-E., Saedler, H. and Sommer, H. (1992) EMBO J. **11**, 251–263
13. Simanis, V. and Nurse, P. (1986) Cell **45**, 261–268
14. Hunt, T. (1991) Semin. Cell Biol. **2**, 213–222
15. Sherr, C.J. (1993) Cell **73**, 1059–1065
16. Feilotter, H., Lingner, C., Rowley, R. and Young, P.G. (1992) Biochem. Cell Biol. **70**, 954–971
17. Norbury, C. and Nurse, P. (1992) Annu. Rev. Biochem. **61**, 440–470
18. Dalton, S. (1992) EMBO J. **11**, 1797–1804
19. Matsushime, H., Ewen, M.E., Strom, D.K., Kato, J.-Y., Hanks, S.K., Roussel, M.F. and Sherr, C.J. (1992) Cell **71**, 323–334
20. Pines, J. and Hunter, T. (1989) Cell **58**, 833-846
21. Schwob, E. and Nasmyth, K. (1993) Genes Dev. **7**, 1160–1175
22. Welsh, P.J. and Wang, J.Y.L. (1992) Proc. Natl. Acad. Sci. U.S.A. **89**, 3093–3097
23. McKinney, J.D. and Heintz, N. (1991) Trends Biochem. Sci. **16**, 430–435
24. Reed, S. I. (1992) Annu. Rev. Cell Biol. **8**, 529–561
25. Reference deleted.
26. Tyers, M., Tokiwa, G. and Fulcher, B. (1993) EMBO J. **12**, 1955–1968
27. Osmani, S.A., May, G.S and Morris, N.R. (1987) J. Cell Biol. **104**, 1495–1504
28. Fobert, P., Coen, E., Murphy G. and Doonan J.H. (1994) EMBO J. **13**, 616–624
29. Lehner, C.F. and O'Farrell, P.H. (1990) Cell **61**, 535–547
30. Whitfield, W.G.F., Gonzalez, C., Maldonada-Codina, G. and Glover, D.M. (1990) EMBO J. **9**, 2563–2572
31. Marzluff, W.F. and Pandey, N.B. (1988) Trends Biochem. Sci. **13**, 49–52
32. Nakayama, T. and Iwabuchi, M. (1993) Crit. Rev. Plant Sci. **12**, 97–110
33. Nurse, P. and Bisset, (1981) Nature (London) **292**, 558–560
34. Fang, F. and Newport, J.W. (1991) Cell **66**, 781–742
35. Fobert, P., Gaudin, V., Lunness, P., Coen, E. and Doonan, J.H. (1996) Plant Cell **8**, 1465–1476
35a. Lyndon, R.F. (1990) Plant Development: The Cellular basis. Unwin Hyman, London
36. Steeves, T.A. and Sussex, I.M. (1989) Patterns in Plant Development, 2nd edn, Cambridge University Press, Cambridge
37. Tilney-Bassett, R.A.E. (1986) Plant Chimeras. Edward Arnold, London
38. Berger, C.A., Witkus, E.R. and Joseph, T.C. (1951) Caryologia **4**, 110–114
39. Harte, C. (1974) Handbook of Genetics (King, R.C., ed.), pp. 315–331, Plenum Press, London
40. Zhang, H., Scofield, G., Fobert, P. and Doonan, J.H. (1996) J. Microscopy **181**, 186–194
41. Arundhati, A., Fielder, H., Traas, J.A., Zhang, H., Lunness, P.A. and Doonan, J.H. (1995) Plant J. **7**, 823–834
42. Hemerly, A., Bergounioux, C., Van Montagu, M., Inzé, D. and Ferreira, P. (1992) Proc. Natl. Acad. Sci. U.S.A. **89**, 3295–3299
43. Martinez, M.C., Jørgensen, J.-E., Lawton, M.A., Lamb, C.J. and Doerner, P.W. (1992) Proc. Natl. Acad. Sci. U.S.A. **89**, 7360–7364

44.   Soni, R., Carmichael, J.P., Shah, Z.H. and Murray, J.A.H. (1995) Plant Cell 7, 85–103
45.   Simon, R., Carpenter, R., Doyle, S. and Coen, E. (1994) Cell 78, 99–107

# Control of Nod-signal-induced cell divisions during root nodule formation

**Attila Fehér\*†, Michael Schultze\* and Eva Kondorosi\*‡**

\*Institut des Sciences Végétales, CNRS, F-91198 Gif-sur-Yvette, Cedex, France, and †Institute of Plant Biology, Biological Research Center, HAS, H-6701 Szeged, P.O. Box 521, Hungary

## Introduction

Control of cell division in eukaryotes can be considered as the integration of two main regulatory systems: (i) control of cell cycle progression through well-defined phases ($G_1$, S, $G_2$, M), and (ii) regulation of transitions between proliferative and resting states of the cells dependent on the initiation of developmental patterns. In plants, several components of the cell cycle machinery have already been identified [e.g. cyclins and cyclin-dependent kinases (CDKs)] (see L. De Veylder et al., chapter 1; M. Ito, chapter 8). However, the integration of cell cycle control with transduction of external signals is still poorly understood. Plant hormones have long been recognized as endogenous regulators of cell division (for reviews, see [1,2]), but due to their multiple effects on plant physiology and development [3], it is difficult to clarify their direct mode of action on cell cycle control. Transduction of external signal molecules triggering unique developmental programmes, such as the rhizobial Nod factors which initiate nodule organogenesis in leguminous plants, might serve as prototypes for mitogen-signalling cascades in plants.

This review describes the structure of the Nod factor, the basis of host specificity, and cell cycle activation in *Medicago* roots and cell cultures. Owing to the positional determination of root cortical cells for Nod-factor-induced cell division, the metabolic stage of the plant, endogenous morphogen gradients and possible interplay between phytohormones and Nod factors will be discussed in connection with cell susceptibility and competence. Based on the mitogenic effects of Nod-factor derivatives in non-legumes, the possible role of these signal molecules as general plant growth regulators will also be considered.

‡*To whom correspondence should be addressed*

## *Rhizobium*–legume symbiosis

Rhizobia, belonging to the genera *Rhizobium*, *Bradyrhizobium* and *Azorhizobium*, are soil bacteria which establish symbiosis with leguminous plants when the combined nitrogen in the soil is limited. The *Rhizobium*–legume symbiosis results in the development of nitrogen-fixing root nodules (reviewed in [4–8]). The structure of root nodules is highly organized. It is composed of three major zones: (i) the meristem, (ii) the invasion zone and (iii) the symbiotic zone where the bacteria are converted into nitrogen-fixing bacteroids capable of reducing the atmospheric nitrogen into ammonia. The invasion and symbiotic zones are surrounded by the nodule parenchyma, which includes the vascular bundles whose peripheral location is a characteristic feature of nodules as compared with the central vascular system of lateral roots. The nodule is surrounded by the nodule endodermis and the nodule cortex, respectively. The nodule endodermis and parenchyma, which serve as oxygen barriers, are implicated in the generation of the microaerobic conditions required for nitrogen fixation.

Development of nitrogen-fixing symbiosis is a complex process composed of multiple stages which are controlled by both symbiotic partners, and which rely on a molecular 'cross-talk' between rhizobia and legume plants (for review, see [9]). Reciprocal signallings control the expression of different sets of genes in a temporally and spatially regulated manner, and determine the host specificity of symbiosis. Most *Rhizobium* species interact with a limited range of host plants: for example, *R. meliloti* nodulates *Medicago*, *Melilotus* and *Trigonella* species, whereas *R. leguminosarum* bv. *viciae* interacts with *Pisum*, *Vicia, Lens* and *Lathyrus*. However, *Rhizobium* strains also exist with a broader host range. *Rhizobium sp.* NGR234, for example, nodulates more than 60 legume genera.

The induction of nodule development, representing the initial stage of symbiosis, is based on a two-way molecular signal exchange (as reviewed in [10–14]). Flavonoid signal molecules exuded by the host plant root induce the expression of the *Rhizobium* nodulation genes in conjunction with the bacterial activator NodD protein. The interaction between the flavonoids and the NodD proteins represents the first barrier of *Rhizobium*–legume symbiosis concerning host specificity. A given flavonoid in combination with different NodD proteins can act either as an inducer or inhibitor of *nod* genes. In *R. meliloti,* more than 20 nodulation genes have been identified (for review, see [13]). Mutations in these genes either inhibit or delay nodulation of the host plant. The nodulation genes code for the synthesis and excretion of the *Rhizobium* Nod signals that are the key determinants of host specificity and act as host-specific morphogens capable of reprogramming differentiated root cortical cells for nodule development. Purified Nod factors applied on the host plant roots at a concentration of $10^{-9}$ M can induce nodule organogenesis [15,16]. This concentration is at least two orders of magnitude lower than the active concentration of exogenously applied phytohormones capable of inducing root cortical cell divisions, like auxin or cytokinin [17,18].

## Nodule organogenesis

Nodule organogenesis itself consists of several stages (for reviews, see [7,8,13]). The first interaction between rhizobia and their host plants occurs at the emerging root-hair zone, behind the root apical meristem. This region is not only the most susceptible to *Rhizobium* infection, but is also the region where induction of mitotic activity in the root cortex occurs opposite to the protoxylem poles. In this zone, attachment of rhizobia to the root-hair tips results in hair curling and entrapment of bacteria. Rhizobia enter the root-hair cells via a newly formed tube structure, the infection thread, which transports the multiplying bacteria towards the root cortex. Simultaneously with the formation of the infection threads, cell divisions start in the cortex, and the cell division centre develops into the nodule primordium. Depending on the host plant, cell division may occur either in the outer or in the inner cortex which leads to the development of one of the two major nodule types, the 'determinate' or 'indeterminate' nodules, respectively. The round-shaped 'determinate' nodules are characteristic to tropical legumes like soybean (*Glycine max*), whereas the 'indeterminate' nodules develop in temperate legumes such as alfalfa (*Medicago sativa*), and their elongated form is due to the persistence of the apical meristem.

The first divisions of cortical cells are anticlinal. After the first divisions, the mitotic activity spreads over the neighbouring cells and results in the development of the nodule primordium. Owing to its polarized development, the apical part becomes the nodule meristem, ensuring the outgrowth of the nodule structure from the root, whereas cells behind the meristem lose their division activity and become infected by rhizobia. Differentiation of the nodule cells, as well as that of the bacteria into bacteroids, leads to the formation of different nodule zones and fully functioning nitrogen-fixing root nodules.

Though nodules are unique plant organs, their development shares certain common features with that of the lateral roots [8]. For example, both organs are predominantly initiated in cells located opposite to the protoxylem poles, but lateral roots originate from the pericycle, while nodule organogenesis starts with cortical cell divisions. Moreover, certain Nod-factor-inducible early nodulin (*enod*) genes, thought to be specific for nodule organogenesis, are also expressed during lateral root development, suggesting involvement of common endogenous plant factors in both developmental processes. In spite of these similarities, however, only nodule, but not lateral, root organogenesis is induced by the Nod factors.

## Effects of Nod factors

Nod factors elicit similar plant responses like rhizobia, and also in a host-specific manner (for review, see 13,14]). Exogenously applied Nod factors induce root-hair deformation within a few hours at concentrations as low as $10^{-12}$ M [16,19,20]. At concentrations of $10^{-9}$ M or higher, they induce cortical cell divisions which, in certain legumes such as *Medicago*, develop into empty nodules, lacking the invasion and symbiotic zones, but accumulating a large number of amyloplasts in the central nodule zone [15]. This indicates that, at least in this legume, Nod factors are the only external signal molecules required for nodule organogenesis. Under the limitation of combined nitrogen, certain genotypes of *Medicago* have the capability to form nodules spontaneously (in the absence of *R. meliloti* or Nod factors). This shows that all steps of the nodule developmental pathway are encoded by the plant, and Nod factors are needed for efficient triggering of this programme [21].

The molecular events in nodule organogenesis, and the mechanism by which the Nod factors induce this pathway, are still unknown. Based on the highly host-specific activity of Nod factors, they are likely perceived by a receptor present in the plasma membrane, or inside the cell. Although a Nod-factor receptor has not been identified yet, a low-affinity Nod-factor binding site was found in *Medicago* root protein extracts [22]. It is not clear how Nod factors elicit, nearly at the same time, responses both in the root hairs and in the cortical cells. Given that a single *Rhizobium* species normally produces several related Nod-signal molecules, root hairs and cortical cells may differ in the Nod factor receptors and their susceptibility towards the various molecules [23]. Nod factors that were the most active in root-hair deformation, however, were also the most potent mitogens. The differential responses of cells towards Nod factors in the different cell layers might be due to their distinct developmental programmes, and their responsiveness to Nod signals might be determined by their sensitivities towards phytohormones and their positions in the plant root. Root-hair cells could generate a secondary signal in response to Nod factors which would elicit the cortical cell responses. Although this hypothesis cannot be ruled out, rapid activation of the cell cycle in the cortical cells, resulting in cell division 12–24 h after *Rhizobium* infection, points to a direct action of the Nod factor [24,25].

The Nod signal transduction pathway is unknown. Until now, only a few *enod* genes have been isolated which are induced or stimulated in the root and in the course of nodule development (as reviewed in [13,26,27]). The functions of these genes remain to be determined. At present, putative functions are provided, based on the sequence analysis and *in situ* hybridization studies. The best-known examples of *enod* genes are *enod12* [28–32] and *enod40* [31–36]. The *enod12* genes encode proline-rich putative cell-wall proteins. In *M. sativa*, there are two *Msenod12* genes: *Msenod12A*, which is expressed in the dividing cortical cells, and *Msenod12B*, which is induced in the root-hair and epidermal cells [37]. The *enod40* gene is induced by Nod factors in the root pericycle and in the dividing cortical cells. Surprisingly, the *enod40* genes of different legume plants have an open-reading-frame coding only for a 12–13-amino-acid-long oligopeptide, though its transcript is a 700–800-nucleotide-long mRNA [32–36]. The mRNA has a highly stable secondary structure, suggesting that it acts as a riboregulator [36]. At present, it is unclear whether the *enod40* function is provided by the oligopeptide, by the highly conserved mRNA, or whether both the oligopeptide and the mRNA contribute to the action of *enod40*.

During the last few years, besides the isolation of *enod* genes, much effort has been made to elucidate cell cycle activation by the Nod factor. Most of these studies have been carried out on *Medicago*. One of the reasons for the selection of the *R. meliloti–Medicago* system was the detailed knowledge on *R. meliloti,* with respect to nodulation, and the availability of a range of purified Nod factors varying in their mitogenic activity, also including inactive derivatives. The other reason was the advanced cell cycle research in *Medicago* which allowed the characterization of cell cycle phases, monitoring the cell cycle progression. The cell division cycle is controlled in all eukaryotes by conserved mechanisms involving the sequential formation, activation, and subsequent inactivation, of a series of cyclin and CDK complexes (see V. Sundaresan and J. Colasanti, chapter 3). The availability of cell cycle marker genes from *Medicago* allowed us to address the question: Which *Medicago* cells are susceptible to the mitogenic action of the Nod factor?

## The bacterial nodulation signal: structure and host specificity

The Nod signals consist of a tetra- or penta-meric $\beta$-1,4-linked *N*-acetylglucosamine backbone which is acylated with unsaturated fatty acids on the non-reducing sugar residue [19]. Usually, the core Nod-factor structure is further decorated by various substitutions on the terminal residues in the different *Rhizobium* species that produce a family of related molecules (for review, see [38]). Identification of the different Nod signal molecules, and measurement of the biological activities of the purified molecules, have revealed that host specificity is influenced essentially by all structural features of the Nod factors. Structural variation occurs with respect to the length of the chito-oligosaccharide chain, the length of the fatty acid chain and the degree of its unsaturation, and the combination of various substituents at either end of the oligosaccharide backbone (Fig. 1). A minimum of one type of decoration, either at the reducing or the non-reducing terminus, is found on the core lipo-chito-oligosaccharide (LCO) structure, whereas the majority of rhizobia produce Nod factors carrying substituents at both termini. The structure of the substituents can be strikingly different. Carbon C-6 of the reducing-end sugar can be decorated by a sulphate or *O*-acetyl group, by differently substituted fucosyl residues, or by D-arabinose. The reducing-end sugar can carry an *O*-acetyl group at the C-6 position, and *O*-carbamoyl groups at C-3, C-4 or C-6. In addition, *N*-methylation at the non-reducing terminus is observed frequently.

For example, the *R. meliloti* Nod factors are composed of tetra- and penta-meric oligochitin backbones that are acylated most frequently with a $C16:2\Delta^{2,9}$ fatty acid, 6-*O*-acetylated at the non-reducing terminus, and sulphated at C-6 of the reducing terminus (Fig. 1) [19,20,39,40]. In contrast, Nod factors of *R.*

*leguminosarum* bv. *viciae* do not have any substituent groups on the reducing *N*-acetylglucosamine residue, but they are 6-*O*-acetylated at the non-reducing terminus and carry either a $C18:4\Delta^{2,4,6,11}$ or a $C18:1\Delta^{11}$ fatty-acyl chain [41] (Fig. 1). An absence of the sulphate group renders the *R. meliloti* Nod factors inactive on its host plant *Medicago*, but active on *Vicia*, the host plant of *R. leguminosarum* bv. *viciae* [16]. Thus, the presence or absence of the sulphate modification is a major determinant of host specificity.

The influence of modifications at the reducing-end is less striking in other systems. For example, lack of the 2-*O*-methylfucosyl group of *Bradyrhizobium japonicum* Nod factors (Fig. 1) reduces the nodulation capacity only slightly [42]. Substitutions on the non-reducing terminus, e.g. the *O*-acetyl group, also contribute to Nod signal activity and may influence host specificity [23,43]. Moreover, Nod signal activity depends on the length of the oligosaccharide chain [20,44], as well as on the nature of the acyl chain [15,41,45]. Although it is likely that the Nod-factor structure is specifically recognized by host plant receptors [23,46], some modifications may increase the stability of Nod factors against degradation by hydrolytic enzymes in the rhizosphere [47,48], and might thereby enhance their apparent morphogenic activity [49].

**Fig. I.**          **Structure of nodulation factors as exemplified for three *Rhizobium* species**

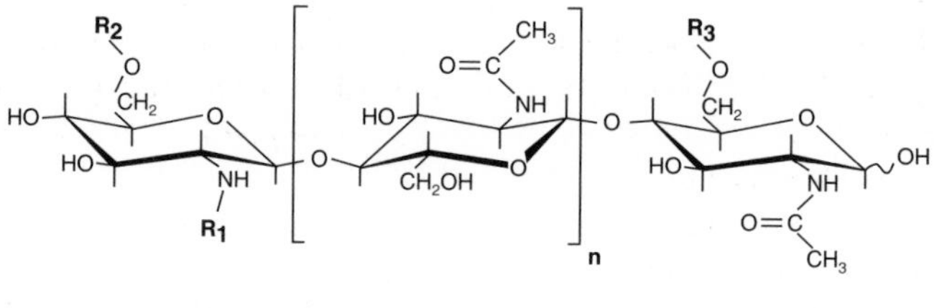

| Rhizobium sp. | n | R$_1$ | R$_2$ | R$_3$ |
|---|---|---|---|---|
| R. meliloti | 2,3 | $C16:2\Delta^{2,9}$ $C16:2\Delta^{2,4,9}$ $C16:1\Delta^{9}$ | H, acetyl | sulphate |
| R. leguminosarum bv. viciae | 2,3 | $C18:4\Delta^{2,4,6,11}$ $C18:4\Delta^{11}$ | acetyl | H |
| B. japonicum | 3 | $C18:4\Delta^{11}$ | H, acetyl | 2-*O*-methyl-fucose |

*Only the most abundant molecular species are indicated. For a more detailed description of Nod factors of different Rhizobium species, see [38].*

# The mitogenic activity of Nod factors on plant cells

One of the earliest morphological changes in the host plant root in response to rhizobial infection or Nod factors is the division of cortical cells in the emerging root-hair zone, opposite to the protoxylem poles. Moreover, purified Nod signal molecules proved to be potent mitogens, not only in legume roots, but also in partially quiescent cell cultures of alfalfa [50], whereas synthetic LCO molecules, unmodified on the reducing terminus, induced cell division in tobacco leaf protoplasts [51], indicating that non-legumes also possess a functional perception system for these molecules.

## Cell cycle activation in Nod-factor-susceptible *Medicago* cells

Differentiated cortical cells in legume roots can have 2C or 4C nuclear DNA contents. Cells with 2C DNA content are arrested in the $G_1$ phase, whereas the 4C DNA content can represent either cells arrested in the $G_2$ phase, or cells that are about to become endopolyploid. Division of the root cortical cells begins 12–24 h after inoculation with *Rhizobium*, or application of Nod factors on the root. This rapid response raised the hypothesis that Nod factor activates $G_2$-arrested cortical cells which are able to enter mitosis rapidly [52].

This hypothesis was recently investigated in *Medicago* cell cultures [50], as well as in intact roots [53]. In spite of the differences between the two experimental systems, the results were similar and gave complementary information. In the Nod-factor-responsive *Medicago* cell cultures, the plant material was not limited. Thus, it was possible to perform biochemical analysis and monitor the expression of various cell cycle genes. In roots, investigation of the spatial and temporal expression of cell cycle marker genes by *in situ* hybridization revealed responsiveness of cortical cells to *R. meliloti* or Nod factor.

The mitogenic effect of *R. meliloti* Nod factors was investigated in a microcallus suspension culture of the *M. sativa* line RA3 that contained both quiescent and dividing cells [50]. Addition of NodRm-IV(C16:2,S) to the suspension culture, at a concentration of $10^{-9}$ M, resulted in highly stimulated expression of *cdc2Ms* within 2 h in comparison to the control cultures. The thymidine incorporation and p34[cdc2]-related histone H1 kinase activities were higher in the Nod-factor-treated suspensions than in the control cultures. However, the Nod-factor treatment did not lead to faster cycling of the cells, but instead suggested the recruitment of quiescent cells for division. Flow-cytometric analysis, and *in situ* localization of the S-phase-specific histone H3 mRNA, revealed that the number of cells in S phase had already increased 5 h after the addition of Nod factor, and before the onset of cell division. These findings indicated that Nod factors most likely activate the cell cycle machinery in the quiescent cells at the $G_0$–$G_1$/S cell cycle transition, contrary to the previous assumption [52]. Induction of cell division necessitated the active Nod-factor structure; saturation of the acyl chain rendered NodRm-IV(C16:0,S) a 1000-fold less active than NodRm-IV(C16:2,S), whereas the non-sulphated Nod factor, NodRm-IV(C16:2), was inactive. In the microcallus suspensions, the Nod factor was required for the $G_0$–$G_1$/S transition; however, it remained an open question whether it was also needed for progression through the $G_2$-M phases.

The expression of cell cycle genes (marking specific stages of the cell cycle) was investigated via *in situ* hybridization in *Medicago* roots after spot inoculation with *R. meliloti* or purified Nod factors [53]. It has been postulated that expression of *cdc2* in non-dividing plant cells contributes to their division potential (see L. De Veylder et al., chapter 1; [54]). However, Nod-signal-susceptible cells did not have a higher *cdc2* expression level than the non-susceptible cells, and expression of *cdc2Ms* was induced in those cortical cells that divided after inoculation with *R. meliloti*. Surprisingly, the S-phase-specific histone H4 gene was induced not only in the inner cortex, but also in the non-dividing outer cortical cells opposite to the protoxylem poles. The induced expression of this gene before the first cell division pointed towards the susceptibility of $G_0$-arrested cells to Nod factors, and confirmed that activation of the cell cycle in the root cortical cells started by entering S phase before mitosis. This is a similar pattern to that described in the *Medicago* suspension cells. Expression of *cycMs2* encoding a mitotic cyclin was detected only in the inner cortical cells. Thus, progression of the cell cycle in the outer cortical cells was halted in the S phase, contrary to the inner cortical cells which had the capability to enter the $G_2$/M phases and divide. The permissive factors which lead to the completion of the cell cycle in these cells are unknown. One can speculate that the differences in cell division potential of the inner and outer cortical cells rely on transverse gradients of endogenous plant morphogens in the root (see J.P. Renaudin et al., chapter 4).

To further characterize the effect of the Nod signal on cell cycle regulation, cell cycle genes expressed at the $G_0$–$G_1$/S cell cycle phase transition could be used as markers. In animals, different types of cyclins regulate the $G_1$-phase progression. The D-type cyclins act primarily as growth factor sensors and mediators of information on the external environment towards the cell cycle control system [55]. Their plant homologues (the so-called δ-cyclins) have been recently isolated from *Arabidopsis* [56] and *Medicago* (*cycMs4*) [57], and it is presumed that they carry out similar functions as the D-type cyclins.

Another alfalfa cyclin gene (*cycMs3*) can be activated rapidly by mitogen signals (auxin plus cytokinin supplied together) in quiescent leaf cells preceding by several hours the expression of *cycMs4* [58] . Since Nod factors seem to activate the cell cycle in the $G_0$-quiescent cells, and the *cycMs3* function appears to be necessary for the quiescent cells to re-enter the cell cycle, expression of *cycMs3* might be induced by the Nod factor before *cycMs4*. *cycMs3* could be a functional homologue of the yeast *CLN3* cyclin gene, directing environmental information towards the cell cycle machinery [58]. It remains to be elucidated whether such cyclin genes are involved in Nod-factor-induced cell cycle reactivation.

## Biological effects of Nod factors on non-legumes: mitogenic activity of LCOs is not restricted to nodulation

In spite of the highly host-specific effect of Nod factors on legume plants, recent data indicate that Nod factors, or their derivatives, might be general plant growth regulators. The first observation was made on a temperature-sensitive mutant carrot line deficient in the transition from the globular to the heart stage of somatic embryos. This mutant could be rescued at the non-permissive temperature by the *R. leguminosarum* bv. *viciae* Nod factor NodRlv-V(Ac,C18:4) [59]. Chitopentaose or chitotetraose, as well as the conventional phytohormones 2,4-dichlorophenoxyacetic acid (2,4-D) and 6-benzylaminopurine (6-BAP), were unable to overcome the arrest.

Mitogenic activity of synthetic LCOs was demonstrated on tobacco protoplasts [51]. Addition of synthetic LCO molecules, consisting of a chitotetraose backbone *N*-acylated with either saturated or unsaturated C18 fatty acids, alleviated the requirement of protoplasts for both auxin and cytokinin for cell growth and division. LCOs substituted with C18:1 *trans*-fatty acid activated maximal rates of cell division at concentrations as low as $10^{-15}$ M, whereas LCOs carrying a C18:1 *cis*-fatty acid (like the *Rhizobium* Nod factors) displayed weaker mitogenic activity, and induced division of tobacco protoplasts at concentrations of $10^{-10}$ M and higher. LCOs with a C18:1 *trans*-fatty acid induced the phytohormone-responsive (-90-base-pair) region of the cauliflower mosaic virus 35S RNA promoter, similarly to auxin, suggesting that LCOs have the ability to substitute for phytohormones. The hormonal effect of LCOs could not only be achieved by an altered phytohormone production, but also by induction of genes that mediate auxin-responsive cell division, thus uncoupling cell division from auxin action. Recently, an auxin-inducible gene, *axi1,* has been characterized from tobacco, that directed auxin-independent growth of tobacco protoplasts [60]. In the absence of auxin, LCOs elicited transcriptional activation of the *axi1* gene that might be, at least partially, responsible for the LCO-induced, auxin-independent growth in tobacco protoplasts [51].

Mitogenic activity of LCOs and auxin on tobacco protoplasts suggests that the signal transduction pathways mediated by these molecules shares common steps in triggering the cell cycle in plants. It is tempting to speculate that LCO-related molecules, acting as universal endogenous plant growth regulators, might exist in plants.

## Factors affecting the competence of root cortical cells for Nod-signal-induced cell division

Root nodule organogenesis is restricted to the emerging root-hair zone. However, rhizobia also interact with the mature root hairs and induce different types of root-hair deformation, but neither root-hair curling nor nodule primordium formation in this root region [61]. Likewise, Nod factors induce root-hair deformation and expression of the root-hair-specific *Mtenod12* gene, not only in the mitotically competent region, but all along the root, starting just behind the growing root tip and extending throughout the zone of root-hair emergence and maturation [62]. This suggests that perception of the Nod signal is not restricted to the emerging root-hair zone. However, only cortical cells in this region can gain transient competence for cell division. Moreover, within this susceptible region, cortical cell divisions are initiated predominantly opposite the protoxylem poles. Positional information determining the sensitivity of cortical cells towards mitogenic stimulation might rely on endogenous gradients of plant morphogens (Fig. 2). The best candidates for influencing cell division competence are obviously the phyto-hormones, especially auxin and cytokinin, whose ratio and endogenous concentration might vary within the different root zones. In addition, the metabolic status of the plant can influence the control of cell competence, since active photosynthesis and limitation of combined nitrogen are prerequisites for nodule primordium formation.

## Morphogen gradients in the plant roots as possible sources for positional information determining cortical cell competence

Since auxins are synthesized predominantly in the shoot apex and young leaves, whereas cytokinins are produced in the root apex, inverse vertical gradients of these hormones exist in the plant root (Fig. 2) [63,64]. These gradients were shown to determine the site of lateral root primordium formation in pea roots [63,64], which, in certain respects, is similar to the initiation of the nodule meristem [8]. In the elongation zone of the pea root, the cell division is minimal due to the high concentrations of cytokinins provided by the root tip [63,64]. However, even in this zone, division of pericycle cells can be prematurely induced by removal of the root tip. In the intact root, lateral root initiation was highly stimulated by auxins supplied exogenously, and inhibited by the presence of cytokinins [65]. Possibly, similar regulation could be responsible for restricting the nodule primordium formation to the emerging root-hair zone (Fig. 2). However, lateral root primordia are initiated closer to the root apex in a region that is separate from that where nodules are formed [66].

Besides the evident vertical gradients of auxins and cytokinins, transverse morphogen gradients also contribute to the cell division potential of the root cortical cells. Auxins and cytokinins, transported by the phloem and xylem respectively, are highly diffusible molecules, and their concentration may decrease towards the root-hair cells and/or may alter hormone sensitivity of cells in the different cell layers.

In addition to auxins and cytokinins, uridine has been identified recently as a cell division factor in pea roots [67]. The excretion of an unknown cell division factor from the central cylinder at the protoxylem poles was first postulated by Libbenga et al. [17], based on experiments with root cortical explants of pea. If the excised cortical tissues were cultured on synthetic media supplemented with auxin

**Fig. 2.**       **Morphogen gradients in the root implicated in Nod signal-induced cell division**

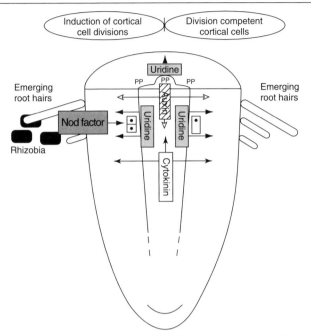

*Vertical and transverse gradients of auxins and cytokinins as well as radial gradient of the cell division factor uridine excreted at the protoxylem poles (PP) from the central cylinder are likely determining division competence of cortical cells in the emerging root-hair zone (left part). Competent cells can be induced to divide by the rhizobial Nod factor (or a secondary signal generated by the Nod factor in the plant root) (right part).*

and cytokinin, cell division foci appeared predominantly opposite to the protoxylem poles. Additional application of crude alcoholic extracts from the central cylinder or the cortex resulted in divisions throughout the cortex. These results were interpreted as evidence for the presence of a radial gradient of an unknown cell division factor in the pea root cortex. This so-called 'stele factor' has been recently purified and identified as uridine-inducing cell divisions in pea root segments at nanomolar concentrations if supplied exogenously [67]. Uridine probably plays a role in restricting the competence for cell cycle reactivation only to certain root cortical cells that are opposite the protoxylem ridges of the central root cylinder (Fig. 2).

The transverse morphogen gradients in the legume roots might also explain the differential response of cortical cells to the Nod signal. Why the outer cortical cells, that re-enter the cell cycle, are arrested before mitosis, and why the inner cortical cells can progress through the whole cell cycle and finally form the primordia of 'indeterminate' nodules [53] is unknown (see also V. Sundaresan and J. Colasanti, chapter 3). However, one can speculate that in tropical legumes, where cell divisions start in the outer cortex in response to the Nod signal, the morphogen gradients are optimal in the outer cortical cell layers for cell division competence.

## Metabolic signals interfering with cortical cell competence: nitrogen regulation of nodulation

The limitation of combined nitrogen for the plant host is the major prerequisite for Rhizobium—legume symbiosis. The presence of combined nitrogen inhibits symbiosis at multiple stages (for review, see [68,69]), and prevents the Nod-factor-induced plant responses, including root-hair deformation and cortical cell division. Formation of spontaneous nodules is also inhibited by the presence of nitrate. Symbiotic nitrogen-fixation requires much energy, and the nodules are storage organs for amyloplasts. Thus, their development requires active photosynthesis. A central role for cytokinins has been proposed in the regulation of carbon/nitrogen metabolism in plants [70–73], and it was hypothesized that metabolic status determines the sensitivity of plant tissues towards growth regulators and serves as a basis for the phenotypic plasticity of plants [74]. Interplays between the endogenous morphogen gradients, hormone sensitivity of root cortical cells and metabolic signals can be hypothesized during root nodule induction.

# Induction of cortical cell divisions by plant hormones

While it is likely that the endogenous morphogen gradients determine division competence of inner cortical cells, this competence, in the absence of Nod factors, can rarely result in cell divisions and only in certain genotypes of *Medicago*. For the efficient stimulation of the nodule developmental pathway, the Nod factors are essential. On the other hand, exogenous application of plant hormones or hormone transport inhibitors had various effects on legume roots, and mimicked, to a certain extent, the mitogenic effect of Nod factors by eliciting the formation of nodule-like structures (also reviewed in [75]). In most cases, the effect of auxins and cytokinins has been investigated, since these phytohormones play major roles in the regulation of various organogenic processes in plants [3] and are required for maintenance of plant cell divisions *in vitro* (especially the auxins).

## Auxins and auxin transport inhibitors

In cortical explants of pea roots, exogenous application of auxin did not result in divisions of cortical cells [17]. In alfalfa roots, addition of $N$-naphthaleneacetic acid (NAA), or 2,4-dichlorophenoxy acetic acid (2,4-D), at a concentration of 10 mM, resulted in the generation of lateral root primordia and anticlinally dividing cells along the root inner cortex [18]. These auxin-induced morphological changes were distinct from those triggered by the Nod factor with respect to the pattern of cortical cell division and amyloplast deposition, and were independent of the nitrate concentration in the medium.

It is generally accepted that auxin is the main phytohormonal inducer of *cdc2* expression (reviewed in [76,77]). Up-regulation of *cdc2* genes by auxin treatment has been shown in pea [78], soybean [79] and *Arabidopsis* [80] roots. In pea roots, expression of *cdc2* was stimulated by auxin within 10 min, suggesting a direct effect of auxin on *cdc2* expression [78]. In soybean, expression of the root-specific *cdc2* gene was induced by rhizobia, as well as by auxin, but not by cytokinin [79]. In cultured alfalfa cells, both auxin [81] and the *R. meliloti* Nod factor [50] increased the steady-state levels of *cdc2* transcripts within 1–3 h.

Although the expression of the *cdc2* gene is indispensable, it is probably not the only requirement for activation of the cell cycle machinery. Clearly, *cdc2* genes are mainly expressed in proliferating tissues in a developmentally regulated way, and they can be stimulated by mitogenic signals in differentiated cells. However, their expression is not always linked to cell division, and reflects, in such cases, the competence of the cells for mitogenic stimulation [54].

Polar auxin transport inhibitors [2,3,5-tri-iodobenzoic acid, TIBA; or *N*-(1-naphthyl)phthalamic acid, NPA] have been reported to elicit development of pseudo-nodules on alfalfa roots [82]. These organs, however, were in many respects distinct from the *Rhizobium-* and Nod-factor-induced nodules. At concentrations of 10 mM, both NPA and TIBA elicited complex, mainly periclinal, cell division patterns in several tissue layers along the root, and the pseudonodules consisted of, mainly, endodermis and vascular knots [18].

## Cytokinins

Several experimental data suggest that cytokinins might be involved more directly in nodule organogenesis. One of the first indications came from complementation of a nodulation-defective *R. meliloti* mutant, with the *tzs* gene of *Agrobacterium tumefaciens* coding for *trans*-zeatin secretion [83]. In addition, exogenously supplied benzyladenine elicited rapid expression of the Nod-factor-inducible early nodulin genes, *enod12* and *enod40* [75], and triggered cortical cell divisions in *Medicago* roots in a similar manner to the *R. meliloti* Nod factor [18]. Cell divisions were mainly anticlinal, and restricted to the inner root cortex opposite protoxylem poles in the emerging root-hair zone. Moreover, cortical cell divisions occurred only in the absence of nitrate, required active photosynthesis, and starch granules accumulated along the root. These observations suggest an interplay between the Nod factor and, at least, some elements of a cytokinin-responsive signalling pathway in nodule induction. It is unclear whether the Nod factor affects directly the cytokinin production or alters the cytokinin/auxin ratio and hormone sensitivity of the cells, or whether cytokinins and Nod factors share common steps in the stimulation of cell division. Dehydrodiconiferyl alcohol glucosides were recently identified as possible mediators of the cell-division-promoting effects of cytokinins [84–86]. It is unknown, however, whether Nod factors also have an effect on the synthesis of these compounds. At present, *enod40* seems to be the best candidate for mediating the hormonal effect of the Nod factor. Leaf protoplasts of transgenic tobacco plants expressing the *enod40* gene were able to divide at high auxin concentrations that inhibited the growth of wild-type protoplasts [87]. It has been speculated that *enod40* expression affects either the cytokinin/auxin balance or hormonal sensitivity of cells.

How may the cytokinin-like effect of the Nod factor activate the cell cycle? In certain tissues requiring both auxin and cytokinin for cell division (e.g. in tobacco pith tissue), cytokinin was reported to be necessary for the Cdc2 kinase activity [78]; when *cdc2* expression was induced by auxin alone, the kinase remained inactive. It was shown that induction of cell divisions also required a cytokinin-dependent factor. Recently, a root-specific $G_1$–cyclin cDNA clone (*cyclin-δ3*) was isolated from *Arabidopsis* [56], which is strongly inducible by cytokinins, indicating a new putative level of the hormonal regulation of the plant cell cycle machinery at the $G_0$–$G_1$/S transition point. This cyclin could be the cytokinin-dependent component of Cdc2 kinase activation observed in tobacco pith tissue [56]. Perhaps the Nod-factor-induced cell cycle activation is also mediated by cytokinin-responsive $G_1$-type cyclins in *Medicago* roots. The expression of these cyclin genes may be induced by the Nod factor directly or indirectly, either through the alteration of the cytokinin/auxin ratio, or the activation of genes mediating hormonal effects.

## Summary and perspectives

At present, very little is known about the molecular mechanisms of Nod signal action on cell cycle reactivation of root cortical cells upon infection by rhizobia. Most of our present knowledge is based on using cell cycle marker genes as molecular probes in the analysis of Nod-factor-induced gene expression in roots and cell cultures, or on studying the effects of exogenously applied plant growth regulators mimicking the action of Nod factors during nodule induction. On the basis of this limited information, only very general and speculative models can be suggested for the early regulatory steps in the induction of nodule organogenesis.

We propose that cortical cell division, in response to the external nodulation signal, is the result of an interplay among several regulatory signals determining, on the one hand, cortical cell competence and, on the other hand, triggering the cell division of the competent cells (Fig. 3). Cell competence is dependent on endogenous cell division factors, and on the metabolic state of the plant, as efficient nodulation requires active photosynthesis and nitrogen limitation. Metabolic signals, directly or indirectly, may change the internal hormone balance in the root and/or sensitize certain cells towards plant growth regulators. A vertical and oppositely oriented gradient of cytokinin and auxin, as well as radial gradients of these morphogens and that of uridine diffusing from the protoxylem poles in the

**Fig. 3.        Possible interplays of external and internal factors resulting in competence and Nod-signal-induced cell division of root cortical cells**

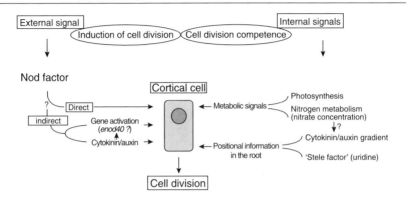

root, might be the crucial factors limiting cell division competence to a few cells in the root inner cortex. At present, it is not known whether the Nod factor itself or a secondary signal, generated in the plant root, triggers division of the inner cortical cells. Preliminary data indicate that de-acylated derivatives of the *R. leguminosarum* bv. *viciae* Nod factors, containing the *O*-acetylated sugar backbone, introduced by microbombardment into the inner cortical cells of *Vicia* roots, are capable of inducing cell division and nodule primordium formation [88]. Thus, it is possible that the fatty acyl substitution of the Nod factors is required for their integration into the plasma membrane and/or for their interaction with the putative receptor(s), whereas, intracellularly, the de-acylated molecules would act as mitogens. The Nod signal transduction pathway most likely results directly in the activation of the cell cycle by generating transcription factors that activate expression of cell cycle genes, like $G_1$-type cyclins and their CDK counterparts. However, induction of certain early nodulin genes, such as *enod40*, or hormone-responsive genes, such as *axi1*, might mediate indirectly hormonal effects or, conversely, decouple cell cycle activation from the hormonal trigger. It cannot be excluded either that the mitogenic effect of Nod factor is exerted through changes in the hormone (auxin/cytokinin) balance or sensitivity in the root, which also affect expression of hormone-regulated genes. The purified Nod-factor molecules exhibit auxin-like activities in certain systems (induction of cell division in tobacco leaf protoplasts, plasma membrane depolarization in alfalfa root hairs and transcriptional induction of *cdc2* genes in several systems). However, the role of cytokinin is certainly more important during nodule initiation as being capable of mimicking Nod-factor action on legume roots if applied at high concentrations exogenously. Perhaps the Nod signal transduction pathway at certain points converges to a cytokinin-controlled signalling pathway. It is unclear whether the cytokinin-like effect of the Nod factor involves perturbation of local cytokinin/auxin balance in the root cortex, or whether it is achieved *via* expression of genes, such as *enod4O*. These are inducible by both Nod factors and cytokinins. Concerning the different auxin- and cytokinin-like effects of the Nod-factor molecules in different plant systems, one can hypothesize that they may depend on the target cells or reflect different hormonal sensitivity of the plant materials.

In the further elucidation of the detailed mechanism of mitogenic Nod-factor signalling, different approaches can be applied. Undoubtedly, one of the most promising approaches is the analysis of legume nodulation mutants that are defective in different stages of nodule organogenesis ([89–93]; for review, see [94]). Pharmacological techniques based on activation and inhibition of signal transduction components in transgenic plants carrying promoter-marker gene fusions can provide useful information on Nod signalling towards early-nodulin gene or cell cycle gene expression. New data can be expected for the mechanism of signal transduction and the mitogenic effect of LCO molecules by the utilization of the recently established Nod-factor-responsive cell culture systems (tobacco leaf protoplasts, alfalfa microcallus culture). The elucidation of the role of uridine as a cell division factor in leguminous roots is undoubtedly required. Studying possible Nod-signal-dependent expression of recently isolated plant genes that are quite likely to play a role in $G_0$–$G_1$/S phase transition (e.g. plant $G_1$-type cyclins) can also provide important additional information on cell cycle regulation during nodule induction. New sensitive techniques, such as the differential display reverse transcription polymerase chain reaction technique, might be useful for the isolation of genes activated transcriptionally at very early stages (even within minutes) after treatment of roots with Nod factor. The isolation of further plant genes, such as those related to cytokinin signalling, mitogenic signal transduction, cell cycle regulation and reactivation, or lateral root initiation, can be expected in the near future. They may prove to be very important in the understanding of the early steps in nodule primordium formation, as well as for the integration of external nodulation signal transduction with the cell cycle control system.

*The authors thank Drs Adam Kondorosi, Denes Dudits and Anibal Lodeiro for the critical reading of the manuscript and for their valuable comments and discussions. Attila Fehér was supported by a long term EMBO postdoctoral fellowship (ALTF 315-1994).*

## References

1.    Dudits, D., Bögre, L., Bako, L., Dedeoglu, D., Magyar, Z., Kapros, T., Felföldi, F. and Györgyey, J. (1993) in Molecular and Cell Biology of the Plant Cell Cycle (Ormrod, J.C. and Francis, D., eds.), pp. 111–131, Kluwer, Dordrecht, The Netherlands
2.    Ferreira, P., Hemerly, A., Van Montagu, M. and Inzé, D. (1994) Plant Mol. Biol. **26**, 1289–1303

3. Davies, P.J. (1988) Plant Hormones and their Role in Plant Growth and Development, Kluwer, Dordrecht, The Netherlands
4. Dart, P. (1977) in A Treatise on Dinitrogen Fixation (Hardy, R.W.F. and Silver, W.S., eds.), pp. 367–472, Wiley, New York
5. Newcomb, W. (1981) in International Review of Cytology (Suppl. 13), Biology of the *Rhizobiaceae* (Giles, K.L. and Atherly, A.G., eds.), pp. 247–298, Academic Press, New York
6. Sprent, J.I. (1989) New Phytol. **111**, 129–153
7. Brewin, N.J. (1991) Annu. Rev. Cell Biol. **7**, 191–226
8. Hirsch, A.M. (1992) New Phytol. **122**, 211–237
9. Franssen, H.J., Vijn, I., Yang, W.C. and Bisseling, T. (1992) Plant Mol. Biol. **19**, 89–107
10. Fisher, R.F. and Long, S.R. (1992) Nature (London) **357**, 655–660
11. Dénarié, J., Debelle, F. and Rosenberg, C. (1992) Annu. Rev. Microbiol. **46**, 497–531
12. Downie, J.A. (1994) Trends Microbiol. **2**, 318–324
13. Schultze, M., Kondorosi, É., Ratet, P., Buiré, M. and Kondorosi, Á. (1994) Int. Rev. Cytol. **156**, 1–75
14. Spaink, H.P. (1995) Annu. Rev. Phytopathol. **33**, 345–368
15. Truchet, G., Roche, P., Lerouge, P., Vasse, J., Camut, S., De Billy, F., Promé, J.C. and Dénarié, J. (1991) Nature (London) **351**, 670–673
16. Roche, P., Debellé, F., Maillet, F., Lerouge, P., Faucher, C., Truchet, G., Dénarié, J. and Promé, J.C. (1991) Cell **67**, 1131–1143
17. Libbenga, K.R., Van Iren, F., Bogers, R.J. and Schraag-Lamers, M.F. (1973) Planta **114**, 29–39
18. Bauer, P., Ratet, P., Crespi, M., Schultze, M. and Kondorosi, A. (1996) Plant J., **10**, 91–105
19. Lerouge, P., Roche, P., Faucher, C., Maillet, F., Truchet, G., Promé, J.C. and Dénarié, J. (1990) Nature **344**, 781–784
20. Schultze, M., Quiclet-Sire, B., Kondorosi, E., Virelizier, H., Glushka, J.N., Endre, G., Géro, S.D. and Kondorosi, A. (1992) Proc. Natl. Acad. Sci. U.S.A. **89**, 192–196
21. Truchet, G., Barker, D.G., Camut, S., De Billy, F., Vasse, J. and Huguet, T. (1989) Mol. Gen. Genet. **219**, 65–68
22. Bono, J.J., Riond, J., Nicolaou, K.C., Bockovich, N.J., Estevez, V.A., Cullimore, J.V. and Ranjeva, R. (1995) Plant J. **7**, 253–260
23. Ardourel, M., Demont, N., Debellé, F.D., Maillet, F., De Billy, F., Promé, J.C., Dénarié, J. and Truchet, G. (1994) Plant Cell **6**, 1357–1374
24. Calvert, H.E., Pence, M.K., Pierce, M., Malik, N.S.A. and Bauer, W.D. (1984) Can. J. Bot. **62**, 2375–2384
25. Dudley, M.E., Jacobs, T.W. and Long, S.R. (1987) Planta **171**, 289–301
26. Nap, J.-P. and Bisseling, T. (1990) Physiol. Plant. **79**, 407–414
27. Munoz, J.A., Palomares, A.J. and Ratet, P. (1996) World J. Microbiol. Biotechnol. **12**, 189–202
28. Scheres, B., Van de Wiel, C., Zalensky, A., Horvath, B., Spaink, H., Van Eck, H., Zwartkruis, F., Wolters, A.M., Gloudemans, T., Van Kammen, A. and Bisseling, T. (1990) Cell **60**, 281–294
29. Pichon, M., Journet, E.P., Dedieu, A., De Billy, F., Truchet, G. and Barker, D.G. (1992) Plant Cell **4**, 1199–1211
30. Allison, L.A., Kiss, G.B., Bauer, P., Poiret, M., Pierre, M., Savouré, A., Kondorosi, E. and Kondorosi, A. (1993) Plant Mol. Biol. **21**, 375–380
31. Vijn, I., Martinez-Abarca, F., Yang, W.C., Das Neves, L., Van Brussel, A., Van Kammen, A. and Bisseling, T. (1995) Plant J. **8**, 111–119
32. Vijn, I., Yang, W.C., Pallisgard, N., Jensen, E.O., Vankammen, A. and Bisseling, T. (1995) Plant Mol. Biol. **28**, 1111–1119
33. Kouchi, H. and Hata, S. (1993) Mol. Gen. Genet. **238**, 106–119
34. Yang, W.C., Katinakis, P., Hendriks, P., Smolders, A., De Vries, F., Spee, J., Van Kammen, A., Bisseling, T. and Franssen, H. (1993) Plant J. **3**, 573–585
35. Hirsch, A.M., Asad, S., Fang, Y., Wycoff, K. and Löbler, M. (1993) in New Horizons in Nitrogen Fixation (Palacios, R., Mora, J. and Newton, W.E., eds.), pp. 291–296, Kluwer, Dordrecht, The Netherlands
36. Crespi, M.D., Jurkevitch, E., Poiret, M., D'Aubenton-Carafa, Y., Petrovics, G., Kondorosi, E. and Kondorosi, A. (1994) EMBO J. **13**, 5099–5112
37. Bauer, P., Crespi, M., Szécsi, J., Allison, L.A., Schultze, M., Ratet, P., Kondorosi, E. and Kondorosi, A. (1994) Plant Physiol. **105**, 585–592

38.  Schultze, M. and Kondorosi, A. (1996) World J. Microbiol. Biotechnol. **12**, 137–149
39.  Roche, P., Lerouge, P., Ponthus, C. and Prome, J.C. (1991) J. Biol. Chem. **266**, 10933–10940
40.  Demont, N., Ardourel, M., Maillet, F., Promé, D., Ferro, M., Promé, J.C. and Dénarié, J. (1994) EMBO J. **13**, 2139–2149
41.  Spaink, H.P., Sheeley, D.M., Van Brussel, A.A.N., Glushka, J., York, W.S., Tak, T., Geiger, O., Kennedy, E.P., Reinhold, V.N. and Lugtenberg, B.J.J. (1991) Nature (London) **354**, 125–130
42.  Stacey, G., Luka, S., Sanjuan, J., Banfalvi, Z., Nieuwkoop, A.J., Chun, J.Y., Forsberg, L.S. and Carlson, R.W. (1994) J. Bacteriol. **176**, 620–633
43.  Ardourel, M., Lortet, G., Maillet, F., Roche, P., Truchet, G., Promé, J.C. and Rosenberg, C. (1995) Mol. Microbiol. **17**, 687–699
44.  Stokkermans, T.J.W., Ikeshita, S., Cohn, J., Carlson, R.W., Stacey, G., Ogawa, T. and Peters, N.K. (1995) Plant Physiol. **108**, 1587–1595
45.  Spaink, H.P., Bloemberg, G.V., Van Brussel, A.A.N., Lugtenberg, B.J.J., Van der Drift, K.M.G.M., Haverkamp, J. and Thomas-Oates, J.E. (1995) Mol. Plant-Microbe Interact. **8**, 155–164
46.  Felle, H.H., Kondorosi, E., Kondorosi, A. and Schultze, M. (1995) Plant J. **7**, 939–947
47.  Staehelin, C., Schultze, M., Kondorosi, E., Mellor, R.B., Boller, T. and Kondorosi, A. (1994) Plant J. **5**, 319–330
48.  Staehelin, C., Schultze, M., Kondorosi, E. and Kondorosi, A. (1995) Plant Physiol. **108**, 1607–1614
49.  Schultze, M. and Kondorosi, A. (1995) Trends Microbiol. **3**, 370–372
50.  Savouré, A., Magyar, Z., Pierre, M., Brown, S., Schultze, M., Dudits, D., Kondorosi, A. and Kondorosi, E. (1994) EMBO J. **13**, 1093–1102
51.  Röhrig, H., Schmidt, J., Walden, R., Czaja, I., Miklasevics, E., Wieneke, U., Schell, J. and John, M. (1995) Science **269**, 841–843
52.  Verma, D.P.S. (1992) Plant Cell **4**, 373–382
53.  Yang, W.C., De Blank, C., Meskiene, I., Hirt, H., Bakker, J., Van Kammen, A., Franssen, H. and Bisseling, T. (1994) Plant Cell **6**, 1415–1426
54.  Hemerly, A.S., Ferreira, P. J., de Almeida Engler, J., Van Montagu, M., Engler, G. and Inzé, D. (1993) Plant Cell **5**, 1711–1723
55.  Sherr, C.J. (1995) Trends Biochem. Sci. **20**, 187–190
56.  Soni, R., Carmichael, J.P., Shah, Z.H. and Murray, J.A.H. (1995) Plant Cell **7**, 85–103
57.  Dahl, M., Meskiene, I., Bögre, L., Cam Ha, D.T., Swoboda, I., Hubmann, R., Hirt, H. and Heberle-Bors, E. (1995) Plant Cell **7**, 1847–1857
58.  Meskiene, I., Bögre, L., Dahl, M., Pirck, M., Cam Ha, D.T., Swoboda, I., Heberle-Bors, E., Ammerer, G. and Hirt, H. (1995) Plant Cell **7**, 759–771
59.  De Jong, A.J., Heidstra, R., Spaink, H.P., Hartog, M.V., Meijer, E.A., Hendriks, T., Lo Schiavo, F., Terzi, M., Bisseling, T., Van Kammen, A. and De Vries, S.C. (1993) Plant Cell **5**, 615–620
60.  Walden, R., Hayashi, H., Lubenow, H., Czaja, I. and Schell, J. (1994) EMBO J. **13**, 4729–4736
61.  Wood, S.M. and Newcomb, W. (1989) Can. J. Bot. **67**, 3108–3122
62.  Journet, E.P., Pichon, M., Dedieu, A., De Billy, F., Truchet, G. and Barker, D.G. (1994) Plant J. **6**, 241–249
63.  Torrey, J.G. (1976) Annu. Rev. Plant Physiol. **27**, 435–459
64.  Wightman, F. and Thimann, K.V. (1980) Physiol. Plant. **49**, 13–20
65.  Wightman, F., Schneider, A.E. and Thimann, K.V. (1980) Physiol. Plant. **49**, 304–314
66.  Libbenga, K.R. and Harkes, P.A.A. (1973) Planta **114**, 17–28
67.  Smit, G., De Koster, C.C., Schripsema, J., Spaink, H.P., Van Brussel, A.A.N. and Kijne, J.W. (1995) Plant Mol. Biol. **29**, 869–873
68.  Streeter, J. (1989) CRS Crit. Rev. Plant Sci. **7**, 1–23
69.  Carroll, B.J. and Mathews, A. (1990) in Molecular Biology of Symbiotic Nitrogen Fixation (Gresshoff, P.M., ed.), pp. 159–180, CRC Press, Boca Raton
70.  Chen, C., Jin, G., Andersen, B.R. and Ertl, J.R. (1993) Aust. J. Plant Physiol. **20**, 609–619
71.  Vincentz, M., Moureaux, T., Leydecker, M.T., Vaucheret, H. and Caboche, M. (1993) Plant J. **3**, 315–324
72.  Brzobohaty, B., Moore, I. and Palme, K. (1994) Plant Mol. Biol. **26**, 1483–1497
73.  Faure, J.D., Jullien, M. and Caboche, M. (1994) Plant J. **5**, 481–491

74. Trewavas, A.J. (1986) in Plasticity in Plants (Jennings, D.H. and Trewavas, A.J., eds.), pp. 31–77, Company of Biologists Ltd, Cambridge
75. Hirsch, A.M. and Fang, Y.W. (1994) Plant Mol. Biol. **26**, 5–9
76. Murray, J.A.H. (1994) Plant Mol. Biol. **26**, 1–3
77. Jacobs, T.W. (1995) Annu. Rev. Plant Physiol. **46**, 317–339
78. John, P.C.L., Zhang, K., Dong, C., Diederich, L. and Wightman, F. (1993) Aust. J. Plant Physiol. **20**, 503–526
79. Miao, G.H., Hong, Z. and Verma, D.P.S. (1993) Proc. Natl. Acad. Sci. U.S.A. **90**, 943–947
80. Martinez, M.C., Jorgensen, J.-E., Lawton, M.A., Lamb, C.J. and Doerner, P.W. (1992) Proc. Natl. Acad. Sci. U.S.A. **89**, 7360–7364
81. Hirt, H., Páy, A., Györgyey, J., Bacó, L., Németh, K., Bögre, L., Schweyen, R.J., Heberle-Bors, E. and Dudits, D. (1991) Proc. Natl. Acad. Sci. U.S.A. **88**, 1636–1640
82. Hirsch, A.M., Bhuvaneswari, T.V., Torrey, J.G. and Bisseling, T. (1989) Proc. Natl. Acad. Sci. U.S.A. **86**, 1244–1248
83. Cooper, J.B. and Long, S.R. (1994) Plant Cell **6**, 215–225
84. Binns, A.N., Chen, R.H., Wood, H.N. and Lynn, D.G. (1987) Proc. Natl. Acad. Sci. U.S.A. **84**, 980–984
85. Orr, J.D. and Lynn, D.G. (1992) Plant Physiol. **98**, 343–352
86. Black, R.C., Binns, A.N., Chang, C.-F. and Lynn, D.G. (1994) Plant Physiol. **105**, 989–998
87. Mylona, P., Pawlowski, K. and Bisseling, T. (1995) Plant Cell **7**, 869–885
88. Spaink, H.P., Bloemberg, G.V., Wijfjes, A.H.M., Ritsema, T., Geiger, O., Lopez-Lara, I.M., Harteveld, M., Kefetzopoulos, D., van Brussel, A.A.N., Kijne, J.W. et al. (1994) in Advances in Molecular Genetics of Plant-Microbe Interactions (Daniels, J., Downie, J.A. and Osbourn, A.E., eds.), pp. 91–98, Kluwer, Dordrecht
89. Dudley, M.E. and Long, S.R. (1989) Plant Cell **1**, 65–72
90. Utrup, L.J., Cary, A.J. and Norris, J.H. (1993) Plant Physiol. **103**, 925–932
91. Pedalino, M. and Kipenolt, J. (1993) J. Exp. Bot. **44**, 1007–1014
92. Pedalino, M., Kipenolt, J., Frusciante, L. and Monti, L. (1993) J. Exp. Bot. **44**, 1015–1020
93. Sagan, M., Huguet, T., Barker, D. and Duc, G. (1993) Plant Sci. **95**, 55–66
94. Caetano-Anollés, G. and Gresshoff, P.M. (1991) Annu. Rev. Microbiol. **45**, 345–382

# 12

# DNA replication initiation and mitosis induction in eukaryotes: the role of MCM and Cdc25 proteins

**Paolo A. Sabelli\*‡, Shirley R. Burgess†, Anil K. Kush\* and Peter R. Shewry†**

Institute of Molecular & Cell Biology, National University of Singapore, 10 Kent Ridge Crescent, Singapore 0511, Republic of Singapore, and †IACR-Long Ashton Research Station, Department of Agricultural Sciences, University of Bristol, Bristol BS18 9AF, U.K.

## Introduction

The cell division cycle consists of three major sequential processes which allow cells to proliferate. These are: (i) the replication of the genome and other structures (S phase), (ii) the segregation of chromosomes to two daughter nuclei (M phase or mitosis), and (iii) cytokinesis. Less dramatic, but still very important, processes take place during the intervals between mitosis/cytokinesis and S phase (the $G_1$ phase) and between S phase and mitosis (the $G_2$ phase), during which cells prepare to execute S phase and mitosis, respectively. Over the past few years, tremendous progress has been made in our understanding of the biochemistry and molecular biology of cell cycle regulation, mainly as a result of the earlier isolation of cell cycle mutants in yeast. One lesson that we have learned is that key cell cycle enzymes have been highly conserved in evolution. It is well established that the regulation of the assembly and activity of complexes between cyclin-dependent kinases (CDKs) and regulatory proteins known as cyclins is crucial to virtually every aspect of cell cycle control in all eukaryotic cells (reviewed in [1,2]). Plant homologues to CDKs and cyclins have been isolated and characterized from numerous species. The roles of these molecules in the plant cell cycle are reviewed in greater detail in several other chapters of this volume. In the present chapter, we focus on two key processes for the onset of S phase and mitosis: the initiation of DNA replication and the role played by minichromosome maintenance (MCM) proteins, and the activation of the Cdc2–cyclin B complex (the M-phase-promoting factor or MPF) by the Cdc25 protein phosphatase at the G2/M phase transition. Although the molecular analysis

‡*To whom correspondence should be addressed*

of the plant cell cycle has mainly focused on CDKs and cyclins, recent results, obtained in our and other laboratories, have provided interesting information on other cell cycle genes, thereby expanding considerably our knowledge of cell cycle control in plants. Thus, the current understanding of DNA replication initiation and MPF activation, both in lower and higher eukaryotes, will be reviewed, and results which have recently been obtained in higher plants will be presented and discussed.

## Initiation of DNA replication in eukaryotes

Three important aspects of DNA replication are specific to eukaryotes (reviewed in [3,4]). First, eukaryotes have many origins of DNA replication (from a few hundred in yeast to many thousands in higher eukaryotes). Secondly, DNA replication is asynchronous, as different regions of the genome are replicated at different times during S phase. A reproducible pattern of early- and late-replicating DNA has been observed, although this pattern may change during development. Thirdly, all DNA is normally replicated exactly once, and only once, in each somatic cycle. Thus, multiple initiations at one origin or incomplete DNA replication are incompatible with proper chromatid segregation at mitosis and cell viability, and are, therefore, prevented from occurring. The complexity of DNA replication introduces several problems for eukaryotic cells. Cells must be able to activate multiple sets of replication units, termed replicons, in a co-ordinated manner in S phase, and must be able to distinguish fired/replicated origins from non-fired/replicated origins. Cells appear to accomplish the above tasks mainly by tightly controlling the temporal and spatial regulation of DNA replication initiation.

Early experiments on animal cell fusion have shown that $G_1$ and $G_2$ nuclei differ in their ability to pass through S phase [5]. In fact, $G_1$ nuclei can initiate DNA replication in response to cytoplasmic signals, whereas $G_2$ nuclei can re-initiate DNA synthesis only after passage through mitosis. Subsequent experiments on replicated *Xenopus* sperm nuclear DNA revealed that, to enter a second round of replication, nuclei either had to go through mitosis, during which the nuclear envelope breaks down and re-assembles, or be permeabilized with non-ionic detergents. These observations, in addition to highlighting the dependence of S phase on mitosis, demonstrated that an intact nuclear membrane is required to prevent re-initiation of DNA replication and led to the 'licensing factor' hypothesis of DNA replication [6]. According to this hypothesis, some essential replication factor(s), unable to cross the nuclear membrane, would enter the nucleus at mitosis, when the nuclear membrane breaks down, and bind to the chromatin. At the $G_1$/S phase transition, such factor(s) would activate replication origins, but would be inactivated or eliminated from the nucleus during S phase. Only a subsequent round of mitosis would allow chromatin to bind fresh licensing factor. Further studies on permeabilization and repair of HeLa nuclei in *Xenopus* egg extracts have provided additional evidence for this hypothesis. In fact, the nuclear membrane prevents re-initiation of DNA replication by excluding an essential positive factor, rather than by retaining a negative inhibitor [7]. In addition, *Xenopus* extracts, treated with protein kinase inhibitors, are functionally devoid of licensing factor, indicating that protein phosphorylation is also required for the activation of licensing factor at mitosis [8]. As described below, genetic analyses of DNA replication origins and the proteins that interact with them in yeast, and biochemical studies in vertebrate cells, have recently converged on the identification of the MCM family of initiation proteins as components of the licensing factor activity of DNA replication. Although the licensing model seems correct, the exact mechanism for the temporal and spatial controls of DNA replication initiation is not known. However, both *cis* sequences and *trans* factors involved in this control are being identified and may reveal new insights into the regulation of eukaryotic DNA replication in the near future. In the following sections, the roles played by such *cis* sequences and *trans* factors will be discussed. However, for an extensive coverage of DNA replication control in eukaryotes, the reader is referred to several excellent reviews (see [3,4,9–15]). The absence of viable cell cycle mutants and problems in synchronizing cell division, together with the large genome size, have hampered the study of DNA replication in plants. Although detailed biochemical studies of replication enzymes have been carried out (reviewed in [15,16]), little is known of the molecular basis of DNA replication in higher plants. The recent identification of *MCM* homologues in plants [17, 17a] is encouraging for further progress in plant DNA replication studies.

## Origins of DNA replication

DNA replication initiates at chomosomal regions termed origins. Defined chromosomal regions required for origin function are known as replicators. It is thought that specific DNA–initiation protein interactions at origins result in the localized unwinding and melting of double-stranded DNA, with the resulting single-stranded DNA becoming available for the loading of replication enzymes, which synthesize the complementary DNA strand as the replication forks are extended bidirectionally. Eukaryotic origins of DNA replication have been best studied in the yeast *Saccharomyces cerevisiae*. Their identity has been elucidated mainly by studies on yeast autonomously replicating sequences (ARSs) which are chromosomal fragments that, when cloned into plasmids depleted of their own replication origin, function as extrachromosomal replication origins allowing plasmid stability [18]. In *S. cerevisiae*, replication origins are relatively short sequences of DNA (150–200 bp) rich in adenosine and thymidine, a base composition that may help to explain why origins represent the sites of initial DNA unwinding ([19], reviewed in [9]). Clearly, ARS elements are essential for chromosomal origin function, as all known chromosomal origins are associated with ARSs. However, only a subset of ARSs are active as chromosomal origins. Several reasons may account for this apparent discrepancy, including the influence of the chromosomal context in which origins reside, which could affect their activity, and the interference between neighbouring origins on the chromosome [20]. Recent studies have shown that ARSs have a modular structure with several functionally important domains (Fig. 1) [21,22]. Domain A is highly conserved, essential (although not sufficient for ARS activity), and is centred on the ARS core sequence (ACS) matching the 11-mer consensus oligonucleotide 5'-(A/T)TTTA (T/C)(A/G)TTT(A/T)-3'. The ACS represents the binding site for an initiation protein complex (see below). The B domain, which is also necessary for ARS activity, is located in a region which generally extends 50–150 bp to the 3' end of the T-rich core strand. Domain B is far less conserved among different ARSs, but, in the ARS1 origin, it comprises three functionally important subdomains: B1, B2 and B3. Subdomain B1 is involved in binding to the same initiation complex which interacts with the ACS; subdomain B2 is characterized by a low thermal stability of DNA and appears to function as a DNA-unwinding element; and subdomain B3 seems to stimulate DNA replication in a manner analogous to that of enhancer elements in gene transcription (see below). An increasing amount of evidence suggests that replicators are also present in the fission yeast *Schizosaccharomyces pombe*, although they have not been defined at the level required to identify essential DNA sequences [14].

In higher eukaryotes, many studies have identified loci at which replication is initiated but the lack of a genetic method to identify origins has hampered the identification of replicators. In metazoans, DNA synthesis is initiated at specific DNA sites, not randomly. Although there have been several convincing demonstrations of the presence of replicators (reviewed in [3,9,10,14,23]), animal replicators may be less specific or may function differently from those of prokaryotes and yeast. Mapping by two-dimensional gel electrophoresis has revealed that initiation sites of replication origins in animals are distributed over broad regions, spanning from several to many kilobase pairs, termed initiation zones. Such regions comprise a primary initiation locus flanked by numerous secondary initiation loci [3,10,23]. Thus, an apparent paradox is present in animal cells: replicators have been identified and shown to co-localize with the initiation sites of DNA synthesis, yet, in numerous instances, initiation may take place at multiple sites in a seemingly random fashion. For example, this occurs in the rapidly cycling cells of the *Xenopus* and *Drosophila* embryos, or when naked DNA is introduced into *Xenopus* eggs or egg extracts [3,10,14,23].

What about DNA replication origins in plants? There are numerous reports of DNA sequences from higher plants which have ASA activity in yeast cells [18,24–27]. However, only one *bona fide* chromosomal replication origin has been identified in the spacer region of pea ribosomal DNA [28,29]. Plant ARS sequences appear to have a similar organization to yeast and animal ARSs with an A/T-rich DNA sequence closely matching the yeast ACS. However, they are generally longer (200–1000 bp) than *S. cerevisiae* ARSs, and belong to different classes of sequences. Plant genomes display a similar organization of DNA replication to other higher eukaryotes with groups of adjacent simultaneously

**Fig. I** **The activation pathway leading to DNA replication initiation in *S. cerevisiae***

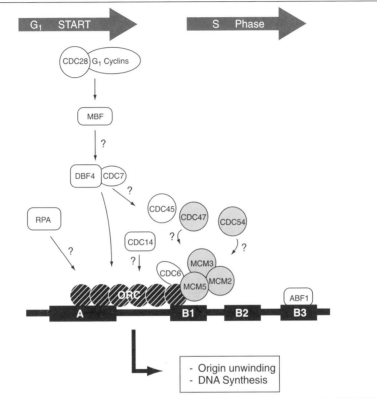

At 'START', the CDC28 kinase activity is required for entry into S phase, and to trigger expression of many MBF-transactivated S-phase-specific genes. Downstream in the pathway, a large number of proteins is involved in the interaction with, and activation of, replication origins (represented by the ARS I origin, black bar). These proteins include the DBF4–CDC7 kinase complex, CDC6, CDC14, the ORC complex, RPA, and members of the MCM protein family (MCM2, MCM3, MCM5, CDC47, and CDC54) (indicated by shaded circles). As a result, double-stranded DNA unwinds at origins and DNA synthesis is initiated. Although genetic analysis has shown these proteins to be required for the initiation of DNA replication, the exact roles and substrates of many of them are not clear.

active replicons forming replicon clusters (reviewed in [15]). Analysis of DNA replication by fibre autoradiography revealed that DNA synthesis advances bidirectionally within replicons. Replicon clusters which are active at the same time throughout the genome are organized in replicon families. As in metazoans, the molecular characterization of plant chromosomal origins has proved difficult.

## Proteins interacting with replication origins

Apart from viral DNA replication, interactions between eukaryotic initiation proteins and replication origins have been best understood and characterized in *S. cerevisiae*, mainly due to the availability of yeast mutants with defects in DNA replication, and of genetically characterized ARS elements. Many genes and proteins involved in the early stages of DNA synthesis have been identified and the emerging scenario is of a relatively large number of molecules that interact genetically or physically, with each other and with replication origins, to trigger DNA synthesis (Fig. 1). Bell and Stillman identified a complex of six essential proteins, called the origin replication complex (ORC), which binds to the ACS in an ATP-dependent manner [22]. ORC binds to double-stranded ACS with high specificity and mutations in the ACS destroy this binding. Subsequent experiments have shown that ORC probably binds to ARS throughout the cell cycle and it is, therefore, unlikely that ORC binding alone is sufficient to activate ARS at the $G_1/S$ phase transition. However, the binding specificity of ORC changes during the cell cycle, both *in vitro* and *in vivo*, as revealed by the ORC footprint of DNAseI protection, suggesting that ORC either undergoes post-translational modification, or interacts with one or more other proteins in a cell cycle regulated manner, and that such events affect its binding properties to ARS [30,31]. All the ORC subunits have recently being cloned, and two of them, ORC2 and ORC6, possess potential sites for phosphorylation by the CDC28 kinase [32,33]. Although the regulation of ORC–ARS interaction by protein phosphorylation is an attractive hypothesis, it remains to be proved experimentally. Another subunit, ORC5, has a potential ATP-binding site consistent with a possible direct binding to ARS [34]. ORC and ARS have been shown to interact genetically with a number of proteins including CDC6, CDC14, MCM5/CDC46, and the DBF4–CDC7 complex [34–36]. *cdc6* mutants, like *cdc14* mutants, display plasmid instability that can be suppressed by additional ARS elements. ORC and CDC6 form a pre-replicative complex at individual replicators and appear to co-operate to determine the frequency of initiation of DNA replication in the genome [36]. CDC14 is likely to be a protein phosphatase, and its interaction with ORC further suggests that the phosphorylation state of ORC may be important for its function [34]. *CDC7* encodes a serine/threonine protein kinase, whose activity peaks at the $G_1/S$ phase boundary, and whose execution point is the last genetically defined step before DNA replication. CDC7 kinase activity requires binding to DBF4, a protein which also binds to the ACS of ARS [35]. Thus, DBF4 could both activate CDC7 and recruit it to ARS, whereas CDC7 may phosphorylate other replication proteins such as RPA, CDC6, MCM5/CDC46 and ORC. Other proteins which interact with replication origins are ABF1 and RPA. ABF1 is a multifunctional DNA-binding protein involved in transcription and replication which binds to the ARS subdomain B3 and probably stimulates replication through its transcriptional activation domain [9]. RPA is a three-subunit, single-stranded DNA-binding protein required for DNA unwinding (reviewed in [4,7]). In vertebrates, RPA assembles post-mitotically into many pre-replication centres on the decondensing chromosomes which serve, upon formation of the nuclear membrane, as replication centres for the initiation of DNA synthesis [37]. RPA undergoes complex phosphorylation during the cell cycle of yeast, *Xenopus* and human. Generally, it is dephosphorylated in $G_1$, but phosphorylated in S phase and mitosis, though the exact identity of the kinase(s) involved in such phosphorylations is not known (reviewed in [4]).

Is the yeast model applicable to DNA replication initiation of higher eukaryotes and plants in particular? As discussed in other sections of this chapter, there is increasing evidence for DNA replication initiation in higher eukaryotes involving specific interactions between replicators and initiation proteins. For example, ARS elements have been identified in both animal and plant genomes. A homologue of the *S. cerevisiae CDC7* gene has recently been cloned from the distantly related *S. pombe* [38], suggesting that the CDC7 protein may be ubiquitous in eukaryotes. Furthermore, the MCM family of proteins is highly conserved in evolution and its members have been found in a range of eukaryotes including plants (see below). Finally, there is increasing confidence in ORC being conserved in eukaryotes, and large protein complexes, possibly interacting with origins of replication, have been detected and partially purified from plants ( P. Sabelli, unpublished work; [39]). These considerations suggest that many, if not most, key elements of the *S. cerevisiae* model for the initiation of DNA synthesis may be conserved in all eukaryotes.

## MCM protein family

Among the proteins which are required for DNA replication initiation, and which interact with ARSs, the MCM family of proteins is highly conserved in evolution and has recently received much attention. MCM proteins were first discovered in the analysis of yeast mutants characterized by a pre-mitotic cell cycle arrest with a concomitant increase in minichromosome loss and recombination [40]. These mutants, termed *mcm* for minichromosome maintenance, are unable to maintain ARS-containing minichromosomes, and the intensity of such defects varies with the ARS sequence present in the minichromosome, suggesting a direct interaction of MCM proteins with replication origins. Genetic experiments have shown these proteins to be essential, non-redundant and to interact with each other and other replication proteins (reviewed in [41]). MCM2 and MCM3 are required for the initiation of DNA synthesis at chromosomal origins and are localized in the nucleus from late mitosis to the onset of DNA replication early in S phase. In addition, a fraction of nuclear MCM2 and MCM3 is tightly bound to chromatin [40,42,43]. MCM5 is identical to CDC46, a protein with the same cell cycle nuclear distribution [44,45]. MCM5/CDC46 interacts genetically with ORC6, the smallest of the ACS-binding proteins comprising ORC, and with CDC45, CDC54, CDC47 and probably MCM2 and MCM3 [34,46]. Cloning of CDC47 and CDC54 has revealed that these proteins also belong to the MCM protein family [47]. The properties of the MCM proteins of *S. cerevisiae*, including their genetic interaction with ARSs and with other replication proteins such as ORC, the $G_1$-specific nuclear localization, the tight binding to chromatin, and their requirement for initiation of DNA synthesis at chromosomal origins, are all consistent with a role analogous to that of the licensing factor of DNA replication postulated in vertebrate cells [41]. However, mitosis in yeast cells lacks the breakdown and re-assembly of the nuclear membrane and, therefore, other mechanisms must be sought to explain the typical pre-S-phase nuclear localization of MCM proteins. MCM3 is unique among budding yeast MCM proteins as it has a simian virus 40 (SV40)-type nuclear localization signal (NLS) at its C-terminus, which also overlaps several putative CDC28 phosphorylation sites [41]. MCM3 might be responsible for targeting a multisubunit complex of MCM (and possibly other) proteins to the nucleus late in mitosis, though experimental data to support this idea are currently lacking. Thus, MCM3 may function as a molecular shuttle for an initiation protein complex to gain access to the nucleus. The presence of CDC28 phosphorylation sites overlapping the NLS suggests that CDC28 kinase activity during mitosis might be responsible for inducing conformational changes which, in turn, activate the NLS and target the MCM3-driven complex to the nucleus. MCM2 has an essential zinc-finger motif suggesting a direct interaction with DNA at ARS, while MCM5/CDC46 interacts with ORC6. These interactions are consistent with the observed changes in ARS-binding specificity of ORC during the cell cycle [30,31].

Homologues of the *S. cerevisiae* MCM family have also been identified in a variety of eukaryotes, including *S. pombe*, *Xenopus*, *Drosophila*, mouse, human and higher plants (Fig. 2 and 3). The Cdc*19*, Cdc*21*, Nda*1*, Nda*4*, and Mis*5* proteins are required for DNA synthesis in *S. pombe* [48–51]. In humans, purification of DNA polα-primase complex has revealed that the accessory protein P1 is an MCM homologue [52]. A family of P1 homologues has subsequently been discovered in both human and mouse [53–55]. Consistent with an essential role in DNA replication, microinjection of antibodies against such proteins blocks DNA synthesis [54]. In addition, the human MCM3 and CDC46 homologues form a stable complex in HeLa cells [56], a behaviour that may be shared by other P1 proteins in mammals [57]. In yeast, the disappearance of MCM proteins from the nucleus during S phase correlates with chromatin regions which have undergone initiation of DNA replication [41]. In contrast, mammalian MCM homologues are localized in the nucleus throughout interphase [54,55,58,59]. The P1 protein is associated with replicative chromatin, and its intranuclear localization shifts during S phase as different regions of the genome are replicated at different times [55,58]. The phosphorylation state of P1 proteins changes during the shifts. Two forms of nuclear P1 proteins have been identified: a hypophosphorylated form, associated with nuclear structure of pre-replication chromatin, and a hyperphosphorylated form, loosely bound to the nucleus and co-localized with chromatin regions which have undergone replication [55,59]. Although the persistence of P1 proteins in the nucleus throughout interphase in mammals indicates the presence of regulatory pathways differing from that in yeast, overwhelming evidence indicates that MCM

## Fig. 2          Alignment of the central domains of the MCM protein family

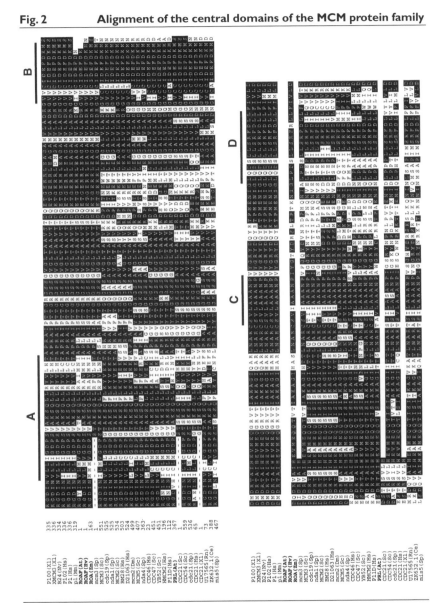

Black boxes indicate residues that match the consensus. Dashes indicate gaps which have been introduced to maximize the alignment. Four highly-conserved subdomains, also present in a large number of nucleic acid-dependent ATPases, are indicated by 'A', 'B', 'C', and 'D'. For sequence names, species, and references, see the legend of Fig. 3. The plant amino acid sequences are indicated by bold letters.

**Fig. 3**              **Phylogenetic tree of MCM-related proteins**

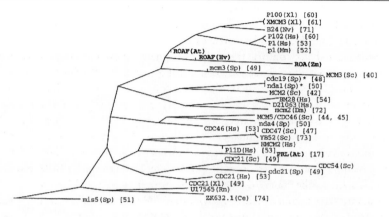

The initials of the species of origin are given in parentheses and are as follows: At, Arabidopsis thaliana; Ce, Caernohabditis elegans; Dm, Drosophila melanogaster; Hs, Homo sapiens; Hv, Hordeum vulgare; Mm, Mus musculus; Nv, Notophtalmus viridescens; Rn, Rattus norvegicus; Sc, Saccharomyces cerevisiae; Sp, Schizosaccharomyces pombe; Zm, Zea mays; Xl, Xenopus laevis. References are indicated in square brackets. D21063(Hs), HMCM2(Hs), CDC54(Sc), and U17565(Rn) are unpublished sequences and have D21063, D28480, U14731, and U17565 database accession numbers, respectively. Plant MCM-like sequences are indicated by bold letters.

proteins play a conserved pivotal role for the initiation of DNA synthesis. In fact, recent exciting results have identified MCM homologues as components of licensing activity of DNA replication in vertebrate cells [60–62]. Permeabilized HeLa $G_2$ nuclei, which are able to replicate in *Xenopus* egg extracts, are unable to do so in extracts depleted of the *Xenopus* MCM3 homologue and other associated MCM-related proteins. The re-addition of the purified MCM3-homologue is sufficient to support DNA replication [60,61]. Also, MCM3 homologues, and other MCM proteins, can confer licensing activity to chromatin added to mitotic *Xenopus* egg extracts which have been treated with kinase inhibitors and, therefore, are unable to initiate DNA replication [62]. Although the above experiments demonstrate a direct involvement of MCM proteins in licensing DNA replication, MCM3 homologues can enter $G_1$ nuclei in the absence of nuclear membrane breakdown, a behaviour that is inconsistent with the definition of licensing factor [62]. However, the properties of other MCM3-interacting proteins, yet to be characterized, may fully satisfy the licensing factor hypothesis.

A highly conserved region can be identified within the central domain of MCM proteins (Fig. 2), which contains an NTP-binding domain comprising a set of conserved motifs also present in a variety of nucleic acid-dependent ATPases (A, B, C, and D motifs) [63]. These motifs form a slightly modified version of the Walker-type purine NTP-binding pattern that consists of two distinct motifs, A and B [64]. Although, in MCM-like proteins, the A motif has the Gly in the G-K-S/T peptide substituted by Ala (Ser in Cdc21 homologues, and Thr in the *Arabidopsis* MCM3 homologue, ROAF), secondary-structure predictions indicate that the typical β-strand-loop-α-helix conformation, the phosphate-binding loop, in the A motif could accommodate various sequence substitutions without affecting the phosphate moiety of the NTP substrate [63]. The presence of an ATP-binding-like domain is consistent with a potential helicase activity required for replication origin unwinding. However, experimental data to support this are not available, presumably because recombinant or purified MCM homologues may display helicase activity only when associated in multimeric complexes. A 'DEAD'-box-like signature, identified in motif B (Asp-236–Asp-239 in the maize ROA protein), in which Ala in the standard 'DEAD' signature is substituted by Phe, may be relevant for the presumed helicase activity of MCM proteins [53]. In fact, 'DEAD' box motifs are found in enzymes involved in DNA unwinding [65]. Additional interesting motifs include the 'psycho' motif, which is potentially involved in protein–protein interactions required for the initiation of DNA replication. Such a motif is present in a subset of MCM proteins (MCM2, BM28 and D21063), and also in the SV40 T antigen and bacterial Purα and Purβ proteins [66]. Two other conserved regions have been found in the MCM family, although no function has been attributed to them (reviewed in [41]). As mentioned earlier, MCM2 contains an essential zinc-finger, which is also present in the closely related Cdc19 and Nda1 proteins from fission yeast [42,48,50]. Furthermore, the Cdc21 protein may have a metal-chelating motif in a similar position [49]. Such motifs may mediate interaction with DNA at origins. In addition to the CDC28 phosphorylation sites overlapping the MCM3 NLS, there are many other potential sites for phosphorylation by CDKs, protein kinase C, and casein kinase II present in MCM proteins, though their biological relevance has yet to be established.

Using a polymerase chain reaction (PCR)-based approach, we have isolated a 2.0 kb cDNA encoding an MCM3 homologue from maize root tips, and also related cDNA fragments from barley and *Arabidopsis* (Fig. 2) [121]. Another MCM-related *Arabidopsis* gene, *PROLIFERA* (*PRL*), has been isolated by transposon tagging [17], but it clearly represents a distinct *MCM* homologue, being more closely related to CDC47-like proteins (Fig. 2 and 3). Gene copy number reconstruction experiments indicate the presence of 2–4 copies of closely related *ROA* sequences in the maize genome, but also some less closely related sequences (not shown). Thus, MCM proteins are also conserved in higher plants, most likely encoded by multigene families.

Information on the expression patterns of MCM-related genes is fragmentary. The human *P1* mRNA peaks during S phase of serum-stimulated fibroblasts [52], while the levels of *cdc19* and *cdc21* mRNAs appear to be constant throughout the cell cycle of fission yeast [48,49]. In budding yeast, *MCM5/CDC46* and *CDC47* mRNAs peak late in mitosis, though protein levels are constant during

**Fig. 4**        *In situ* hybridization to transversal sections of the maize root apex

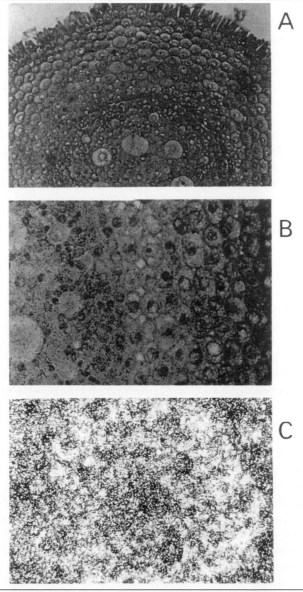

*A shows the cellular patterning in the root apex. B and C show a higher magnification of a similar section as in A labelled with an antisense ROA riboprobe. A and B were photographed under bright-field illumination, C under dark-field illumination. The patchy pattern of labelling is evident.*

the cell cycle [43,44,47]. In higher plants, expression analysis by Northern blot [121], *in situ* hybridization (Fig. 4), and β-glucuronidase (GUS)-staining [17], have demonstrated that MCM homologues are expressed in actively cycling cells and tissues. Although post-translational regulatory processes are likely, such cell-proliferation-dependent transcriptional control is consistent with that of most *cdk* and cyclin plant genes (reviewed in [2,67]). In addition, *in situ* analysis of mRNA expression has revealed a patchy pattern of labelling in the root meristem of maize, indicative of temporally regulated expression during the cell cycle in asynchronously cycling cells (Fig. 4) [68,69] (see J.H. Doonan, chapter 10). In animal cells, the expression of MCM homologues is also regulated during development [70]. Interestingly, the promoter of the murine *P1* gene contains two putative binding sites for E2F [55], known to transactivate animal genes in a cell cycle specific fashion, analogous to MBF in budding yeast and Dsc1 in fission yeast. A deeper understanding of the cell cycle dependent transcriptional control of the *ROA* gene will require the characterization of promoter sequences.

## Regulation of DNA replication initiation

DNA replication is a complex process which needs to be co-ordinated with the other cell cycle events. There is a large body of evidence indicating that the coupling of DNA replication to the cell cycle is mediated by CDKs (reviewed in [12]). For example, DNA synthesis is inhibited in $G_2$ and early M phase when CDK activity is high, and its mitotic elimination correlates with the resumption of licensing activity. Several mutations in the *cdc2* and *cdc13* (which encodes a B-type cyclin) genes result in endoreduplication in *S. pombe* [75,76]. There is clear evidence that some chromatin modifications are required in late mitosis to make nuclei competent to initiate DNA replication. These modifications are likely to include alteration of chromatin conformation and assembly of a pre-replicative complex at origins which may include ORC, CDC6, CDC14, RPA, and MCM proteins. In addition, CDKs inhibit the assembly of pre-replicative centres in *Xenopus* extracts [77]. Replication-competent origins would be activated after passage through 'START', possibly by post-translational modifications of origin-bound complexes by the addition/modification of proteins in the complex. The CDC7 kinase is a likely candidate as an effector of these modifications. These events constitute the S-phase control over DNA replication initiation, and are also under the ultimate control of CDKs. In fact, it is well established that CDK activity at 'START' in yeast leads to the transcription of several DNA synthesis genes. In mammals, the sequential activation of different CDKs is important for S-phase progression. In higher eukaryotes, cyclins and CDKs have been localized within replication foci [78]. Also, both in yeast and higher eukaryotes, CDK inhibitors block the initiation of DNA synthesis [79].

Once replication forks are formed and DNA synthesis is initiated, other modifications in the interaction between origins and replication complexes prevent further initiation and constitute a block to re-replication of the DNA. Dissociation of CDC6 and MCM proteins from the complex, probably as a result of phosphorylation, is likely to represent such a block. Also, phosphorylation of RPA is likely to induce conformational changes inhibiting replication initiation at origins. Studies in animal cells have shown that the nuclear membrane plays a key role in the once-per-cell-cycle control of DNA replication and that the nuclear/cytoplasmic compartmentalization of licensing factor activity appears to be crucial for this control. Plant cells frequently display a polyploid DNA content due to endoreduplication events in which there are two or more rounds of DNA synthesis, each of which is not followed by mitosis [80,81]. In some tissues, such as the seed endosperm and the embryo suspensor, endopolyploid cells represent the vast majority. A similar situation is also found in the *Drosophila* larva. This apparent failure to enforce the once-per-cell-cycle control over the initiation of DNA replication in endopolyploid cells remains largely unexplained. It would be interesting to study the behaviour of *MCM* genes during the endocycles of plant cells.

Although the identity and regulation of replication origins are not clear in higher eukaryotes, a large body of evidence suggests that the structural conformation of chromatin and nuclear scaffold may play very important roles in the control of initiation of DNA replication. For example, nuclear assembly and functional nuclear pores are required for the replication of DNA introduced into *Xenopus* eggs or egg extracts (reviewed in [4]).

There is an indication that DNA replication and gene transcription are processes which reciprocally affect each other (reviewed in [14]). For example, ORC has been shown to be involved in gene silencing, though its exact role is not clear [34]. Transcriptionally active euchromatin is usually replicated early in S phase, while transcriptionally inactive heterochromatin is replicated late. As active regions of chromatin are relatively decondensed to allow transcription, they may also allow the action of initiation/replication proteins. Thus, the structure of chromatin and nuclear matrix may play crucial roles in the co-ordinate control of DNA replication initiation and transcription. This possibility is supported by the observation that the transition from random to non-random initiation of DNA replication in the genome of *Xenopus* embryos coincides with the stage at which the pattern of transcriptionally active chromatin becomes more specifically determined [82]. There is increasing evidence for an important role played by the nucleoskeleton in specifying the sites of DNA replication initiation [83,84]. Stable replicon clusters have been observed through several cell cycles in pea root tip cells, suggesting that they may be held together by the nuclear matrix [85].

The analysis of tumour virus DNA replication (e.g. SV40) has greatly contributed to our present knowledge of the control of DNA synthesis initiation in animal cells. These viruses create a cellular environment permissive for their replication by the interaction of a virally encoded oncoprotein with members of the retinoblastoma (Rb) protein family. Upon binding to the viral protein, the Rb protein is unable to sequester transcription factors essential for S-phase entry, and, as a result, cellular DNA synthesis is activated. Interaction between the viral and Rb proteins is mediated by the lxcxe motif present in the viral protein. Interestingly, this Rb-binding motif has recently been identified in the Rep protein of wheat dwarf geminivirus and shown to be essential for viral replication [86]. Such a finding suggests that Rb-like proteins may exist in plants. This possibility is also supported by the presence of the lxcxe motif in plant D-type cyclins, although its functional relevance remains to be demonstrated [87].

## The Cdc25 phosphatase activates the M-phase-promoting factor at the $G_2$/M transition

The transition from $G_2$ to M phase depends on the activity of the MPF complex between a 34 kDa catalytic subunit Cdc2 protein kinase and a 45–60 kDa regulatory subunit, cyclin B (Fig. 5). This complex is responsible for driving cells into mitosis, and perhaps is the most highly conserved cell cycle controlling element in eukaryotes (reviewed in [1]). During interphase, Cdc2 is phosphorylated at Thr-161 in higher eukaryotes (and the equivalent Thr-167 in yeast) by a CDK-activating kinase. This is a stimulating phosphorylation, required for CDK kinase activity. However, Cdc2 is also phosphorylated at Thr-14 and Tyr-15 within its kinase catalytic domain. This is an inhibitory phosphorylation, which results in the inactivation of the Cdc2–cyclin B complex. The Wee1 kinase (together with the Mik1 kinase in fission yeast) is responsible for Tyr-15 phosphorylation, while a distinct kinase phosphorylates Thr-14. At the onset of M phase, Cdc2 becomes rapidly dephosphorylated at both Thr-14 and Tyr-15, while Thr-161/Thr-167 remains phosphorylated. As a result, MPF is fully active, and cells are driven into mitosis. These regulatory pathways that target Cdc2 are extremely important for the timing of the induction of mitosis and for cell cycle progression (reviewed in [88]). They are also probably involved in mediating the coupling of M phase to the completion of DNA replication and repair. The dephosphorylation of Thr-14 and Tyr-15 is carried out by the Cdc25 phosphatase, which was originally identified in fission yeast ([89–94], reviewed in [95]). This dual activity is rare among protein phosphatases, which usually belong to only one of the two classes of phosphatases. Also, a striking feature of Cdc25 phosphatases is their extremely high specificity for MPF, unlike the substrate promiscuity of most other protein phosphatases and kinases [95]. Yeast cells contain only one Cdc25 protein, and this appears to be a dosage-dependent inducer of mitosis. In fission yeast, Cdc25 (and Wee1) appears to be involved in the checkpoint control on the size at which cell division takes place (see D. Francis, chapter 9).

Homologues of Cdc25 have been identified in a range of species (Fig. 6 and 7). Several examples of functional complementation of *cdc25* mutants by Cdc25-like proteins from higher eukaryotes have been reported. Cdc25 homologues share obvious sequence homology, mainly in their C-terminal catalytic domains, which include the conserved active site of protein phosphatases (IVxHCxxxxxR), the so-

**Fig. 5**        **Mechanism of Cdc25 action and regulation at the G₂/M transition in vertebrates**

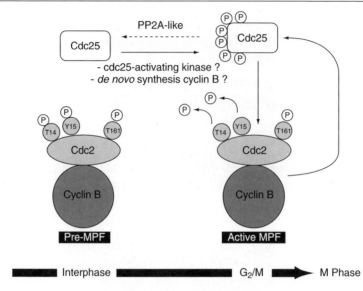

In late interphase, Cdc2 is associated with cyclin B to form pre-MPF. Cdc2 is phosphorylated at Thr-14, Tyr-15 (inhibitory phosphorylation) and at Thr-161 (activating phosphorylation). The interphase form of Cdc25 is non-phosphorylated and characterized by a low phosphatase activity. At the G₂/M transition, Cdc25 becomes phosphorylated at the non-catalytic N-terminus, probably by one or several activating kinase(s) which appear to include the MPM-2 epitope kinase. As a result, Cdc25 dephosphorylates the Cdc2 subunit of pre-MPF at Tyr-15 (and probably also at Thr-14), generating active MPF. The MPF kinase phosphorylates a range of substrates driving cells into mitosis. Recent evidence indicates that active MPF also phosphorylates Cdc25 which, in turn, activates more MPF. This Cdc25–MPF activation loop explains the sudden burst of active MPF at the onset of M phase, and the self-perpetuating activation of Cdc25 observed in higher eukaryotes.

called HC motif, despite there being no other obvious sequence similarity [96] (Fig. 6). The cysteine residue in the HC motif is believed to be required for the formation of an unstable thiophosphate bond. Interestingly, not all Cdc25-like proteins cloned so far possess this cysteine residue, while all have the HXE motif. This may indicate differences in their activities or substrate specificities.

In vertebrates, multiple *CDC25* genes have been identified, which may have distinct functions, probably activating different CDKs at different cell cycle phases [92]. For example, the CDC25a protein is predominantly expressed in late $G_1$ in mammals and its immunodepletion blocks cell cycle progression from $G_1$ into S phase [97].

**Fig. 6**     **Multiple sequence alignment of the conserved C-terminal domains of Cdc25-related proteins from various species**

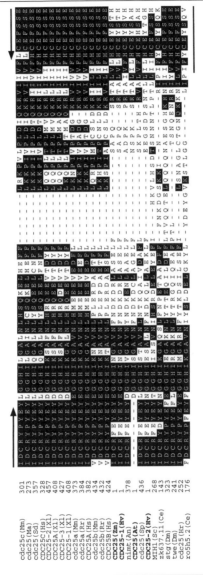

*Black boxes indicate identical residues to the consensus. Dashes indicate gaps which have been introduced to maximize the alignment. The arrows indicate the locations of the oligonucleotide primers used for PCR amplification. The conserved catalytic domain is underlined. For sequence names, species of origin, and references, see the legend of Fig. 7. The names of Cdc25-related proteins from plants are shown in bold letters.*

## Regulation of Cdc25

What controls Cdc25 activity? While the pathway which regulates Cdc25 in yeast has not been identified, there is increasing evidence that Cdc25 in higher eukaryotes is also activated post-translationally at mitosis by phosphorylation (at the non-catalytic N-terminus domain) by MPF ([94,98–100], reviewed in [88]). A basal level of Cdc25 phosphatase activity is observed throughout S phase, but it is near the onset of mitosis that Cdc25 becomes phosphorylated and highly active. This results in the dephosphorylation of Thr-14 and Tyr-15 of Cdc2, which activates MPF. Active MPF phosphorylates and activates more Cdc25 which, in turn, produces more active MPF (Fig. 5). This loop mechanism of activation is consistent with the rapid activation of MPF at the onset of mitosis, and the observed self-amplification of active MPF. Recently, a novel conserved domain in cyclin B (the P box) has been identified, which mediates its binding to Cdc2 and activates the low-activity interphase form of Cdc25 [101,102]. Although the above model provides an explanation for the rapid increase in the reciprocal activation of Cdc25 and MPF, the initial triggering kinase of Cdc25 (and presumably a counter-acting phosphatase) remains to be clearly identified. Evidence for such a kinase has been obtained from *Xenopus* extracts depleted of cyclin B and with inhibition of protein phosphatase 2A (PP2A) by okadaic acid, in which Cdc25 is efficiently phosphorylated [94,103]. Recent work has shown that Cdc25 is phosphorylated by the MPM-2 epitope (ME) kinase. This enzyme forms a phosphopeptide epitope in many proteins at mitosis, which can be recognised by the MPM-2 monoclonal antibody [104]. These results suggest that ME kinase and PP2A may be the stimulatory kinase and inhibitory phosphatase respectively of the low-activity form of Cdc25 during interphase.

     *cdc25* genes, homologous to the mitosis-inducer *cdc25* gene of *S. pombe*, are specifically transcribed at the $G_2/M$ phase transition. Promoter analysis of the human *CDC25C* gene has revealed that its cell cycle regulated expression is mainly due to cell cycle dependent association (in $G_0/G_1$) and dissociation (in S phase/$G_2$) of a repressor protein(s) with defined *cis* promoter sequences [105,106]. In addition, developmentally regulated transcription also appears to be important for Cdc25 activity [107–110]. Furthermore, association in complexes with other proteins may represent another level of Cdc25 regulation. It has recently been shown in humans that CDC25A (and probably CDC25B), but not CDC25C, form a complex with the RAF1 kinase, which is involved in the transduction of mitogenic signals. In addition, RAF1 phosphorylates CDC25 *in vitro* and stimulates its phosphatase activity [111]. Furthermore, over-expression *in vivo* of CDC25A and CDC25B (but not CDC25C) promotes oncogenic transformation, further suggesting that CDC25 phosphatases may be potential oncogenes [112]. Proteins belonging to the 14-3-3 family also form complexes with CDC25, though the biological relevance of this is not known [113].

## The search for plant *cdc25* homologues

Do plants have *cdc25* homologous genes? We attempted to isolate DNA sequences related to *cdc25* using PCR with degenerate primers, on the basis of the conserved C-terminal sequences in Cdc25-related proteins. A potentially interesting band was obtained from the amplification of genomic DNA from both maize and barley. These PCR fragments were subcloned into M13 phagemid vectors and several individual clones were sequenced for each cloning experiment. Three sequences, two from barley (*HVCDC25-1* and *HVCDC25-2*), and one from maize (*ZMC DC25*), showed homology to the *cdc25* gene family (Fig. 6 and 7). *ZMCDC25* and *HVCDC25-1* are 148 bp long and 93% identical. They encode polypeptides of 49 amino acids which are 96% identical. When compared with the sequences of the *cdc25* gene and gene product from *S. pombe*, they are about 59% similar. In contrast, *HVCDC25-2* is 155 bp long encoding a 51-residue polypeptide. Its nucleotide sequence is 53% similar to that of fission yeast *cdc25*, and the encoded protein is 46% identical with Cdc25; figures which are considerably lower than those of *ZMCDC25* and *HVCDC25-1*. In fact, *HVCDC25-2* is only 65–68% identical with *ZMCDC25* and *HVCDC25-1*, and only 65% identical at the amino acid level. Thus, while *ZMCDC25* and *HVCDC25-1* probably correspond to homologous genes, *HVCDC25-2* is likely to represent a distinct *cdc25* homologue from barley. The presence of at least two barley *cdc25*-like sequences is consistent

---

**Fig. 7**          **Phylogenetic tree of the Cdc25 protein family**

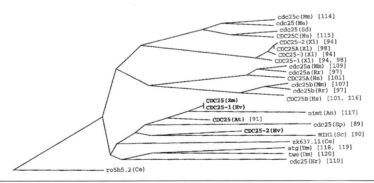

*The initials of the species are given in parentheses and are as follows: At, Arabidopsis thaliana; An, Aspergillus nidulans; Ce, Caenorhabditis elegans; Dm, Drosophila melanogaster; Hr, Helobdella robusta; Hs, Homo sapiens; Hv, Hordeum vulgare; Ma, Merocricetus auratus; Mm, Mus musculus; Rr, Rattus rattus; Sc, Saccharomyces cerevisiae; Sp, Schizosaccharomyces pombe; Sd, Scrofa domestica; Zm, Zea mays; Xl, Xenopus laevis. References are indicated in square brackets. cdc25(Ma), cdc25(Sd), zk637.11(Ce) and ro5h5.2(Ce) are unpublished sequences and have D10878, S43123, P30634 and Z48795 database accession numbers, respectively. The names of cdc25-related sequences from plants are shown in bold letters.*

with the presence of multiple *cdc25* homologues in other eukaryotes, suggesting differences in the functions of different cdc25 family members. Another plant Cdc25-like sequence from *Arabidopsis* has been reported, although a different combination of PCR primers was used [91].

Using a cDNA fraction obtained from maize root apices as a template for PCR amplification of *cdc25*-like sequences, a fragment of 148 bp was obtained which, upon sequencing, was confirmed to be identical to *ZMCDC25*. Although at least one gene related to *cdc25* appears to be expressed in maize root tips, the isolation of the corresponding transcript from a maize root tip cDNA library has been difficult, presumably because of the low abundancy of the mRNA. While the maize and barley fragments hybridize to each other under stringent conditions, under more permissive conditions, the barley *cdc25*-related fragment (from which both *HVCDC25-1* and *HVCDC25-2* have been isolated) cross-hybridizes to the region of fission yeast *cdc25* which encodes the conserved catalytic domain (not shown). However, Southern blot experiments with maize and barley genomic DNAs, using the maize and barley PCR fragments as probes, showed that *cdc25*-related sequences are present in the genomes of these two cereals, and that the two probes hybridized more strongly to homologous rather than heterologous DNA (Fig. 8). The control probe from the fission yeast *cdc25* gene (kindly provided by Prof. Paul Nurse, Imperial Cancer Research Fund, London) did not give any specific hybridization to plant DNA. Although both *S. cerevisiae* and *S. pombe* cells have only one Cdc25 protein, animal cells appear to contain two or more Cdc25-like proteins, which has been proposed to reflect a more complex pattern of cell cycle regulation in higher eukaryotes [92,97,109]. The PCR amplification of two *cdc25*-homologous sequences from barley suggests that this may also be the case in plants. During the course of this study, no hybridization was detected in Northern blot experiments between the *ZMCDC25* probe and transcripts in RNA fractions extracted from root apices and from other tissues/organs, presumably because expression of maize *cdc25*-related sequences is very low, even in dividing tissues. A possible implication of this is that *cdc25*-related genes are expressed only transiently during the cell cycle, which would make it difficult to detect transcripts using conventional techniques, even in proliferating tissues. This would also, in part, explain the difficulty in isolating *cdc25*-homologous cDNAs from plants. Although the molecular cloning of full-length plant *cdc25* genes remains an open problem, the independent isolation of PCR-sequences from monocotyledonous and dicotyledonous plants using different primer combinations suggests they are present, probably as small gene families. The expression of the *cdc25* gene from *S.pombe* in tobacco, which results in abnormal phenotypes, also suggests the presence of a Cdc25-dependent pathway in higher plants ([121], see D. Francis, Chapter 9).

**Fig. 8**   **The *cdc25*-related sequences from maize and barley hybridize to homologous genomic fragments**

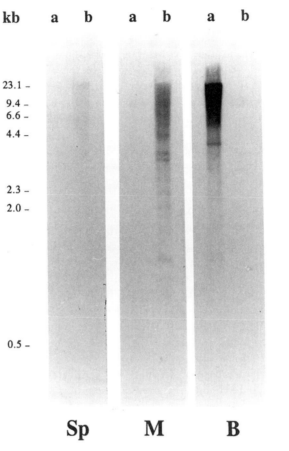

Approximately 4 μg of barley genomic DNA (lane a) and 10 μg of maize genomic DNA (lane b) were digested with EcoRI, fractionated by agarose gel electrophoresis, blotted on to a nylon membrane and hybridized with the ~150 bp PCR-amplified fragments of the cdc25 gene from S. pombe (Sp), maize (M) and barley (B). Final wash was at 60°C in 0.1 × SSC buffer for 1 h, and exposure was for 5 days.

## Conclusions and perspectives

In recent years, there has been remarkable progress in the study of the initiation of DNA replication and it has become one of the prominent fields in cell cycle research. In the present chapter, we have reviewed recent findings on the regulation of eukaryotic DNA replication, emphasizing the role played by MCM proteins. DNA replication appears to be regulated at multiple levels in eukaryotes. In addition to the activity of replication factors, the influence of the structure and conformation of chromatin and nuclear matrix, and of transcriptional activity, is becoming increasingly evident.

It is clearly established that in yeast, as in viruses and bacteria, DNA synthesis is initiated at replication origins as the result of specific interactions between initiation proteins and DNA sequences. In higher eukaryotes, including higher plants, the identification and characterization of defined cellular *cis* sequences required for DNA replication, and interacting proteins has been difficult. The lack of mutants in DNA replication initiation, the large genome size, the confinement of DNA replication to small cell populations in multicellular organisms, and difficulties in cell synchronization are some of the problems encountered. However, there is increasing evidence for the existence of replicators in the genome of higher eukaryotes, though their function and regulation may be different from yeast. What is emerging is that the regulation of DNA replication initiation in eukaryotes is highly conserved. Certain initiation proteins appear to be bound to replication origins throughout the cell cycle (i.e. ORC). Specific sets of replication origins are made competent for replication late in M phase by the assembly of additional protein in complexes, probably driven by CDKs. After passage through 'START' in late $G_1$, further modification of such complexes, probably by protein phosphorylation and/or interaction with other proteins, leads to DNA unwinding and initiation of DNA synthesis. CDKs seem to play an important role also in this process. After firing, origins are prevented from re-firing during S phase by other modifications of the initiation complexes, such as the dissociation of key proteins, probably brought about by protein phosphorylation. Thus, DNA replication is initiated once, and only once, between two rounds of mitosis. Passage through mitosis is required for the origin's competence to support DNA synthesis. In this context, the MCM protein family plays a crucial role, and their recent identification in plants is encouraging for further insights in this area of research.

Much remains to be learned about DNA replication in plants. Important questions concern the role played by specific CDKs and cyclins in the activation of DNA replication and the identification of their substrates, in the identification of replicators and initiation proteins, and how DNA replication is dependent on mitosis in the somatic cell cycle. During plant development, DNA synthesis is often uncoupled from mitosis, resulting in endopolyploid cells. While the biological significance of this is not known, plant endoreplication cycles may represent a unique opportunity to unravel the dependence of S phase on mitosis in normal cell cycles. Also, a deeper understanding of how plant-specific mitogenic signals trigger DNA synthesis is required. For example, it has long been known that auxin and cytokinin induce differentiated cells to enter S phase of the cell cycle. However, it has been difficult to distinguish the influence that phytohormones specifically have on DNA replication from a general induction of cell proliferation.

We have also reviewed Cdc25 phosphatases and their roles in cell cycle progression, mainly in the activation of MPF at the $G_2/M$ phase transition. In contrast to yeast, higher eukaryote cells have multiple Cdc25 proteins which are also differentially expressed during the cell cycle. Thus, Cdc25 proteins may play additional important roles in the cell cycle, and recent evidence points to a potential oncogenic role. Whether higher plants have *cdc25* genes is a subject of lively debate, particularly in view of the fact that no dephosphorylation of tyrosine residues has yet been demonstrated in plants. Early attempts to clone plant *cdc25* homologues by heterologous library screening have failed, but this is not surprising considering the high degree of sequence divergence within the *cdc25* gene family. Using a PCR-based approach, sequences encoding the conserved Cdc25 C-terminus catalytic domain have been cloned from both monocotyledonous and dicotyledonous plants. Two distinct *cdc25*-like sequences have been identified in barley, suggesting that plants, in addition to animals, may possess multiple Cdc25 phosphatases. However, the detection of the corresponding transcripts has proved difficult, presumably due to the narrow window of expression during the cell cycle. This also complicates the cloning of *cdc25* cDNAs from plants, although the final demonstration of the existence of *cdc25* genes in higher plants will require the isolation of full-length cDNA or genomic clones. The availability of PCR sequences represents a promising starting point in this direction.

*The data presented here were obtained in part with the support of a grant from the UK AFRC Plant Molecular Biology I Programme to IACR, Bristol (Ref. No. PG206/517) and from the Singaporean National Science & Technology Board to IMCB, Singapore. IACR receives grant-aided support from the BBSRC (U.K.).*

## References

1. Nurse, P. (1990) Nature (London) **344**, 503–508
2. Jacobs, T.W. (1995) Annu. Rev. Plant Physiol. Plant Mol. Biol. **46**, 317–339
3. Coverly, D. and Laskey, R.A. (1994) Annu. Rev. Biochem. **63**, 745–776
4. Blow, J.J. (1995) in Cell Cycle Control (Hutchinson, C. and Glover, D.M., eds.), pp. 177–205, IRL Press, Oxford
5. Rao, P.N. and Johnson, R.N. (1970) Nature (London) **225**, 159–164
6. Blow, J.J. and Laskey, R.A. (1988) Nature (London) **332**, 546–548
7. Coverly, D., Downes, C.S., Romanowski, P. and Laskey, R.A. (1993) J. Cell Biol. **122**, 985–992
8. Blow, J.J. (1993) J. Cell Biol. **122**, 993–1002
9. Rowley, A., Dowell, S.J. and Diffley, J.F.X. (1994) Biochim. Biophys. Acta **1217**, 239–256
10. DePamphilis, M.L. (1993) Annu. Rev. Biochem. **62**, 29–63
11. Heichman, K.A. and Roberts, J.M. (1994) Cell **79**, 557–562
12. Su, T.T., Follette, P.J. and O'Farrell, P.H. (1995) Cell **81**, 825–828
13. Li, J.J. (1995) Curr. Biol. **5**, 472–475
14. Huberman, J.A. (1995) Cell **82**, 535–542
15. Bryant, J.A. and Dunham, V.L. (1988) Oxford Surveys Plant Mol. Cell Biol. **5**, 23–55
16. Aves, S.J. and Bryant, J.A. (1993) in Molecular and Cell Biology of the Plant Cell Cycle (Ormrod, J.C. and Francis, D., eds.), pp. 45–56, Kluwer Academic Publishers, Amsterdam

17. Springer, P.S., McCombie, W.R., Sundarasen, V. and Martienssen, R.A. (1995) Science **268**, 877–880
17a Sabelli, P.A., Burgess, S.R., Kush, A.K., Young, M.R. and Shewry, P.R. (1996) Mol. Gen. Genet. **252**, 125–136
18. Stichcomb, D.T., Thomas, M., Kelly, J., Selker, E. and Davis, R.W. (1980) Proc. Natl. Acad. Sci. U.S.A. **77**, 4559–4563
19. Newlon, C.S. and Theis, J.F. (1993) Curr Opin. Genet. Dev. **3**, 752–758
20. Marahrens, Y. and Stillman, B. (1994) EMBO J. **13**, 3395–3400
21. Marahrens, Y. and Stillman, B. (1992) Science **255**, 817–823
22. Bell, S.P. and Stillman, B. (1992) Nature (London) **367**, 128–134
23. Hamlin, J.C. and Dijkwel, P.A. (1995) Curr. Opin. Genet. Dev. **5**, 153–161
24. Berlani, R.E., Davis, R.W. and Walbot, V. (1988) Plant Mol. Biol. **11**, 161–172
25. Berlani, R.E., Walbot, V. and Davis, R.W. (1988) Plant Mol. Biol. **11**, 173–182
26. Ohtani, T., Uchimiya, H., Kato, A., Harada, H. and Sugita, M. (1984) Mol. Gen. Genet. **195**, 1–4
27. Eckdahal, T.T., Bennetzen, J.L. and Anderson, J.N. (1989) Plant Mol. Biol. **12**, 507–516
28. van't Hof, J. and Lamm, S.S. (1991) EMBO J. **10**, 1949–1953
29. van't Hof, J. and Lamm, S.S. (1992) Plant Mol. Biol. **20**, 377–382
30. Diffley, J.F.X. and Cocker, J.H. (1992) Nature (London) **357**, 169–172
31. Diffley, J.F.X., Cocker, J.H., Dowell, S.J. and Rowley, A. (1994) Cell **78**, 303–316
32. Bell, S.P., Kobayashi, R. and Stillman, B. (1993) Science **262**, 1844–1849
33. Li, J.J. and Herskowitz, I. (1993) Science **262**, 1870–1874
34. Loo, S., Fox, C.A., Rine, J., Kobayashi, R., Stillman, B. and Bell, S.P. (1995) Mol. Biol. Cell **6**, 741–756
35. Dowell, S.J., Romanowski, P. and Diffley, J.F.X. (1994) Science **265**, 1243–1246
36. Liang, C., Weinreich, M. and Stillman, B. (1995) Cell **81**, 667–676
37. Adachi, Y. and Laemmly, U.K. (1992) J. Cell Biol. **119**, 1–15
38. Masai, H., Miyake, T. and Arai, K. (1995) EMBO J. **14**, 3094–3104
39. Bryant, J.A., Fitchett, P.N., Hughes, S.G. and Sibson, D.R. (1992) J. Exp. Bot. **43**, 31–40
40. Gibson, S.I., Surosky, R.T. and Tye, B.-K. (1990) Mol. Cell. Biol. **10**, 5707–5720
41. Tye, B.-K. (1994) Trends Cell Biol. **4**, 160–166
42. Yan, H., Gibson, S. and Tye, B.K. (1991) Genes Dev. **5**, 944–957
43. Yan, H., Merchant, A.M. and Tye, B.-K. (1993) Genes Dev. **7**, 2149–2160
44. Hennessy, K.M., Clark, C.D. and Botstein, D. (1990) Genes Dev. **4**, 2252–2263
45. Chen, Y., Hennessy, K.M., Botstein, D. and Tye, B.-K. (1992) Proc. Natl. Acad. Sci. U.S.A. **89**, 10459–10463
46. Hennessy, K.M., Lee, A., Chen, E. and Botstein, D. (1991) Genes Dev. **5**, 958–969
47. Dalton, S. and Whitbread, L. (1995) Proc. Natl. Acad. Sci. U.S.A. **92**, 2514–2518
48. Forsburg, S.L. and Nurse, P. (1994) J. Cell Sci. **107**, 2779–2788
49. Coxon, A., Maundrell, K. and Kearsey, S.E. (1992) Nucleic Acids Res. **20**, 5571–5577
50. Miyake, S., Okishio, N., Samejima, I., Hiraoka, Y., Toda, T., Saittoh, I. and Yanagida, M. (1993) Mol. Biol. Cell **4**, 1003–1015
51. Takahashi, K., Yamada, H. and Yanagida, M. (1994) Mol. Biol. Cell **5**, 1145–1158
52. Thommes, P., Fett, R., Schray, B., Burkhart, R., Barnes, M., Kennedy, C., Brown, N.C. and Knippers, R. (1992) Nucleic Acids Res. **20**, 1069–1074
53. Hu, B., Burkhart, R., Schulte, D., Mushahl, C. and Knippers, R. (1993) Nucleic Acids Res. **21**, 5289–5293
54. Todorov, I.T., Peppertok, R., Philipova, R.N., Kearsey, S.E., Ansorge, W. and Werner, D. (1994) J. Cell Sci. **107**, 253–265
55. Kimura, H., Nozaki, N. and Sugimoto, K. (1994) EMBO J. **13**, 4311–4320
56. Burkhart, R., Schulte, D., Hu, B., Musahl, C., Gohring, F. and Knippers, R. (1995) Eur. J. Biochem. **228**, 431–438
57. Someya, A., Shioda, M. and Okuyama, A. (1995) Biochem. Biophys. Res. Commun. **209**, 823–831
58. Starborg, M., Brundell, E., Gell, K., Larsson, C., White, I., Daneholt, B. and Hoog, C. (1995) J. Cell Sci. **108**, 927–934

59. Schulte, D., Burkhart, R., Mushal, C., Hu, B., Schlatterer, C., Hameister, H. and Knippers, R. (1995) J. Cell Sci. **108**, 1381–1389
60. Kubota, Y., Satoru, M., Nishimoto, S., Takisawa, H. and Nojima, H. (1995) Cell **81**, 601–609
61. Madine, M.A., Khoo, C.-Y., Millis, A.D. and Laskey, R.A. (1995) Nature (London) **375**, 421–424
62. Chong, J.P.J., Mahbubani, H.M., Khoo, C.-Y. and Blow, J.J. (1995) Nature (London) **375**, 418–421
63. Koonin, E.V. (1993) Nucleic Acid Res. **21**, 2541–2547
64. Walker, J.E., Saraste, M., Runswick, M.J. and Gay, N.J. (1982) EMBO J. **1**, 945–951
65. Schmid, S.R. and Linder, P. (1992) Mol. Microbiol. **6**, 283–292
66. Ma, Z.W., Bergemann, A.D. and Johnson, E.M. (1994) Gene **149**, 311–314
67. Ferreira, P., Hemery, A., Van Montagu, M. and Inzé, D. (1994) Plant Mol. Biol. **26**, 1289–1303
68. Ruiz-Avila, L., Burgess, S.R., Stiefel, V., Ludevid, M.D. and Puigdomenech, P. (1992) Proc. Natl. Acad. Sci. U.S.A. **89**, 2414–2418
69. Fobert, P.R., Coen, E.C., Murphy, G.J.P. and Doonan, J.H. (1994) EMBO J. **13**, 616–624
70. Starborg, M. and Hoog, C. (1995) Eur. J. Cell Biol. **68**, 206–210
71. Bucci, S., Ragghianti, M., Nardi, I., Bellini, M., Mancino, G. and Lacroix, J.C. (1993) Int. J. Dev. Biol. **37**, 509–517
72. Treisman, J.E., Follette, P.J., O'Farrell, P.H. and Rubin, G.M. (1995) Genes Dev. **9**, 1709–1715
73. Busserau, F., Mallet, L., Gaillon, L. and Jacquet, M. (1993) Yeast **9**, 797–806
74. Wilson, R., Ainscough, R., Anderson, K., Baynes, C., Berks, M., Bonfield, J., Burton, J., Connell, M., Copsey, T., Cooper, J., Coulson, A., Craxton, M., Dear, S., Du, Z., Burbin, R., Favello, A., Fraser, A., Fulton, L., Gardner, A., Green, P., Hawkins, T., Hillier, L., Jier, M., Johnston, L., Jones, M., Kershaw, J., Kirsten, J., Laisster, N., Latreille, P., Lightning, J., Lloyd, C., Mortimore, B., O'Callaghan, M., Parsons, J., Percy, C., Rifken, L., Roopra, A., Saunders, D., Shownkeen, R., Sims, M., Smaldon, N., Smith, A., Smith, M., Sonnhammer, E., Staden, R., Sulston, J., Thierry-Mieg, J., Thomas, K., Vaudin, M., Vaughan, K., Waterson, R., Watson, A., Weinstock, L., Wilkinson-Sproat, J. and Wohldman, P. (1994) Nature (London) **368**, 32–38
75. Broek, D., Bartlett, R., Crawford, K. and Nurse, P. (1991) Nature (London) **349**, 388–393
76. Hayles, J., Fisher, D., Woollard, A. and Nurse, P. (1994) Cell **78**, 813–822
77. Adachi, Y. and Laemmli, U.K. (1994) EMBO J. **13**, 4153–4164
78. Cardoso, M.C., Leonhardt, H. and Nadal-Ginard, B. (1993) Cell **74**, 979–992
79. Strausfeld, U.P., Howell, M., Rempel, R., Maller, J.L., Hunt, T. and Blow, J. (1994) Curr. Biol. **4**, 876–883
80. D'Amato, F. (1977) in Developmental and Cell Biology series (Abercrombie, M., Newth, D.R. and Torrey, J.G., eds.), pp. 102–134, Cambridge University Press, Cambridge
81. D'Amato, F. (1984) in Embryology of Angiosperms (Johri, B.M, ed.), pp. 519–566, Springer-Verlag, New York
82. Hyrien, O., Maric, C. and Mechali, M. (1995) Science **270**, 994–997
83. Hozak, P., Hassan, A.B., Jackson, D.A. and Cook, P.R. (1993) Cell **73**, 361–373
84. Gilbert, D.M., Miyazawa, H. and DePamphilis, M. (1995) Mol. Cell. Biol. **15**, 2942–2954
85. Sparvoli, E., Levi, M. and Rossi, E. (1994) J. Cell Sci. **107**, 3097–3103
86. Xie, Q., Suarez-Lopez, P. and Gutierrez, C. (1995) EMBO J. **14**, 4073–4082
87. Soni, R., Carmichael, J.P., Shah, Z.H. and Murrayt, J.A.H. (1995) Plant Cell **7**, 85–103
88. Dunphy, W.G. (1994) Trends Cell Biol. **4**, 202–207
89. Russell, P. and Nurse, P. (1986) Cell **45**, 145–153
90. Russell, P., Moreno, S. and Reed, S.I. (1989) Cell **57**, 295–303
91. Millar, J.B.A., McGowan, C.H., Lenaers, G., Jones, R. and Russell, P. (1991) EMBO J. **10**, 4301–4309
92. Sebastian, B., Kakizuka, A. and Hunter, T. (1993) Proc. Natl. Acad. Sci. U.S.A. **90**, 3521–3524
93. Dunphy, W.G. and Kumagai, A. (1991) Cell **67**, 189–196
94. Kumagay, A. and Dunphy, W.G. (1992) Cell **70**, 139–151
95. Millar, J.B.A. and Russell, P. (1992) Cell **68**, 407–410
96. Moreno, S. and Nurse, P. (1991) Nature (London) **351**, 194
97. Jinno, S., Suto, K., Nagata, A., Igarashi, M., Kanaoka, Y., Nojima, H. and Okayama, H. (1994) EMBO J. **13**, 1549–1556
98. Izumi, T., Walker, D. and Maller, J.L. (1992) Mol. Biol. Cell **3**, 927–939

99.   Hoffmann, I., Clarke, P.R., Marcote, M.J., Karsenti, E. and Draetta, G. (1993) EMBO J. **12**, 53–63

100.  Strausfeld, U., Fernandez, A., Capony, J.-P., Girard, F., Lautredou, N., Derancourt, J., Labbe, J.-C. and Lamb, N.J.C. (1994) J. Biol. Chem. **269**, 5989–6000

101.  Galaktionov, K. and Beach, D. (1991) Cell **67**, 1181–1194

102.  Zheng, X.-F. and Ruderman, J.V. (1993) Cell **75**, 155–164

103.  Izumi, T. and Maller, J.L. (1995) Mol. Biol. Cell **6**, 215–226

104.  Kuang, J., Ashorn, C.L., Gonzalez-Kuyvenhoven, M. and Penkala, J.E. (1994) Mol. Biol. Cell **5**, 135–145

105.  Lucibello, F.C., Truss, M., Zwicker, J., Ehlert, F., Beato, M. and Muller, R. (1995) EMBO J. **14**, 132–142

106.  Zwicker, J., Lucibello, F.C., Wolfrain, L.A., Gross, C., Truss, M., Engeland, K. and Muller, R. (1995) EMBO J. **14**, 4514–4522

107.  Kakizuka, A., Sebastian, B., Borgmeyer, U., Hermans-Borgmeyer, I., Bolado, J., Hunter, T., Hoekstra, M.F. and Evans, R.M. (1992) Genes Dev. **6**, 578–590

108.  Edgar, B.A., Lehman, D.A. and O'Farrell, P.H. (1994) Development **120**, 3131–3143

109.  Wickramasinghe, D., Becker, S., Ernst, M., Resnick, J.L., Centanni, J.M., Tessarollo, L., Grabel, L. and Donovan, P.J. (1995) Development **121**, 2047–2056

110.  Bissen, S.T. (1995) Development **121**, 3035–3043

111.  Galaktionov, K., Jessus, C. and Beach, D. (1995) Genes Dev. **9**, 1046–1058

112.  Galaktionov, K., Lee, A.K., Eckstein, J., Draetta, G., Meckler, J., Loda, M. and Beach, D. (1995) Science **269**, 1575–1577

113.  Conklin, D.S., Galaktionov, K. and Beach, D. (1995) Proc. Natl. Acad. Sci. U.S.A. **92**, 7892–7896

114.  Nargi, J. and Woodford-Thomas, T.A. (1994) Immunogenetics **39**, 99–108

115.  Sadhu, K., Red, S.I., Richardson, H. and Russell, P. (1990) Proc. Natl. Acad. Sci. U.S.A. **87**, 5139–5143

116.  Nagata, A., Igarashi, M., Jinno, S., Suto, K. and Okayama, H. (1991) New. Biol. **3**, 959–968

117.  O'Connell, M.J., Osmani, A.H., Morris, N.R. and Osmani, S.A. (1992) EMBO J. **11**, 2139–2149

118.  Edgar, B. and O'Farrell, P.H. (1989) Cell **57**, 177–187

119.  Jimenez, J., Alphey, L., Nurse, P. and Glover, D.M. (1990) EMBO J. **9**, 3565–3571

120.  Alphey, L., Jimenez, J., White-Cooper, H., Dawson, I., Nurse, P. and Glover, D.M. (1992) Cell **69**, 977–988

121.  Bell, M.H., Halford, N.G., Ormrod, J.C. and Francis, D. (1993) Plant. Mol. Biol. **23**, 445–451

# 13

# Histone gene expression

## Nicole Chaubet and Claude Gigot*

Institut de Biologie Moléculaire des Plantes, CNRS, Université Louis Pasteur, 12, rue du Général Zimmer, 67084 Strasbourg Cedex, France

## Introduction

The co-ordinated activation and arrest of various meristems are responsible for controlling the major morphogenetic changes that occur during plant development. Therefore, identifying genes which control the progression of meristematic cells through the cell cycle, and elucidating the mechanisms which induce cell cycle-specific gene expression, is of major interest.

A very detailed picture of the events regulating cell cycle progression in eukaryotes has recently emerged from a multiplicity of results obtained in yeast [1,2] and in animals [3,4]. In both systems, the transition from $G_1$ to S phase has been shown to be a crucial checkpoint of cell cycle control, where a cohort of proteins necessary for DNA replication and chromatin reconstruction are synthesized ([5]; see P. Sabelli et al., chapter 12). Histones represent the largest group among these proteins, the massive synthesis of which is necessitated during S phase to associate with the nascent DNA. Thus, the genes encoding the five major classes of replication-dependent histones—H1, H2A, H2B, H3 and H4—represent an excellent model to study the mechanisms triggering co-ordinated gene expression at this very crucial checkpoint of the cell cycle.

In animals, the biosynthesis of histones is regulated at several levels, and by multiple, and rather complex, mechanisms [6]. The typical replication-dependent genes encoding histones are exceptional among class II genes in that they are devoid of introns, and are transcribed into non-polyadenylated mRNAs. Their steady-state amounts are regulated at three different levels: (i) initiation of transcription, (ii) 3′-processing of the pre-mRNAs and (iii) degradation of the mature mRNAs. Positive elements involved in transcriptional regulation during the cell cycle have been defined in the promoter regions of some animal histone genes [7,8], and both positive and negative regulatory elements have been found in the promoter regions of the yeast H2A and H2B genes [8]. Intriguingly, in the animal kingdom, the elements responsible for cyclic transcription are strikingly variable among species, and even among the five classes of histone genes within the same species [8]. These cis-elements are generally sufficient to confer cell-cycle-specific expression by themselves, but account only for an approximate 5-fold increase in the steady-state

*To whom correspondence should be addressed

level of histone mRNAs during S phase. Increases in both the post-transcriptional 3'-processing and in the mRNA half-life, involving a characteristic hairpin structure in the 3'-untranslated region (UTR) of the pre-mRNAs, result in an additional 5–10-fold increase in the amounts of histone mRNAs during S phase [9].

In addition to this major type of replication-dependent gene expression, a few histone genes are transcribed constitutively at a low rate. They encode the so-called 'replacement histones' which accumulate in the chromatin of non-proliferating tissues. These genes behave like ordinary protein-coding genes, giving rise to polyadenylated mRNAs, and several of them contain introns. Additional minor types of histone genes exist, encoding developmental or tissue-specific variants [10].

In the plant kingdom, histone genes and cDNAs have been cloned from more than 10 different species [11]. They differ from their animal counterparts by several characteristics. First, each multigene family encoding one particular class of core histones (i.e. H2A, H2B, H3 or H4) is organized into multiple subfamilies with a specific environment [12], which are dispersed on different chromosomes [13], while animal histone genes are often clustered within rather short chromosomal arrays. Secondly, as found in lower eukaryotes and the replication-independent histone genes of animals, they are transcribed into polyadenylated mRNAs with long 3'-UTRs [14–16] lacking the palindromic structure specific to the animal replication-dependent histone genes. Hence, the mechanisms which specifically regulate the levels of animal histone mRNAs during the cell cycle at the level of pre-mRNA processing and stability of mature mRNAs do not operate in plants. This suggests that, essentially, expression of plant histone genes may be regulated at a transcriptional level. Therefore, functional approaches based on promoter analysis in transgenic plants are often used to highlight the elements regulating cyclic plant histone gene expression.

## Expression pattern of plant histone genes

Studies of histone gene expression in plant cells suffered for a while from an absence of a suitable cell-synchronization system. It was therefore restricted for several years to a very general evaluation of organ-specific mRNA steady-state levels, and *in situ* hybridization approaches. Recently, synchronizable alfalfa [17], tobacco BY2 [18] and rice [19] cell suspensions allowed more sophisticated and detailed studies on cell cycle regulated expression, and opened new perspectives in plant cell cycle research.

### Expression in developing plant tissues

Northern analysis during germination of maize seeds showed that histone H2B, H3 and H4 mRNAs accumulate in parallel with the onset of DNA synthesis from approximately 12 h after imbibition, and peak after 4 days [20,21]. Under the same conditions, histone H1 mRNA is somewhat delayed, but afterwards shows the same expression pattern as the core histone genes [22]. Similar observations have been described in germinating wheat [23]. Co-ordinated histone gene expression was also shown in organs from several species, including maize [20–22], alfalfa [24] and *Arabidopsis* [25]. Northern analysis revealed high steady-state levels of histone mRNAs in organs with high mitotic activity, and very low amounts in fully developed organs of adult plants. A more precise expression pattern was inferred from *in situ* hybridization studies, showing that histone mRNAs accumulated in proliferating tissues, such as root tips and apical meristems [26–28]. The spotty expression pattern observed within these tissues was indicative of periodic cell-cycle-coupled transcription. This was elegantly demonstrated by using a double-labelling technique allowing a co-localization of the cells undergoing DNA synthesis during S phase, and those expressing histone mRNA [29]. However, histone gene expression was also noticed in some non-dividing cells or tissues, such as endothecium of the anthers of cereals and endosperm and elongation zones of the roots, some of which have been proposed to undergo rounds of endoduplication without cellular division [26,27,29].

## Expression during the cell cycle

Northern analysis during the cell cycle of alfalfa, rice and tobacco suspension-cultured cells clearly demonstrated a strict temporal correlation between histone gene expression and the S phase of the cell cycle [17,30,31]. The histone mRNA level outside of S phase was estimated to be around 5–10% of the maximum level reached in S phase. However, careful analysis showed that, in contrast to animals, the amount of histone mRNAs was not directly coupled to the DNA synthesis level [31]. Indeed, interrupting ongoing DNA synthesis at mid-S-phase led to the maintainance of histone mRNAs at high levels for at least 12 h, and measurements of both mRNA half-life and transcription rate showed that this phenomenon was not due to a stabilization of the transcripts, but to a sustained transcription rate (N. Chaubet and C. Gigot, unpublished data). This observation suggested that the signals directing the arrest of histone gene transcription might be connected to the natural completion of chromosome replication. Crucially, slowing down DNA replication and, thus, delaying S-phase completion by using low doses of blocking agent led to a similar delay in the disappearance of histone mRNAs.

Interestingly, uncoupled histone gene expression and the rate of DNA synthesis were also noticed at the entry into S phase. In the presence of the DNA polymerase inhibitor, aphidicolin, high levels of histone mRNAs accumulated in the absence of DNA synthesis, suggesting that induction of histone gene expression takes place slightly before the onset of DNA replication. This was also inferred from results of double-labelling experiments *in situ* for histone mRNA and DNA synthesis in pea-root tips [29], and noticed in a yeast *cdc* mutant arrested at a particular point at the $G_1$/S transition (32).

On the other hand, hydroxyurea, which blocks DNA synthesis by inhibiting the synthesis of deoxyribonucleoside triphosphates (dNTPs), and is therefore assumed to act earlier in the cell cycle progression towards S phase, inhibited DNA synthesis and histone gene induction co-ordinately [31]. We conclude that induction of histone gene expression occurs temporally between the two particular stages where these two inhibitors arrest the cell cycle progression around the $G_1$/S transition.

Interestingly, although the amount of histone mRNA does not strictly parallel the rate of DNA synthesis, particular mechanisms exist which prevent an overproduction of histones in the case of interrupted DNA synthesis. These mechanisms act at a translational level by preventing the translation of excess histone mRNAs (N. Chaubet and C. Gigot, unpublished data).

# Functional analysis of histone promoters in transgenic plants

Several systems were tentatively used by different authors to characterize plant histone promoters with respect to their level of activity and/or their specificity of expression in meristems and throughout the cell cycle. Transcriptional fusions between histone promoters and the β-glucuronidase (GUS) coding region were tested in transient assays, or after integration into the plant genome.

## Transient expression

Transient expression studies of *Arabidopsis* and maize histone gene promoters in cultured tobacco mesophyll protoplasts revealed a basal activity of the promoters, but failed to reveal any additional induction in parallel with DNA synthesis after 24 h of culture [33]. This negative result was explained because, in this system, transcription of the transfected constructs takes place only during the first 24 h of culture, before the onset of DNA synthesis. However, other attempts to detect the cell cycle dependent activity of histone promoters by transfecting protoplasts at later stages of cell culture under conditions of active DNA synthesis were also unsuccessful (C. Gigot et al., unpublished data). We postulate that the cell-cycle-dependent activity of histone promoters is only detectable once the transgene has integrated into chromatin. Indeed, mesophyll protoplasts isolated from transgenic tobacco plants and cultivated under the same conditions as in the transient expression studies displayed both a basal, cell cycle independent level of expression, and a replication-dependent expression occurring in parallel with DNA synthesis and inhibited by the DNA synthesis inhibitor, hydroxyurea [17,33].

In spite of this restriction inherent to the system, transient expression in tobacco, rice or wheat protoplasts was often used to reveal regions in the promoters which might play a role in the promoter activity [34–36]. The involvement of these regions in the transcriptional regulation was then further, and more thoroughly, demonstrated by testing the constructs in transgenic plants.

## Expression in transgenic plants

An *Arabidopsis* H4 promoter was studied in both transgenic *Arabidopsis* and tobacco plants [25,33]. Histone-promoter-directed GUS expression was maximal in meristematic tissues, such as shoot and root apices, lateral root primordia, flower buds and carpels, young leaves, axillary buds and stem cambial tissue, and low in mature tissues, such as fully developed leaves and non-cambial stem tissue. The GUS activity in meristems is assumed to correspond to the cell-cycle-dependent activity of the histone promoter, whereas the low expression in non-dividing tissues is assumed to correspond to the replication-independent activity. Notably, *in situ* hybridization studies revealed only some cells containing histone mRNA within a meristem, corresponding to the cells undergoing S phase during the time-course of the experiment. However, histochemical localization of GUS activity led to uniform staining of the whole meristem owing to the longer persistence in the cells of the very stable GUS protein, in comparison to the short-lived histone mRNAs. GUS expression also occurred to some extent in cotyledons at early stages of seedling development, and in the submeristematic part of roots. This expression was attributed to the capacity of the non-dividing cells in these particular tissues to undergo endoreduplication of their DNA. For example, in *Arabidopsis*, ploidy levels up to 32C exist in cotyledons, as well as in many other tissues [37], and DNA endoreduplication was shown in the elongation zone of pea roots [27].

A maize histone H3 promoter was also studied in the two dicot plants mentioned above. In tobacco, although the preferential GUS expression in meristems was clearly apparent, the maize promoter was less efficient than the *Arabidopsis* one in all the tissues [38]. In contrast to the *Arabidopsis* promoter, intense expression was observed in mature pollen. This may be characteristic of monocots (see below), since *in situ* hybridization experiments detected expression of histone genes in mature maize pollen grains (C.Gigot, unpublished data). Intriguingly, the expression pattern of the same promoter in *Arabidopsis* was not exactly the same as in tobacco (C.Gigot, unpublished data). While expression in pollen was as high as in tobacco, expression in the pistil and in root meristems was rather faint and, in lateral root primordia, it was essentially concentrated at the branching of the pericycle. Intense expression was also found in stipules, at the bases of axillary buds and bracts of the inflorescences. This different behaviour between the two types of transgenic plants might reflect quantitative as well as qualitative differences in the histone-specific transcription factors in some tissues which may be of minor importance for detecting expression of a dicot promoter, but of major importance for a monocot one.

Study of a wheat H3 promoter in a transgenic monocot plant, rice, also led to similar results, including expression in mature pollen, but with some particularly surprising sites of expression in non-dividing tissues, such as the root cap and the endothecium of anthers [30]. Activities of alfalfa H3 and wheat H2B promoters were also studied in transgenic tobacco [17,35], and led to similar results as the *Arabidopsis* and maize promoters.

## Expression in synchronized transgenic cells

Measurement of GUS enzymic activity during the cell cycle in partially synchronized rice cells evaded any detection of a change in relation to a particular phase of the cycle, because the GUS protein remains stable even outside the phases when it is expressed [19]. Coupling of transgene expression to the cell cycle in rice or alfalfa cells was thus assessed by measuring the GUS mRNA level by Northern blots, or by S1-nuclease mapping [17,19]. In both cases, histone-promoter-directed expression was restricted to S phase, thus confirming the observations already made on transgenic plants that the promoter alone is sufficient to confer faithful expression to the reporter gene. Slight differences were observed at the end of S phase in these two systems, the decay of GUS mRNA occurring earlier (in rice) or later (in alfalfa) than the decay of the endogenous histone transcripts. A discrepancy between expression of endogenous histone genes and histone-promoter-directed GUS expression was found in the case of rice cells arrested at the $G_1/S$ transition by an aphidicolin treatment before synchronization, where the GUS mRNA level was noticeably lower than that of the endogenous histone transcripts, when compared with the maximal level reached one hour after the release of the drug. This was also observed in synchronized transgenic tobacco cells carrying an *Arabidopsis* promoter/GUS fusion gene (C. Gigot and N. Chaubet, unpublished data). Whether these differences relate to the chromosomal environment of the transgene, or to the absence of the homologous terminators, needs further investigation. In similar experiments, alfalfa cells had been synchronized by a treatment with very low doses of hydroxyurea, and, consequently, arrested at a later point within S phase, thus preventing a similar observation.

## Identification of *cis*-acting elements

*cis*-Elements of plant histone promoters were tentatively identified by two kinds of approaches. The structural approach takes advantage of the binding of *trans*-acting proteinaceous factors to the *cis*-regulatory elements of the promoter. This binding can be revealed *in vitro* by gel-mobility-shift assays, and the DNA targets localized by competition experiments and *in vitro* footprinting. The binding can also be revealed *in vivo*, either by a rough investigation of the chromatin structure of the promoter with nucleases, or by *in vivo* footprinting experiments which enable detection, at the nucleotide level, of the sites of interaction between *trans*-acting factors and DNA regulatory sequences. The functional approach consists of assaying transgenic material for the activity of promoter/GUS fusions carrying deletions or mutations in the hypothetical regulatory regions or sequences of the promoter.

## Structural approach

*In vivo* studies of the chromatin structure of a maize H3 promoter using micrococcal nuclease or DNaseI digestion contributed towards delimiting a promoter region of about 200 bp displaying an increased accessibility to nucleases, and the detection of several hypersensitive sites within this region [38]. These sites were located in the vicinity of three remarkable sequence elements: (i) the universal TATA box, (ii) a highly conserved plant histone gene-specific octameric sequence CGCGGATC present in a direct or reverse orientation, and (iii) a nonamer of consensus sequence CCATCCAAC fairly conserved in plant histone promoters. These observations suggested that transcription factors bound to the promoter on, or close to, these sequence elements.

*In vivo* footprinting experiments with dimethylsulphate (DMS) or DNaseI on maize H3 and H4 promoters visualized the overall surfaces (with DNaseI), as well as the intimate points (with DMS) of interaction between proteins and the *cis*-elements of these two promoters [39]. It confirmed that the octameric and nonameric sequences were targets for *trans*-acting factors *in vivo*. It also revealed other *cis*-elements interacting with proteins which proved, upon sequence comparison, to be common to most of the plant histone promoters (Fig. 1): (i) a conserved CCGTCC sequence located 8–10 bp upstream from the conserved nonamer (NON) and 10–30 bp downstream from the conserved octamer (OCT); (ii) a degenerate copy of the octamer (dOCT) differing by only one bp from the ideal copy (often at the second position), which exists in many plant histone gene promoters about 50 bp up- or down-stream from the ideal copy; (iii) a sequence

---

**Fig. I**      ***Cis*-elements in plant histone promoters**

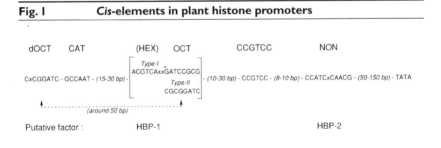

The sequences and abbreviated names of the cis-elements common to the majority of plant histone promoters are indicated, together with their relative positions with respect to each other. The two main forms of the highly conserved plant histone octamer (OCT, types I and II) are shown. The cis-elements can be found in either direct or reverse orientation, except the CCGTCC and nonamer (NON) elements, which are always in direct orientation, and the octamer in the type-I element, which is always in reverse orientation (indicated by a star). The CCAAT and degenerate octamer (dOCT) are represented upstream from the OCT element, but they may be found downstream at equivalent distances. Putative transcription factors, either already isolated or only detected in vitro, are indicated (see text for further details).

analogous to the animal GCCAAT box (CAT) which is also found in many plant histone gene promoters generally 15–30 bp upstream from the octamer; and finally, (iv) a hexameric sequence ACGTCA (HEX), which exists in the promoters of various nuclear, viral and T-DNA genes, and which is present in some of the plant histone gene promoters in close association with a reverse-oriented octamer. The association of these two motifs has been designated 'type-I element' [23]. Consequently, the octamer non-associated to the hexamer is sometimes designated 'type-II element'.

In vivo footprinting experiments on an Arabidopsis H4 promoter, showed that the cis-elements revealed in maize are also binding trans-acting factors in Arabidopsis (C. Gigot et al., unpublished data). Furthermore, the protein footprints detected over the conserved octamer were quite similar in maize and Arabidopsis, suggesting a particular conservation of the trans-acting factor in the plant kingdom, and therefore a crucial function in transcriptional regulation.

Differences in the footprints were observed between dividing and non-dividing cells [39]. The nonamer and CCGTCC sequences only bound proteins in proliferating cells, whereas the hexamer, octamer and CCAAT box were 'constitutively' protected by proteins, whatever the proliferative state of the cells. The footprints observed over these 'constitutive' cis-elements, however, were much stronger in proliferating than in quiescent cells, which may be indicative of protein–protein interactions between the transcription factors they bind and the factors binding to the 'cell-division-inducible' cis-elements.

In vitro binding assays led to the characterization of several nuclear factors binding to particular sequences in wheat H3 or H4 promoters (see M. Iwabuchi et al., chapter 14). Histone-promoter-binding proteins (HBPs)-1a and -1b were shown to bind to the hexameric sequence ACGTCA [40]. cDNAs corresponding to these proteins were isolated by Southwestern screening, and sequence analysis of the corresponding proteins revealed, in both, a basic/leucine-zipper (bZIP) domain important for binding and dimerization [41,42]. Other proteins have been reported to bind to single-stranded DNA in the region around the type-I element, but no cDNA is available yet for these proteins [43]. The nonameric sequence was found to be a target for a protein named HBP-2 [44], but no corresponding cDNA could be isolated. Instead, a cDNA was isolated which codes for a zinc-finger protein (WZF-1) binding to a wheat-H3-promoter-specific CACTC sequence located 6 bp upstream from the nonameric sequence [45]. Whether one, or several of, these proteins is/are involved in the regulation of the S-phase-dependent or -independent activity of histone promoters in vivo remains, as yet, unknown.

In vitro binding assays with nuclear extracts of synchronized tobacco BY2 cells indicate that, in addition to the hexamer and nonamer sequences, the region encompassing the highly conserved plant histone octamer CGCGGATC is also able to bind several proteins, and that some of the proteins binding to the histone promoter are subjected to phosphorylation/dephosphorylation mechanisms at particular phases of the cell cycle (C. Gigot et al., unpublished data).

## Functional approach

We wished to determine which characteristic motifs are candidates for regulating the cell-cycle-dependent or -independent activity of plant histone promoters. Several attempts were undertaken, using various plant materials and different methods, to measure the promoter activity after *in vitro* modification of the promoter, by deletion series or by site-directed mutational changes.

### Transient expression

The effect of disrupting some of the above-cited *cis*-elements in shortened (i.e. deleted just upstream of the approximately 200 bp region containing all the identified remarkable motifs) wheat H3 or maize H4 histone promoters was analysed by transient expression assays in wheat, rice or tobacco protoplasts [34,36,46]. Mutation of the widespread hexamer, or the closely linked reverse octamer, led to an approximate 50% decrease in the wheat promoter activity, but it had no significant effect on the maize promoter. However, it should be pointed out that the maize promoter was studied in protoplasts of a dicot plant (tobacco) whereas the wheat promoter was studied in monocot protoplasts, which might explain the discrepancy between the results. The nonameric sequence was the only motif out of the most conserved ones whose mutation perturbed the promoter activity (5-fold decrease) in the maize promoter/tobacco protoplasts system, but unfortunately the effect of mutating this element within the wheat promoter was not tested.

The effects of 5′-deletions in the promoter regions were also studied using the same systems. Several regions with positive or negative influence were found to exist in the wheat promoter region, far upstream from the minimal region containing the remarkable sequences [47]. Some of these regions had the same effect in monocot- and dicot- protoplasts, but others had either no influence, or opposite effects, on the promoter activity in the tobacco protoplasts system compared with the rice system. Furthermore, some regions were able to function in an orientation-independent manner in the monocot protoplasts, whereas their effect was orientation-dependent in the dicot protoplasts. Interestingly, two redundant *cis*-elements already present in the proximal region of the promoter (a CCAAT box and the degenerate octamer) were also present in the upstream fragment producing the strongest activation.

5′-Deletions were also studied within the short promoter region containing the conserved *cis*-elements [46]. Successive deletions in the 36 bp region containing the hexamer and the closely associated octamer produced fluctuations in the promoter activity, thus indicating an interplay between positive and negative regulatory factors at the level of these two sequences, at least in protoplasts. Further deletion, including the nonamer, led to a marked decrease in the activity, which was in agreement with the result obtained upon base-substitution in the nonamer. Hence, the nonamer element was essential for at least replication-independent expression.

## Expression in transgenic plants

Two plant histone promoters were investigated in detail: (i) an *Arabidopsis* H4 promoter in the corresponding homologous plant [5] and (ii) a wheat histone promoter in transgenic rice [30, 36]. Successive deletions in the 5'-region indicated that, for both promoters, a minimal region of about 200 bp, comprising the mRNA start site and the conserved *cis*-elements, was able to mimic the tissue-specific expression driven by the full-length promoter (except for the expression of the wheat promoter in the root cap, which was due to far upstream regions). The ratio between the promoter activity driven by the short promoter in dividing, and non-dividing, tissues was almost the same as the full-length promoter, indicating that all the elements necessary for specific expression were contained in this short promoter fragment. In the case of the wheat promoter, the proximal promoter region also retained a capacity to direct the expression in the particular non-dividing tissues mentioned above: endothecium of anther, and mature pollen of rice. We conclude that the *cis*-elements responsible for cell-division-dependent and cell-division-independent activity are not physically separated in the promoter sequence. However, for both promoters, upstream sequences were necessary to achieve the final expression level; the activity driven by the 200 bp promoter was about half that of a 1 kb promoter fragment.

Mutation of the wheat hexamer, or octamer, caused a reduction of the promoter activity in stably transformed rice calli, confirming that these two motifs are positive *cis*-elements. However, GUS histochemical assays suggested only the mutation of the octamer reduced the GUS expression level in the root-tip meristem. Neither of these two motifs appeared to play a role in the cell-division-independent expression occurring in anthers. In contrast, these two motifs influenced the promoter activity in transient expression assays which are assumed to measure the replication-independent activity (as discussed above). Hence, we cannot yet conclude about the relative involvement of these two *cis*-elements in the cell-division-dependent, or -independent, expression.

A complete set of mutational changes in all the *cis*-elements revealed by *in vivo* footprinting experiments was performed with the *Arabidopsis* H4 promoter, and analysed in various tissues of transgenic *Arabidopsis* (C. Gigot et al., unpublished data). All five *cis*-elements of this promoter, i.e. octamer, degenerate octamer, nonamer, CCAAT box and CCGTCC motif, behaved as positive *cis*-elements in all the tissues (dividing and non-dividing), but to different extents. The element of greatest importance was the nonamer, whose mutation caused a reduction of the promoter activity to 7% in non-dividing tissues, such as fully expanded rosette leaves. This confirmed the drastic reduction observed upon deletion or mutation of this element in transient assays (see above). However, the reduction was even more severe in many dividing tissues such as flower buds (4%

activity). Moreover, GUS activity in meristems was almost undetectable histochemically, indicating that this element is necessary for specific expression. A similar conclusion could be drawn from the mutation of the closely positioned CCGTCC motif, although the decrease of the promoter activity was less. GUS activity could be detected histochemically with this construct, but with abnormal tissue-specificity, especially in roots, where it was concentrated to vessel junctions. The other *cis*-elements apparently had less influence on the tissue-specificity of the promoter. However, upon octamer disruption, a delay in the GUS expression was noted in meristems of young seedlings, thus suggesting a developmental regulatory function for this motif. The importance of these three *cis*-elements—nonamer, CCGTCC motif and octamer—was further demonstrated by a combination of double mutations, which led to various expression patterns, ranging from an almost constitutive expression in young seedlings and loss of any expression at the later stages, to abnormally strong and exclusive expression in the non-dividing cells of the root cap.

## Tentative model for plant histone gene transcriptional regulation

Although studies on the mechanisms regulating histone gene transcription are still in progress, we have now gained a good knowledge of the *cis*-elements present in histone promoters, and we are starting to get an insight into the relative functional interactions between the different DNA/protein transcriptional complexes. A schematic model of histone promoter architecture, taking into account the results obtained by different approaches, is presented in Fig. 2. The five conserved *cis*-elements, and their respective putative *trans*-acting factors, are indicated. All five *cis*-elements were shown to function synergistically as positive elements to achieve the final expression level in proliferating cells. In addition, some of them also had a predictable influence on the expression pattern of the histone promoter [51].

The CCAAT-like box (CAT) contributed towards an enhanced promoter activity, but had apparently no influence on the specificity. It is thus assumed to play a secondary role, probably by stabilizing the structure of the final multiprotein complex.

The plant histone-specific octamer (OCT) and its degenerate copy (dOCT) are generally separated from each other by a fixed distance and they can substitute for each other, at least in part. Furthermore, according to the observed footprints, they bind identical, or very similar, proteins. It is thus likely that the complexes over these two elements have equivalent functional interactions with those interacting with the other *cis*-elements up- and down-stream. This is the reason why the promoter is schematically represented in Fig. 2 as a bent line, with the two octamers facing each other. The absolute conservation of the octamer in all the plant species analyzed so far, the redundancy of this element in most of the plant

**Fig. 2**          **Tentative model for plant histone transcriptional induction complex**

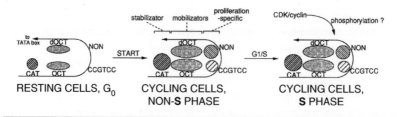

Cis-elements and their occupancy by putative transcription factors are indicated at different steps of the progression of cells from $G_0$ to S phase. It is postulated that proliferation-specific factors bind to the CCGTCC and NON elements, upon mobilization and stabilization by a multicomponent complex sitting on the OCT and CAT cis-elements. At the $G_1/S$ transition, new protein–protein interactions and/or post-translational modifications, such as phosphorylation by a $G_1/S$ phase-specific CDK/cyclin complex, direct an additional increase of the transcriptional level.

histone promoters, and the dramatic effect obtained upon mutation of both copies, suggest that this element plays a key role in histone promoter transcriptional regulation. Given that proteins are loosely bound to these elements in quiescent or blocked cells, but strongly bound in proliferating cells, and that the sequence is necessary to promote expression in newly initiated meristems, we postulate that the role of the octamer is to mobilize the proliferation-specific factors binding to the nonamer and CCGTCC cis-elements.

It is intriguing that two main types of histone promoters exist in plants relative to the octamer: the 'type-I' promoters, in which a reverse copy of the octamer is closely associated to a hexamer ACGTCA that is not specific to plant histone promoters, and the so-called 'type-II' promoters, which do not contain the hexamer (see Fig. 1). The 'type-I' promoters represent about 20% of the histone promoters sequenced up to now, and have been found in several plant species. The octamer and hexamer sequences in the 'type-I' element are so close to each other (2 bp) that it can be postulated that they constitute a single functional block. Since the two types of promoters apparently direct the same kind of expression pattern in plants as well as during the cell cycle, we can therefore speculate on the reasons for this duality in histone promoters, and the respective roles of the octamer in these two types of promoters.

The nonamer (NON) and CCGTCC elements were shown to bind proteins only in proliferating cells. They were also shown to be directly involved in meristem-specific expression as quantitative and/or qualitative modulators. Both results provide evidence that these two elements are targets for proliferation-specific factors. This is in agreement with sequence data, showing that the nonameric element is absent from the promoters of histone genes encoding replacement histone H3 variants which do not show preferential expression in meristems [48,49]. However, the involvement of these elements in the meristematic expression does not necessarily imply their involvement in the S-phase-specific expression owing to the long half-life of the GUS protein. Indeed, it should be kept in mind that histochemically revealed promoter-directed GUS activity in meristems may, or may not, relate to a cell cycle phase-specific expression.

We thus postulate that histone gene transcription is regulated at several levels of induction (Fig. 2):

- the passage from a quiescent, non-proliferative $G_0$ state to a cycling state, with binding of proliferation-specific factors on the nonamer and CCGTCC *cis*-elements,

- the transition from the $G_1$ to the S phase of the cell cycle, probably mediated, as suggested by preliminary results (C. Gigot et al., unpublished data), by new protein–protein interactions, or by post-translational modifications of the prebound proteins.

## Concluding remarks

A great deal of effort has been made during the past decade to investigate plant histone gene expression, especially at the level of histone promoter transcriptional regulation. The results obtained by several groups have now led to a sound knowledge of the whole set of *cis*-elements involved in transcriptional regulation, although their respective contributions in basal, cell-division-induced or S-phase-induced expression still remains unclear. The involvement of each individual *cis*-element in the S-phase-coupled expression is now being studied in synchronized cells, and should help us to understand the connections between induction of histone genes (and potentially, other co-ordinately expressed genes), and the general cell cycle machinery, namely cyclin/cyclin-dependent kinase (CDK) complexes which control the progression through $G_1$ to S phase. In this notable respect, the multicomponent transcription factor HinF-D interacting with multiple recognition motifs in several human histone promoters contains CDC2, cyclin A and a retinoblastoma (Rb)-like protein as associated subunits [50]. The presence of cell-cycle-related mediators in this histone transcription factor thus suggests a direct link between transcriptional regulation of histone expression and the transition from $G_1$ to S phase. It could also help us to understand how the expression of a series of S-phase-specific genes, such as those encoding enzymes involved in DNA replication, is co-ordinately induced during this crucial step of the cell cycle.

Whether similar multiprotein complexes, and mechanisms coupling histone gene transcription to the cell cycle machinery, are also active in plant cells remains to be demonstrated. A major difference is already apparent between plant and animal histone promoters at the DNA sequence level, since several conserved *cis*-elements with given positioning exist in plants whereas no particular sequence is common to animal histone promoters. It will be of major interest to isolate the whole set of plant histone promoter-binding factors previously detected *in vitro* or *in vivo*, and to investigate their post-translational modifications, as well as their interactions between each other and with cell cycle components throughout the cell cycle. Some of these binding factors have been already identified and isolated in pioneering work by M. Iwabuchi and his group, and their potential role in histone transcriptional regulation are described and discussed in detail in the following chapter.

## References

1.  Nurse, P. (1990) Nature (London) **344**, 503–507
2.  Nasmyth, K. (1993) Curr. Opin. Cell Biol. **5**, 166–179
3.  Pines, J. (1993) Trends Biol. Sci. **18**, 195–197
4.  Sherr, C.J. (1993) Cell **73**, 1059–1065
5.  Johnson, L.F. (1992) Curr. Opin. Cell Biol. **4**, 149–154
6.  Schümperli, D. (1988) Trends Genet. **4**, 187–191
7.  Heintz, N. (1991) Biochim. Biophys. Acta **1088**, 327–339
8.  Osley, M.A. (1991) Annu. Rev. Biochem. **57**, 349–374
9.  Birnstiel, M.L., Büsslinger, M. and Strub, K. (1985) Cell **41**, 349–359
10. Old, R.W. and Woodland, H.R. (1984) Cell **38**, 624–626
11. Chaboute, M.E., Chaubet, N., Gigot, C. and Philipps, G. (1993) Biochimie **75**, 523–531
12. Chaubet, N., Philipps, G. and Gigot, C. (1989) Mol. Gen. Genet. **219**, 404–412
13. Chaubet, N., Philipps, G., Gigot, C., Guitton, C., Bouvet, N., Freyssinet, G., Schneerman, M. and Weber, D.F. (1992) Theor. Appl. Genet. **84**, 555–559
14. Chaboute, M.E., Chaubet, N., Clément, B., Gigot, C. and Philipps, G. (1988) Gene **71**, 217–223
15. Chaubet, N., Chaboute, M.E., Clément, B., Ehling, M., Philipps, G. and Gigot, C. (1988) Nucleic Acids Res. **16**, 1295–1304
16. Wu, S.C., Györgyey, J. and Dudits, D. (1989) Nucleic Acids Res. **17**, 3057–3063
17. Kapros, T., Stefanov, I., Magyar, Z., Ocsovsky, I. and Dudits, D. (1993) In Vitro Cell. Dev. Biol. **29P**, 27–32
18. Nagata, T., Nemoto, Y. and Hasezawa, S. (1992) Int. Rev. Cytol. **132**, 1–30
19. Ohtsubo, N., Nakayama, T., Terada, R., Shimamoto, K. and Iwabuchi, M. (1993) Plant Mol. Biol. **23**, 553–565
20. Chaubet, N., Clement, B., Philipps, G. and Gigot, C. (1991) Plant Mol. Biol. **17**, 935–940
21. Joanin, P., Gigot, C. and Philipps, G. (1992) Plant Mol. Biol. **20**, 581–588
22. Razafimahatratra, P., Chaubet, N., Philipps, G. and Gigot, C. (1991) Nucleic Acids Res. **19**, 1491–1496
23. Nakayama, T. and Iwabuchi, M. (1993) Crit. Rev. Plant Sci. **12**, 97–110
24. Wu, S.C., Bögre, L., Vincze, E., Kiss, G.B. and Dudits, D. (1988) Plant Mol. Biol. **11**, 641–649
25. Atanassova, R., Chaubet, N. and Gigot, C. (1992) Plant J. **2**, 291–300
26. Raghavan, V. and Olmedilla, A. (1989) Cell. Differ. Dev. **27**, 183–196
27. Koning, A.J., Tanimoto, E.Y., Kiehne, K., Rost, T. and Comai, L. (1991) Plant Cell **3**, 657–665
28. Köhler, S., Coraggio, I., Becker, D. and Salamini, F. (1992) Planta **186**, 227–235
29. Tanimoto, E.Y., Rost, T.L. and Comai, L. (1993) Plant Physiol. **103**, 1291–1297
30. Terada, R., Nakayama, T., Iwabuchi, M. and Shimamoto, K. (1993) Plant J. **3**, 241–252
31. Reichheld, J.P., Sonobe, S., Clément, B., Chaubet, N. and Gigot, C. (1995) Plant J. **7**, 245–252
32. Matsumoto, S., Yanagida, M. and Nurse, P. (1987) EMBO J. **6**, 1093–1097
33. Lepetit, M., Ehling, M., Chaubet, N. and Gigot, C. (1992) Mol. Gen. Genet. **231**, 276–285
34. Nakayama, T., Sakamoto, A., Yang, P., Minami, M., Fujimoto, Y., Ito, T. and Iwabuchi, M. (1992) FEBS Lett. **300**, 167–170
35. Yang, P., Taoka, K.I., Nakayama, T. and Iwabuchi, M. (1995) Plant Mol. Biol. **28**, 155–172
36. Terada, R., Nakayama, T., Iwabuchi, M. and Shimamoto, K. (1995) Plant Mol. Biol. **27**, 17–26

37. Brown, S.C., Bergounioux, C., Tallet, S. and Marie, D. (1991) in A Laboratory Guide for Cellular and Molecular Plant Biology (Negrutiu, I. and Gharti-Chhetri, G., eds.), pp 326–345, Birkhauser, Basel
38. Brignon, P., Lepetit, M., Gigot, C. and Chaubet, N. (1993) Plant Mol. Biol. **22**, 1007–1015
39. Brignon, P. and Chaubet, N. (1993) Plant J. **4**, 445–457
40. Mikami, K., Sakamoto, A., Takase, H., Tabata, T. and Iwabuchi, M. (1989) Nucleic Acids Res. **17**, 9707–9717
41. Tabata, T., Takase, H., Takayama, S., Mikami, K., Nakatsuka, A., Kawata, T., Nakayama, T., and Iwabuchi, M. (1989) Science **245**, 965–967
42. Tabata, T., Nakayama, T., Mikami, K. and Iwabuchi, M. (1991) EMBO J. **10**, 1459–1467
43. Takase, H., Minami, M. and Iwabuchi, M. (1991) Biochim. Biophys. Res. Commun. **176**, 1593–1600
44. Kawata, T., Nakayama, T., Mikami, K., Tabata, T., Takase, H. and Iwabuchi, M. (1988) FEBS Lett. **239**, 319–323
45. Sakamoto, A., Minami, M., Huh, G.H. and Iwabuchi, M. (1993) Eur. J. Biochem. **217**, 1049–1056
46. Lepetit, M., Ehling, M., Atanassova, R., Chaubet, N. and Gigot, C. (1993) Plant Sci. **89**, 177–184
47. Ito, T., Fujimoto, Y., Nakayama, T. and Iwabuchi, M. (1995) Plant Cell Physiol. **36**, 1281–1289
48. Kapros, T., Bogre, L., Nemeth, K., Bako, L., Gyorgyev, J., Wu, S.C. and Dudits, D. (1992) Plant Physiol. **98**, 621–625
49. Chaubet, N., Clément, B. and Gigot, C. (1992) J. Mol. Biol. **225**, 569–574
50. Van Wijnen, A.J., Aziz, F., Grana, X., De Luca, A., Desai, R.K., Jaarsveld, K., Last, T.J., Soprano, K., Giordano, A., Lian, J.B., Stein, J.L. and Stein, G.S. (1994) Proc. Natl. Acad. Sci. **91**, 12882–12886
51. Chaubet, N., Flenet, M., Clement, B., Brignon, P. and Gigot, C. (1996) Plant J., in the press

# 14

# Transcriptional control of histone genes

**Masaki Iwabuchi\*, Takuya Nakayama and Tetsuo Meshi**

Department of Botany, Graduate School of Science
Kyoto University, Kyoto 606, Japan

## Introduction

Histones, ubiquitous nuclear proteins in eukaryotes, are necessary for assembling DNA into nucleosomes, and they constitute five major classes: four core histones, H2A, H2B, H3, H4, and one linker histone, H1. The genes coding for these proteins exist as a multigene family. Most histones are expressed cell-cycle-dependently, and are coupled to DNA synthesis at the S phase during the cell cycle as a result of transcriptional and post-transcriptional regulation [1–5]. In contrast, there are minor classes of histones, the expression of which is cell-cycle-independent and/or tissue-specific, and these are encoded by genes different from the replication-dependent genes [6]. Therefore, understanding the transcriptional regulation of histone genes, especially the S-phase-specific regulation of transcription, is important to elucidate the means by which cell proliferation is controlled, because the replication-dependent histone genes offer a unique model system for studying the molecular control mechanisms of gene expression during the cell cycle. A number of studies on histone gene expression have been carried out, so far mainly with vertebrates [7,8] and yeast [9–11]. Recently, a considerable amount of information on the regulation of plant histone gene expression has been accumulated, in addition to animal and yeast histone genes [12–26].

In this review, we will describe the transcriptional regulation of plant histone genes, focusing, in particular, on the structures and functions of putative *trans*-acting regulatory factors interacting with the *cis*-acting elements in a wheat H3 gene promoter.

### *Cis*-acting control elements of wheat histone genes

In this section, we will briefly describe *cis*-regulatory elements involved in the S-phase-specific transcription of the wheat H3 gene (TH012). The H3 gene is one of the histone genes that were first isolated from higher plants [27], and it is one of the plant histone genes that have been well investigated with respect to transcriptional regulation. In the proximal region of the H3 gene promoter are several *cis*-acting

\*To whom correspondence should be addressed.

elements, such as the type-I element, the nonamer motif and the ACT box. Of these, the type-I element (CCACGTCANCGATCCGCG), composed of the hexamer sequence and the reverse-oriented plant histone-specific octamer sequence located 2 bp downstream of the hexamer, may be involved in the S-phase-specific transcription of the wheat H3 gene (TH012) ([16,16a]. This suggestion was initially inferred from evidence that in cultured rice cells transformed with H3/β-glucuronidase (GUS) chimaeric genes lacking the upstream sequence of the type-I element, the mRNA transcribed from the chimaeric genes accumulated predominantly at the S phase [16]. Supporting evidence has been obtained from the loss-of-function and gain-of-function experiments using cultured rice cells transformed with the H3/GUS or CaMV 35 S core promoter/GUS chimaeric genes [16a]. The loss-of-function experiments indicated that, when either or both of the hexamer and octamer sequences were point-mutated, the resulting H3/GUS genes, with a mutated type-I element, were no longer transcribed S-phase-specifically. In the gain-of-function experiments, joining three copies of the type-I element sequence upstream of the CaMV 35 S core promoter/GUS gene caused the chimaeric gene to be transcribed S-phase-specifically, despite the cell-cycle-independent activity of the 35 S core promoter.

The hexamer and octamer sequences function synergistically to enhance the transcriptional efficiency in transformed tobacco (TBY-2) cells, when fused upstream of the CaMV 35 S/GUS gene (M. Iwabuchi and H. Kaya et al., unpublished data]. In the upstream region (positions -289 to -246) of the H3 gene, two copies of the type-I-like sequence are located about 100 bp upstream of the type-I element. When a portion containing the type-I element was internally deleted so that the type-I-like sequence replaced the authentic position of the type-I element, these sequences seemed to confer S-phase-specific activity to the H3 promoter in the same way as the type-I element. We conclude that the type-I element is involved in the S-phase-specific transcription of the wheat H3 gene, in addition to having an enhancing effect on promoter activity.

The type-I element has also been found in the regulatory regions of some other plant H3 and H4 genes. Thus, cis-acting elements common to histone gene subtypes seem to exist in higher plants, as well as in yeast H2A and H2B genes [10], although such cis-acting elements have not been found in vertebrate histone genes [8]. It is unknown whether the type-I elements present in the promoters of other plant H2B, H3 and H4 genes are also involved in regulating S-phase-specific transcription. In addition to the type-I element, another cis-regulatory sequence (TCACGCGGATC), named the type-II element, has been found in many plant histone genes, including several wheat histone genes [19]. It is also unknown whether the type-II element has the ability to confer S-phase-specific transcription. All plant H1 and H2B genes isolated to date have the CCAAT motif just 15 bp downstream of the octamer motif ([25], K. Taoka et al., submitted for publication). The CCAAT motif is known to be the cis-acting element of a variety of histone genes, such as human H4 [8], rat testis-specific H2B [28], Xenopus H2B [29] and sea urchin H3 [30] genes, as well as in many eukaryotic protein-coding genes. Functional analyses of the wheat H1 promoter in a transient expression system have suggested that the CCAAT and octamer motifs appear to function together as a modular element. The stretch which features these two elements, named the type-III element, may be implicated in the S-phase-specific transcription of the H1 gene (K. Taoka et al., unpublished data). The type-III element is also found in *Arabidopsis* [31] and tomato [32] H1 genes, and a maize H2B gene [33]. Thus, it is reasonable to conclude that the type-I, -II, and -III elements are deeply involved in transcriptional regulation of plant histone genes.

## *Trans*-acting control factors of wheat H3 gene transcription

Transcriptional activation is primarily mediated through *trans*-acting regulatory proteins that interact with *cis*-acting elements in control regions, such as promoters and enhancers. Such molecular interactions take place through the binding of *trans*-acting factors to specific DNA sequences, and they are often crucial for the initiation of transcription. As for animal and yeast histone genes, cell cycle dependent gene expression is regulated, in part, at the transcriptional level [7,8,10], and S-phase-specific transcriptional regulation of human H1, H2B and H4 genes is achieved through interactions between subtype-specific *cis*-acting elements and sequence-specific *trans*-acting factors [7,8]. For the past 15 years, tremendous effort has been spent on searches for eukaryotic transcription factors, and their structures and functions have been analysed extensively. Consequently, a considerable number of transcription factors and their candidate proteins have been identified in animals and yeast [34,35]. In recent years, similar efforts have been made in higher plants, and the cDNAs and genomic clones coding for plant transcription factors have been isolated [36–38]. The same is true for histone gene transcription factors. In this section, we will describe structural and functional characteristics of several sequence-specific DNA-binding proteins, identified as putative transcription factors of the wheat H3 gene.

### HBP-1 family

DNA-binding proteins, HBP (histone promoter-binding protein)-1a and -1b, were identified as nuclear proteins with Hex motif (CCACGTCA) binding specificity [39–42]. HBP-1a and -1b exhibit distinct DNA-binding properties; the former binds specifically to the Hex motif of the H3 promoter and G-box (CCACGTGG), whereas the latter binds to both the H3 Hex motif and hexamer (ACGTCA)-containing sequences present in nuclear, viral and T-DNA gene promoters [21,40,41]. On the basis of the difference in DNA-binding specificity, two distinct cDNA clones encoding the proteins, named HBP-1a(17) and HBP-1b(c38), were first isolated by the Southwestern screening of a wheat cDNA expression library [43,44]. Sequence analyses of the cDNAs have revealed that HBP-1a(17) and -1b(c38) contain a basic/leucine zipper (bZIP) domain at the C- and N-terminal halves, respectively.

In addition to these two HBP-1 proteins are three other similar bZIP proteins, HBP-1a(c14), -1a(1) and -1b(c1), and cDNA analysis has revealed them to be distinct from HBP-1a(17) and -1b(c38). These five HBP-1 proteins may constitute an HBP-1 family, which is further classified into two subfamilies, HBP-1a and HBP-1b, based on characteristics of their primary structure and DNA-binding specificity [45]. The three members of the HBP-1a subfamily all contain a proline-rich region near the N-terminus, besides the C-terminal bZIP domain. In contrast, the two HBP-1b subfamily members contain well-conserved bZIP domains in the N-terminal half, and a glutamine-rich region near the C-terminus. The bZIP domain comprises two regions; one is rich in basic amino acids for DNA binding, and the other is a region called the leucine-zipper (ZIP), where the leucine residues appear every 7th residue, over 3 to 7 repeat units. The ZIP region is located immediately C-terminal to the basic region, and is required for protein dimerization. The bZIP domain is known as one of the representative characteristics of eukaryotic transcription factors [34,35]. The ZIP region of HBP-1a(17) contains a hexa-leucine structure, and HBP-1b(c38) has a short stretch of tri-leucine. Proline-rich and glutamine-rich regions of several animal and plant transcription factors are known to function as transcription activation domains [46–49].

The five members of the HBP-1 family can bind to the Hex motif, but their DNA-binding specificities and affinities differ [45]. Three of the HBP-1a subfamily members can also bind to the G-box element (CCACGTGG), widely distributed in regulatory sequences, including the promoter regions of UV- and light-responsive plant genes. However, they cannot bind to the TGACGTAA sequence present in the as-1 positive element of the CaMV 35S promoter. HBP-1a subfamily members may recognize the Hex-a motif (CCACGT). The two HBP-1b subfamily members, which may recognize the Hex-b motif (ACGTCA), can bind to the as-1 element, but not to the G-box element. Since the Hex-a and Hex-b motifs share the ACGT core, the sequences flanking the core motif probably play an important role in determining the DNA-binding specificities of the two subfamilies.

Binding of the bZIP-type transcription factors to DNA requires the dimerization through their ZIP domains. With respect to the dimerization of the HBP-1 family members, heterotypic mixing experiments with the bacterially expressed HBP-1 family members have demonstrated that homo- and hetero-dimer formation occurs within the same HBP-1 subfamily, but not between the different subfamilies [44,45]. This result is reasonable, as judged from the overall structures and DNA-binding specificities of these family members.

As for the *trans*-regulatory function of the HBP-1 family members affecting H3 gene expression, some interesting results have been obtained with HBP-1a(17) [49a] and HBP-1b(c38) (T. Ito et al., unpublished data), and in particular, analyses of HBP-1a(17) are much advanced. Primarily, HBP-1a(17) may act as a negative factor to down-regulate H3-promoter activity in transient expression systems. The effect of effector genes capable of over-producing HBP-1a(17), on expression of GUS reporter gene driven by the H3 promoter with the type-I element, was examined in tobacco TBY-2, rice or wheat protoplasts. In connection with the above result, intriguing evidence has been obtained from analyses of the functional domains of HBP-1a(17), suggesting that the proline-rich domain is composed of several modules with *trans*-activating ability. We have arrived at this conclusion from co-transfection experiments employing the GAL4 fusion protein system, in which fusion proteins comprising dissected portions of HBP-1a(17), and the DNA-binding domain of the yeast transcription factor GAL4, were used as effectors. To determine the functional domains, a series of effector genes were constructed. The cDNAs for three distinct portions of HBP-1a(17)— the N-terminal proline-rich region, the central CK region containing sites phosphorylated by *Arabidopsis* casein kinase II, and the C-terminal bZIP region— were joined in-frame to the DNA coding sequence for the GAL4 DNA-binding domain, which was driven by the CaMV 35 S promoter. The reporter gene comprised the GUS coding sequence and wheat H3 minimal promoter, lacking the type-I element but having the GAL4-binding site. Co-transfection experiments with tobacco TBY-2 protoplasts indicated that both proline-rich and bZIP regions tend to negatively regulate the expression of the reporter gene, but the CK region has no effect. This result differs from the report that the proline-rich region of *Arabidopsis* GBF-1 (amino acids 1–110), corresponding to the HBP-1a(17) proline-rich-region, has a *trans*-activating function [48]. Interestingly, when the proline-rich region was divided into the N- and C- terminal halves, each of them was able to activate expression of the reporter gene, and their *trans*-activating efficiencies are 6–8- and 3–4-fold higher in the N- and C- terminal halves respectively, when compared with the control. When the N-terminal half of the proline-rich region was further divided into two sub-regions (amino acids 5–30 and 30–56), each of them still retained approximately half the *trans*-activating ability of the N-terminal half.

Another interesting result was the existence of a repeated sequence of DPVYP, where PVYP corresponds to the sequence of amino acid positions 31–34 in the N-terminal half of the proline-rich region, which exhibits a weak *trans*-activating ability (ca. 2-fold), whereas a repeated sequence of DPGKL encoded by reverse-oriented oligonucleotides has a *trans*-repressing effect. The reason why the entire proline-rich region and HBP-1a(17) suppress the expression of the reporter gene, in spite of the presence of a *trans*-activating potential in both N- and C-terminal halves of the proline-rich region, is still unknown. Plausibly, HBP-1a(17) could be a *trans*-activator, but often exhibits a negative effect on the H3 promoter activity through its interactions with other transcription factors (to be discussed later in this chapter). For example, it is possible that HBP-1a(17) undergoes conformational changes to disclose the proline-rich region at its surface, so as to interact with basal transcription factors or putative co-activators.

**Fig. 1**     Comparison of amino acid sequences between proline-rich
              regions of HBP-1a/GBF-type factors

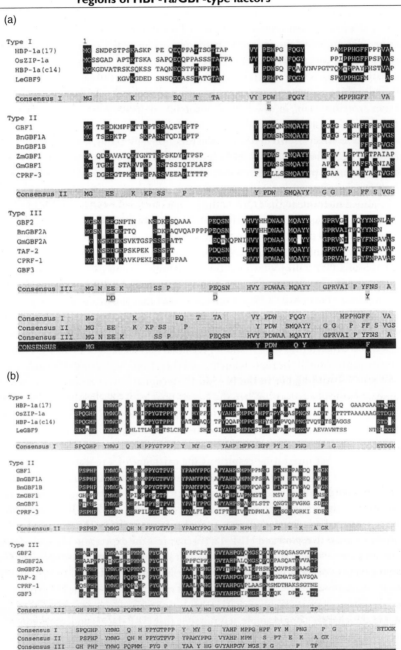

A modular structure of the activation domain has been reported for the acidic domain of the transcription factor VP16, where one of the two subdomains requires further aromatic and hydrophobic amino acids for its activity [50]. Tomato HSF8 and HSF30 also contain two activation domains, in which an aromatic residue (W) is flanked by acidic residues (D/E) [51]. In the VYPEWP (amino acids 32–37) sequence of the N-terminal half of the proline-rich region of HBP-1a(17), aromatic amino acids (Y and W) are flanked by hydrophobic (V) or acidic (E) residues. This sequence is conserved as Y-P-D(or E)-W in the corresponding regions of many HBP-1a/GBF-type bZIP proteins (Figure 1a). Similarly, several other amino acid stretches, containing aromatic residues, are well conserved in the proline-rich region of HBP-1a/GBF-type bZIP proteins (Figures 1a and 1b). These findings suggest that some aromatic residues may be critical for *trans*-activation. If this is the case, then the *trans*-activating function of a repeated sequence of DPVYP can be explained.

Our recent analyses of quantitative alterations of hexamer-specific DNA-binding activities during the cell cycle have revealed the existence of an HBP-1a species with an S-phase-specific binding activity in nuclear extracts obtained from partially synchronized wheat cultured cells (M. Minami et al., unpublished data). This S-phase-specific activity has been shown to be modulated by phoshorylation/dephosphorylation treatments of nuclear extracts. As yet, whether this S-phase-specific activity corresponds to one of the already identified HBP-1a isoforms, or whether it is a novel member of the HBP-1a subfamily, is unknown. In any case, it is likely that this S-phase-specific HBP-1a activity will be deeply implicated in the cell cycle dependent regulation of the H3 gene expression.

## Octamer-binding regulatory factors

DNA-binding proteins, which specifically interact with the plant-histone-specific octamer *cis*-acting sequence (GATCCGCG) in the type-I element of the wheat H3 promoter, have not been identified. With the maize histone H3 gene, DNase-I hypersensitive analyses of the promoter region have suggested the existence of

---

**Fig. I (contd.)**

*(a)*, *Comparison of the N-terminal half of the proline-rich region of HBP-1a(17) and its corresponding regions in other bZIP-type factors. (b), Comparison of the C-terminal half of the proline-rich region of HBP-1a(17) and its corresponding regions in other bZIP-type factors. These factors can be classified into three groups: type I, II and III, according to their amino acid sequence similarities. Spaces are inserted into the sequence for maximum matching alignment. Note that the N-terminal half of the proline-rich regions (A) are less conserved than the C-terminal half of the proline-rich regions (B). Sources (accession numbers) of the sequences: HBP-1a(17), M28704; osZIP-1a, U04295; HBP-1a(c14), D12920; LeGBF9, X7493; GBF1, X63894 (S94514); BnGBF1a, X83922; BnGBF1b, X83921; ZmGBF1, U10270; GmGBF1, L01447; CPRF-3, X58576; GBF2, X63895 (S94516); BnGBF2A, X83920; GmGBF2A, L01448 TAF-2, Z48602; CPRF-1, X58575; GBF3, X63896 (S94518).*

DNA-binding proteins interacting with a relatively long sequence containing the octamer motif [23,24]. More recently, several octamer-motif-specific DNA-binding proteins, named OBRFs (octamer-binding regulatory factors), have been identified by the mobility-shift assay of nuclear extracts from wheat-cultured cells (M. Minami et al., unpublished data). This *in vitro* DNA–protein-binding assay indicated that there are at least three subtypes of OBRF, although their quantity is very low. One of them, OBRF-1, is detectable at a relatively higher level at the S phase, whereas the other two, OBRF-2 and -3, appear to be present constantly throughout the cell cycle. In maize histone H3 and H4 genes, two different octamer-motif-binding proteins seem to exist at the proliferative stage, at which the octamer motif may be crucial in co-ordinating the interplay between upstream *cis*-acting sequence-binding general transcription factors and downstream sequence-specific inducible factors [23]. It is unclear whether an apparently higher level of DNA-binding activity of OBRF-1 at the S phase can be ascribed to the real enhancement of its binding activity, or to an increase in the net amount of active OBRF-1. Considered together with the DNA-binding specificity of OBRF-1 and the S-phase-specific HBP-1a species, described above, we postulate that S-phase-specific transcription of the H3 gene is controlled through co-operative binding of the OBRF members and the HBP-1a subfamily members to their *cis*-acting sequences in the type-I element, and through protein–protein interactions between the two sequence-specific DNA-binding proteins. Since neither cDNAs nor genomic clones for OBRFs have been isolated, the structures and functions of these proteins remain to be proven.

## Histone promoter-binding protein-2

Histone promoter-binding protein-2 (HBP-2) was initially identified by DNaseI footprinting as a wheat DNA-binding protein, which interacts with the region containing the conserved nonamer motif, CATCCAACG, located in the wheat H4 gene (TH011) promoter [52]. The nonamer motif is a *cis*-acting element of the H3 gene [13]. A similar DNA-binding protein has been reported in maize, which interacts with a region containing a sequence resembling the nonamer motif present in the proximal region of the histone H3 gene promoter [23,24]. Until now, no cDNA clone encoding the wheat HBP-2 has been isolated, and therefore no direct evidence exists for HBP-2 as a transcription factor. However, HBP-2 is still a candidate for the nonamer-motif-specific transcription factor.

## Wheat zinc-finger protein-I

Wheat zinc-finger protein-1 (WZF-1) is a DNA-binding protein which was deduced from the analysis of a cDNA, obtained in an attempt to isolate cDNA clones encoding HBP-2 by Southwestern screening of a wheat cDNA expression library [53,53a]. Nucleotide sequence analysis of the cDNA has revealed that it encodes a protein containing two zinc-finger motifs. Unexpectedly, this wheat protein cannot bind to the nonamer motif of the H3 and H4 promoters, but interacts with a novel *cis*-acting sequence, the ACT box (CACTC), present 6 bp upstream of the nonamer motif. Since the ACT box is also found in the WZF-1 gene promoter, and functions as a positive *cis*-acting element in a transient expression system with tobacco protoplasts, WZF-1 should be involved in the transcriptional regulation of wheat H3 and H4 gene expression.

The two zinc-finger motifs of WZF-1 resemble a typical Cys-2/His-2-type zinc finger, but apparently differ from it: WZF-1 zinc-fingers are separated by a long sequence of 49 amino acids, and there are glycine residues in the region corresponding to the recognition helix of the typical Cys-2/His-2 fingers. Such structural features have been reported for petunia transcription factors, EPF1 and EPF2 [54,55]. For both WZF-1 and EPF1, the sequence identities between any two of the finger motifs are greater than 50%, and each carries a short basic amino acid sequence that is regarded as a nuclear localization signal.

Regarding the functional role of WZF-1 as a transcriptional regulator, this wheat protein negatively regulates the activity of the promoters carrying the ACT box in a co-expression system of the effector gene (WZF-1 cDNA driven by the CaMV 35 S promoter) and the reporter gene (the GUS-coding sequence driven by the WZF-1 gene promoter that contains several copies of the ACT box motif). Unfortunately, it is not yet clear how WZF-1 contributes to the expression of wheat H3 and H4 genes, because the entire WZF-1 protein had no effect on the gene expression of chimaeric H3/GUS in co-transfection experiments. However, functional-domain analysis of WZF-1, using the GAL4-based fusion protein system, revealed that the N-terminal region of WZF-1 has a strong repressive potential. These results do not agree with the observation that WZF-1 binds to the ACT box of a postive *cis*-acting element.

To explain this discrepancy, it has been postulated that, depending on changes of intracellular circumstances, WZF-1 could become either a *trans*-activator or a *trans*-repressor through interactions with other transcription factors, or through post-translational modifications, such as phosphorylation. There is evidence for such bifunctional activity of a transcription factor in Krüppel, a *Drosophila* zinc-finger protein [56–59], the mammalian YY-1 zinc-finger protein [60], adenovivus E1A [61], and so on. Thus, one can speculate that WZF-1 regulates wheat H3 and H4 gene transcription as a positive or negative factor at different stages during the cell cycle, although there is no direct evidence that this DNA-binding protein modulates wheat histone gene expression. To elucidate whether bifunctional changes of WZF-1 actually occur, protein–protein interactions between this protein and other transcription factors, and the structural modifications of this protein by phosphorylation or dephosphorylation, must be investigated.

## HBP-1-associated leucine-zipper factor-1

HBP-1-associated leucine-zipper factor-1 (HALF-1) was identified in the process of cloning cDNAs encoding proteins which interact with HBP-1a(17) through protein–protein interactions [61a]. A cDNA clone has been isolated from a wheat cDNA expression library by the far-Western screening method using [32]P-labelled HBP-1a(17) as a probe. Nucleotide sequence analysis of the cDNA clone revealed that HALF-1 contains a typical bZIP motif near the C-terminus. This protein has very high homology to some plant bZIP-type proteins, including HBP-1a subfamily members; this is particularly conspicuous in the basic region. Among these proteins, the most similar protein is EmBP-1b, a DNA-binding protein which interacts with a *cis*-acting element of the wheat Em gene [62].

HALF-1 includes a proline/alanine-rich region at the N-terminus, and the GCB motif commonly existing in some plant bZIP proteins ([61a,63]. A number of serines and acidic residues are also located in the region C-terminal to the proline/alanine-rich region. Interaction of HALF-1 and HBP-1a(17), through their leucine-zipper domains, has been demonstrated *in vivo* using the yeast two-hybrid system. In dimerization experiments *in vitro*, HALF-1 can not only homodimerize, but can also heterodimerize with HBP-1a(17) via the zipper-domain; both the homo- and hetero-dimers bind to the Hex motif of the wheat H3 gene promoter. Thus, HALF-1 can be regarded as a novel member of the HBP-1a subfamily.

Functional studies of HALF-1, using the transient co-expression system of the effector and reporter genes, have shown that overexpression of HALF-1 has no effect on the expression of the reporter GUS gene driven by the wheat H3 promoter, even in the presence of a HBP-1a(17) expression plasmid. On the other hand, in the experiments using the GAL4-based fusion-protein system, the region containing the GCB motif of HALF-1 raised the expression efficiency of the reporter gene by 6–10-fold, when joining GAL4 target sequences upstream of the reporter gene. These results suggest that the entire HALF-1 does not function as a *trans*-regulator under the above conditions, although the GCB-motif-containing region has a *trans*-activation ability. Such a phenomenon is somewhat similar to that of HBP-1a(17), concerning the functional relationship between the entire protein, and its N-terminal proline-rich region.

## Others

Other DNA-binding proteins exist for the wheat H3 and H4 genes. For example, single-stranded DNA-binding proteins (ssDBPs)-1 and -2 have been identified *in vitro* by the mobility-shift assay, and by the methylation-interference assay with wheat germ nuclear extracts [64]. They can bind to single-stranded DNA containing the hexamer and octamer motifs in the H3 and H4 promoters. The former interacts sequence-specifically with the lower strand, whereas the latter binds to the upper strand in the same manner. It is unclear whether these proteins are transcription factors of the H3 and H4 genes. In maize, DNA-binding proteins, which can bind to the CCGTCC and GCCAA sequences in the H3 and H4 gene promoters, were detected by DNaseI footprinting assays [23]. These proteins have not yet been demonstrated to be transcription factors.

# Modification of HBP-1a by phosphorylation

Post-translational modification often results in the modulation of the activities of transcription factors, such as DNA binding, nuclear transport, and activation/repression. There are a large number of reports indicating that transcription factors are functionally modulated by phosphorylation/dephosphorylation in yeast and animals [65–67]. Since gene expression is controlled mostly through signal transduction pathways mediated by many protein kinases and phosphatases, it is important to elucidate functional modulations of transcription factors by phosphorylation/dephosphorylation. In plants, as well as in animals, DNA-binding activity in nuclear extracts is modulated by phosphorylation [68–74]. With respect to purified plant transcription factors, phosphorylation of *Arabidopsis* GBF1 by casein kinase II (CKII) has been demonstrated to raise DNA-binding activity [75].

Phosphorylation of transcription factors of plant histone genes has been investigated with the HBP-1a(17) isoform (T. Meshi et al., unpublished data). This protein can be phosphorylated *in vitro* by certain kinds of kinases contained in the fractions separated from wheat nuclear extracts by Heparin–Sepharose column chromatography. More detailed results have been obtained with three distinct fragments of HBP-1a(17): polypeptides C1, C2 and C6. The N-terminal fragment (polypeptide C1), which contains the proline-rich region, was poorly phosphorylated by kinases contained in the nuclear extract. The central fragment (polypeptide C2) is well phosphorylated by a kinase included in the 0.1–0.35 M KCl fraction of the Heparin–Sepharose column chromatography. It is thought to be CKII from the following observations: first, five possible phosphorylation sites by CKII, with the consensus sequences of S/TxxE/D, are clustered near the N-terminal region of C2 (downstream of the proline-rich region); secondly, C2 is a good substrate for a recombinant *Arabidopsis* CKII; and thirdly, tryptic phosphopeptide-mapping analyses revealed that kinase activity in the 0.35 M KCl fraction phosphorylated the same serine residues in C2 as the *Arabidopsis* CKII did. The significance of this phosphorylation is still unknown, but it seems very likely that phosphorylation by CKII could modulate the activation potential of the proline-rich region and/or the GCB motif, which are, respectively, N- and C-flanking to the CKII phosphorylation sites. It is also possible that phosphorylation modulates the DNA-binding activity of HBP-1a(17) to the cognate sequence motif.

As previously described, the S-phase-specific DNA-binding activity of HBP-1a, present in nuclear extracts from partially synchronized wheat-suspension-cultured cells, was regulated by phosphorylation/dephosphorylation, in which CKII may be involved.

The C-terminal region (polypeptide C6), containing the bZIP domain, was phosphorylated by kinases in the 0.1 M KCl flow-through (FT) fraction, and by kinases in the 0.3 M KCl eluate. The kinase activity in the FT fraction was stimulated approximately 10-fold by calcium. Phosphorylation sites have been determined as three serine residues at positions 261, 265 and 269 in the basic region of the bZIP domain, by tryptic peptide mapping, and phosphoamino acid analyses of C6 and its derivatives with serine-to-alanine changes. Amino-acid substitutions of glutamate residues for serine residues at positions 265 or 269, which mimicked phosphorylated forms of C6, prevented the binding of the mutant C6 protein to the Hex sequence. Thus, it is highly probable that phosphorylation/dephosphorylation of HBP-1a(17) in its basic region is central to the binding of this protein to DNA. These serine residues in this region are highly conserved among plant bZIP-type transcription factors known so far. It is therefore supposed that there is a common regulatory mechanism of plant bZIP proteins, through calcium-mediated phosphorylation of the conserved serine residues. The probable kinase for serine phosphorylation is a calcium-dependent protein kinase, which has been implicated in various calcium-regulated cellular processes in plants, and has been found in the nucleus, as well as in the cytoplasm [76].

## A tentative model for the S-phase-specific transcription of the wheat H3 gene

Focusing mainly on the transcriptional regulation of wheat H3 gene expression, we have described DNA-binding proteins interacting specifically with several *cis*-acting elements involved in H3 gene transcription. Some of them are regarded as putative transcription factors, based on their structural and functional properties analyzed by using their cDNA clones. However, the others have been merely identified as protein–DNA complexes by the mobility-shift assay or the DNaseI footprinting method. Therefore, to date, most of the identified wheat DNA-binding proteins are functionally unknown, other than their binding specificities.

Although study of the transcriptional regulation of plant histone genes is still in progress, we propose a tentative model to explain the S-phase-specific transcriptional regulation of the wheat H3 gene. This will be based upon accumulated evidence on the structures and functions of *cis*-acting elements; in particular, the type-I element, and its sequence-specific DNA-binding proteins (see Fig.2). In this model, special attention is paid to the relationships between HBP-1 family members and OBRFs, both of which are thought to play crucial roles in the cell-cycle-dependent regulation of wheat histone gene expression, because the type-I element can confer S-phase-specific activity of the H3 promoter [19,20]. Another important point of this model is to postulate a cofactor which can bind to both the HBP-1 family members and OBRFs *via* protein–protein interactions. This cofactor could explain the negative effect of the entire HBP-1a(17) by the squelching mechanism, and is useful to interpret how overproduction of HBP-1a(17) containing a *trans*-activating domain results in the down-regulation of activity of the type-I element-containing H3 promoter. To confirm the reliability of this model, the existence of the putative cofactor must first be proven, necessitating structural and functional analyses of this cofactor as well as of HBP-1 family

**Fig. 2**          **A tentative model of the involvement of transcription factors in wheat histone H3 gene expression.**

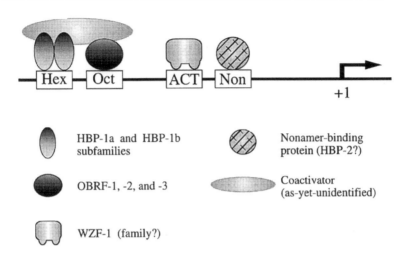

members and OBRFs. In addition, *in vivo* functions of transcription factors will have to be elucidated in transgenic plants, using a homologous gene expression system. Nevertheless, a great deal of effort has been made towards isolation of cDNA clones encoding the putative cofactor, using the yeast two-hybrid system and far-Western screening.

In general, the regulatory region of a given gene consists of multiple *cis*-acting elements. This means that a number of transcription factors are involved in the expression of a specific gene, and they form multi-component complexes on defined regions of the promoter. Thus, in the case of the wheat H3 gene, it is highly probable that the DNA-binding proteins form multi-protein complexes on the regulatory regions, including the type-I element, and that they further interact with core-promoter-specific basal transcription factors *via* protein–protein interactions. In fact, possible multiple complexes of DNA-binding proteins have been reported for the maize H3 or H4 promoter [23].

Are other sequences, besides the type-I element, involved in S-phase-specific transcription? Actually, the type-III element, found in a wheat H1 gene, is supposed to be implicated in S-phase-specific transcription. It seems unlikely that WZF-1 and HBP-2 are directly involved in S-phase-specific regulation, although they are DNA-binding proteins interacting with positive *cis*-acting sequences.

## Concluding remarks

During the past decade, studies on structural analysis of plant histone genes have been steadily advancing, and many *cis*-acting elements involved in transcription have been identified in different histone gene subtypes from various plant species. However, the molecular mechanisms of the transcriptional regulation in plant histone gene expression have scarcely been elucidated. In particular, transcription-factor-mediated gene regulation still remains unknown, because only a small number of sequence-specific DNA-binding proteins have been demonstrated to be *trans*-acting factors.

The most important issue in histone gene expression involves the understanding of the stage-specific gene regulation during cell cycle progression. Judged from the limited evidence obtained so far, it is highly probable that many transcription factors, involved in the regulation of plant histone gene expression, form a network, through both protein–DNA and protein–protein interactions. Therefore, more extensive investigations of transcription factors, as well as *cis*-acting elements, will be needed for elucidating the regulatory mechanisms of S-phase-specific transcription in plant histone gene expression. Since members of the HBP-1a and -1b subfamilies can form heterodimers within each subfamily in all pairwise combinations, heterodimer formation among these bZIP proteins may generate an expanded repertoire of regulatory potential in the expression of not only histone genes, but also other protein-coding genes in plants. In addition, since transcription factors are the final targets in signal transduction pathways, it is of importance to study the post-translational modification of transcription factors, through protein phosphorylation and the regulated expression of the genes coding for the protein factors.

*We thank former and present members of our laboratory for their contri-butions to our histone research project, and for allowing us to cite some of their unpublished data. Thanks are also due to Mrs Yuriko Suzuki for her assistance in preparation of the manuscript. Our work described here has been supported, in part, by Grants-in-Aid for Scientific Research from the Ministry of Education, Science, Sports and Culture of Japan.*

## References

1.   Artishevsky, A., Wooden, S., Sharma, A., Resendez, E., Jr. and Lee, A.S. (1987) Nature (London) **328**, 823–827
2.   DeLisle, A.J., Graves, R.A., Marzluff, W.F. and Johnson, L.F. (1983) Mol. Cell. Biol. **3**, 1920–1929
3.   Harris, M.E., Bohni, R., Schneiderman, M.H., Ramamurthy, L., Schumperli, D. and Marzluff, W.F. (1991) Mol. Cell. Biol. **11**, 2416–2424
4.   Heintz, N., Sive, H.L. and Roeder, R.G. (1983) Mol. Cell. Biol. **3**, 539–550
5.   Plumb, M., Stein, J. and Stein, G. (1983) Nucleic Acids Res. **13**, 2391–2410
6.   Wu, R.S. and Bonner, W.M. (1981) Cell **27**, 321–330

7.  Heintz, N. (1991) Biochim. Biophys. Acta **1088**, 327–339
8.  Stein, G.S., Stein, J.L., Wijnen, A.J. and Lian, J.B. (1992) Curr. Opin. Cell Biol. **4**, 166–173
9.  Lycan, D.E., Osley, M.A. and Hereford, L. (1987) Mol. Cell. Biol. **7**, 614–621
10. Osley, M.A. (1991) Annu. Rev. Biochem. **60**, 187–193
11. Xu, H., Kim, V.-J., Schuster, T. and Grunstein, M. (1992) Mol. Cell. Biol. **12**, 5249–5259
12. Chaubet, N., Clement, B., Philipps, G. and Gigot, C. (1991) Plant Mol. Biol. **17**, 935–940
13. Nakayama, T., Sakamoto, A., Yang, P., Minami, M., Fujimoto, Y., Ito, T. and Iwabuchi, M. (1992) FEBS Lett. **300**, 167–170
14. Atanassova, R., Chaubet, N. and Gigot, C. (1992) Plant J. **2**, 291–300
15. Lepetit, M., Ehling, M., Chaubet, N. and Gigot, C. (1992) Mol. Gen. Genet. **231**, 276–285
16. Ohtsubo, N., Nakayama, T., Terada, R., Shimamoto, K. and Iwabuchi, M. (1993) Plant Mol. Biol. **23**, 553–565
16a. Ohtsubo, N., Nakayama, T., Kaya, H., Terada, R., Shimamoto, K., Meshi, T. and Iwabuchi, M. (1997) Plant J., in the press
17. Terada, R., Nakayama, T., Iwabuchi, M. and Shimamoto, K. (1993) Plant J. **3**, 241–252
18. Terada, R., Nakayama, T., Iwabuchi, M. and Shimamoto, K. (1995) Plant Mol. Biol. **27**, 17–26
19. Mikami, K. and Iwabuchi, M. (1993) in Control of Plant Gene Expression (Verma, D.P.S., ed.), pp. 51–68, CRC Press, Boca Raton
20. Nakayama, T. and Iwabuchi, M. (1993) Crit. Rev. Plant Sci. **12**, 97–110
21. Takase, H. and Iwabuchi, M. (1993) J. Plant Res. (Special Issue) **3**, 37–50
22. Lepetit, M., Ehling, M., Atanassova, R., Chaubet, N. and Gigot, C. (1993) Plant Sci. **89**, 177–184
23. Brignon, P. and Chaubet, N. (1993) Plant J. **4**, 445–457
24. Brignon, P., Lepetit, M., Gigot, C. and Chaubet, N. (1993) Plant Mol. Biol. **22**, 1007–1015
25. Yang, P., Taoka, K., Nakayama, T. and Iwabuchi, M. (1995) Plant Mol. Biol. **28**, 155–172
26. Ito, T., Fujimoto, Y., Nakayama, T. and Iwabuchi, M. (1995) Plant Cell Physiol. **36**, 1281–1289
27. Tabata, T., Fukasawa, M. and Iwabuchi, M. (1984) Mol. Gen. Genet. **196**, 397–400
28. Choi, Y.-C. and Chae, C.-B. (1991) J. Biol. Chem. **266**, 20504–20511
29. Hinkley, C. and Perry, M. (1992) Mol. Cell. Biol. **12**, 4400–4411
30. Diliberto, M., Lai, Z., Fei, H. and Childs, G. (1989) Genes Dev. **3**, 973–985
31. Gantt, J.S. and Lenvik, T.R. (1991) Eur. J. Biochem. **202**, 1029–1039
32. Jayawardene, N. and Riggs, C.D. (1994) Eur. J. Biochem. **223**, 693–699
33. Joanin, P., Gigot, C. and Philipps, G. (1994) Plant Physiol. Biochem. **32**, 693–696
34. Johnson, P.F. and McKnight, S.L. (1989) Annu. Rev. Biochem. **58**, 799–839
35. Pabo, C.O. and Sauer, R. T. (1992) Annu. Rev. Biochem. **61**, 1053–1095
36. Katagiri, F. and Chua, N.-H. (1992) Trends Genet. **8**, 22–27
37. Brunelle, A.N. and Chua, N.-H. (1993) Curr. Opin. Genet. Dev. **3**, 254–258
38. Meshi, T. and Iwabuchi, M. (1995) Plant Cell Physiol. **36**, 1405–1420
39. Mikami, K., Tabata, T., Kawata, T., Nakayama, T. and Iwabuchi, M. (1987) FEBS Lett. **223**, 273–278
40. Mikami, K., Takase, H., Tabata, T. and Iwabuchi, M. (1989) FEBS Lett. **256**, 67–70
41. Mikami, K., Sakamoto, A., Takase, H., Tabata, T. and Iwabuchi, M. (1989) Nucleic Acids Res. **17**, 9707–9717
42. Mikami, K., Nakayama, T., Kawata, T., Tabata, T. and Iwabuchi, M. (1989) Plant Cell Physiol. **30**, 107–119
43. Tabata, T., Takase, H., Takayama, S., Mikami, K., Nakatsuka, A., Kawata, T., Nakayama, T. and Iwabuchi, M. (1989) Science **245**, 965–967
44. Tabata, T., Nakayama, T., Mikami, K. and Iwabuchi, M. (1991) EMBO J. **10**, 1459–1467
45. Mikami, K., Sakamoto, A. and Iwabuchi, M. (1994) J. Biol. Chem. **269**, 9974–9985
46. Mermod, N., O'Neill, E.A., Kelly, T.J. and Tjian, R. (1989) Cell **58**, 741–753
47. Courey, A.J., Holtzman, D.A., Jackson, S.P. and Tjian, R. (1989) Cell **59**, 827–836
48. Schindler, V., Terazaghi, W., Beckmann, H., Kadesch, T. and Cashmore, A.R. (1992) EMBO J. **11**, 1275–1289
49. Katagiri, F., Yamazaki, K., Horikoshi, M., Roeder, R.G. and Chua, N.-H. (1990) Genes Dev. **4**, 1899–1909
49a. Nakayama, T., Okanami, M., Meshi, T. and Iwabuchi, M. (1997) Mol. Gen. Genet. **253**, 553–561

50.  Regier, J.L., Shen, F. and Triezenberg, S.J. (1993) Proc. Natl. Acad. Sci. U.S.A. **90**, 883–887
51.  Treuter, E., Nover, L., Ohme, K. and Scharf, K.-D. (1993) Mol. Gen. Genet. **240**, 113–125
52.  Kawata, T., Nakayama, T., Mikami, K., Tabata, T., Takase, H. and Iwabuchi, M. (1988) FEBS Lett. **239**, 319–323
53.  Sakamoto, A., Minami, M., Huh, G.H. and Iwabuchi, M. (1993) Eur J. Biochem. **217**, 1049–1056
53a. Sakamoto, A., Omirulleh, S., Nakayama, T. and Iwabuchi, M. (1996) Plant Cell Physiol. **37**, 557–562
54.  Takatsuji, H., Mori, M., Benfey, P.N., Ren, L. and Chua, N.-H. (1992) EMBO J. **11**, 241–249
55.  Takatsuji, H., Nakamura, N. and Katsumoto, Y. (1994) Plant Cell **6**, 947–958
56.  Licht, J.D., Grossel, M.J., Figge, J. and Hansen, V.M. (1990) Nature (London) **346**, 76–79
57.  Sauer, F. and Jäckle, H. (1993) Nature (London) **364**, 454–457
58.  Sauer, F., Fondell, J.D., Ohkuma, Y., Roeder, R.G. and Jäckle, H. (1995) Nature (London) **375**, 162–164
59.  Roberts, S.G.E. and Green M.R. (1995) Nature (London) **375**, 105–106
60.  Natesan, S. and Gilman, M.Z. (1993) Genes Dev. **7**, 2497–2509
61.  Bondesson, M., Mannervik, M., Akusjärvi, G. and Svensson, C. (1994) Nucleic Acids Res. **22**, 3053–3060
61a. Okanami, M., Meshi, T., Tamai, H. and Iwabuchi, M. (1996) Genes Cells **1**, 87–99
62.  Guiltinan, M.J., Marcotte, W.R. and Quatrano, R.S. (1990) Science **250**, 267–271
63.  Meier, I. and Gruissem, W. (1994) Nucleic Acids Res. **22**, 470–478
64.  Takase, H., Minami, M. and Iwabuchi, M. (1991) Biochem. Biophys. Res. Commun. **176**, 1593–1600
65.  Hunter, T. and Karin, M. (1992) Cell **70**, 375–387
66.  Hill, C.S. and Treisman, R. (1995) Cell **80**, 199–211
67.  Hunter, T. (1995) Cell **80**, 225–236
68.  Datta, N. and Cashmore, A.R. (1989) Plant Cell **1**, 1069–1077
69.  Harrison, M.J., Lawton, M.A., Lamb, C.J. and Dixon, R.A. (1991) Proc. Natl. Acad. Sci. U.S.A. **88**, 2515–2519
70.  Takase, H., Tabata, T., Mikami, K. and Iwabuchi, M. (1991) Plant Cell Physiol. **32**, 1195–1203
71.  Sarokin, L.P. and Chua, N.-H. (1992) Plant Cell **4**, 473–483
72.  Sun, L., Doxsee, R.A., Harel, E. and Tobin, E.M. (1993) Plant Cell **5**, 109–121
73.  Després, C., Subramaniam, R., Matton, D.P. and Brisson, N. (1995) Plant Cell **7**, 589–598
74.  Sessa, G., Morelli, G. and Ruberti, I. (1993) EMBO J. **12**, 3507–3517
75.  Klimczak, L.J., Schindler, V. and Cashmore, A.R. (1992) Plant Cell **4**, 87–98
76.  Roberts, D.M. and Harmon, A.C. (1992) Annu. Rev. Plant Physiol. Plant Mol. Biol. **43**, 375–414

# 15

# Cell cycle dependent nucleation and assembly of plant microtubular proteins

**\*Marylin Vantard, Virginie Stoppin and Anne-Marie Lambert**

Institut de Biologie Moléculaire des Plantes du C.N.R.S., Université Louis Pasteur, 12, rue du Général Zimmer, 67084 Strasbourg Cedex, France

## Introduction

Microtubules are cytoskeletal polymers involved in a wide variety of cellular mechanisms, such as segregation of chromosomes during mitosis, determination of cell shape, and intracellular transport in eukaryotic cells. Microtubules are tubular structures composed of $\alpha$ and $\beta$ tubulin heterodimers (the basic subunits of the microtubules), and of specific associated proteins [microtubule-associated proteins, (MAPs)]. Intrinsic properties of microtubules include their dynamic instability and polarity [1]. They stochastically alternate between periods of growth or shrinkage by the addition or loss of tubulin subunits, respectively, at their ends. Transition of a microtubule from a growth phase to a shrinking phase is called catastrophe, the opposite transition is called rescue. As microtubules exhibit dynamic instability behaviour, one of their ends (the plus-end) exchanges more tubulin subunits than the other end (the minus-end) [2].

*In vivo*, microtubules are not randomly assembled within the cell, but are distributed in specific arrays. Higher plant cells contain five dynamic microtubular arrays, which are extensively reorganized throughout the cell cycle and during cell differentiation ([3], see Fig. 1). During interphase, cortical microtubules located under the plasmalemma are thought to participate in the deposition and the orientation of the cellulosic microfibrils, whereas nucleus-associated microtubules radiate to the cell cortex. During $G_2/M$ transition, cortical microtubules progressively disappear, simultaneously with the constitution of a new cortical ring of microtubules, the preprophase band (PPB). PPB progressively narrows at the equator, predicting the site of the future division plane, while increased microtubule assembly occurs at the nuclear surface. As the cell progresses into mitosis, microtubules reorganize into an anastral bipolar spindle. In telophase, phragmoplast microtubules participate in the formation of the cell plate during cytokinesis. Hence, during the cell cycle in higher plant cells, the redistribution of microtubules

*To whom correspondence should be addressed*

**Fig. I.**          **Spatial organization of microtubules during the cell cycle in higher plant cells**

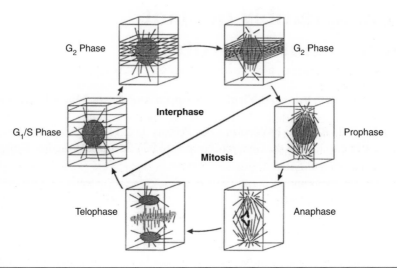

*During $G_1$ and S phases, microtubules are mainly organized in thick bundles beneath the plasma membrane (cortical microtubules). During $G_2$ phase, new microtubules which constitute the preprophase band (PPB) are distributed perpendicular to the spindle axis at the site of the future cell plate, while the number of microtubules emerging from the nuclear surface increases. In late $G_2$, microtubule arrays are gradually replaced by the bipolar mitotic spindle. In prophase, all microtubules are reorganized into the anastral mitotic spindle around the nucleus. In anaphase, the mitotic spindle ensures the accurate segregation of the chromosomes. In telophase, the phragmoplast microtubules are involved in the cell-plate formation.*

in different patterns is closely linked to their acquisition of specialized properties and functions.

The molecular mechanisms which regulate microtubule dynamics in higher plant cells are not yet understood. However, in animal cells, the spatial and temporal distribution of microtubules during interphase and mitosis is controlled mainly by both microtubule-organizing centres (MTOCs) and MAPs, which modify their dynamic properties by specific and transient binding to microtubules. Recent studies suggest strongly that the interactions between MAPs and microtubules are regulated during the cell cycle, and during development [4].

A major purpose of this chapter is to point out molecular mechanisms which may regulate microtubule nucleation, microtubule dynamics and microtubule organization during the cell cycle in higher plant cells. We will restrict our discussion to selected molecular components that we consider to be crucial in understanding the uniqueness of plant microtubule cytoskeleton organization, such as MTOC components and microtubule-interacting proteins.

Recent reviews have been written on the plant cytoskeleton (see [3,5–10]). They provide references and detailed sources of information.

# How, and where, are microtubules nucleated in higher plant cells?

MTOCs are sites within eukaryotic cells where the assembly of microtubules occurs preferentially [11]. MTOCs can be defined by their two major functions: (i) they initiate the assembly of microtubules (a process called nucleation) which are subsequently elongated (a process called elongation); and (ii) they define the polarity of microtubules, the minus-end of the microtubules being anchored at the MTOCs. Consequently, MTOCs are deeply involved in the spatial and temporal organization of the microtubular network *in vivo*. In most eukaryotic cells, this MTOC activity is associated with a discrete organelle from which microtubules emanate. This includes the centrosomes of mammalian cells, composed of two centrioles surrounded by the pericentriolar material responsible for the nucleation of microtubules, and the spindle pole bodies of fungi.

Unlike many animal cells and lower plant cells, higher plant cells lack structurally distinguishable MTOCs. In all higher plant cells examined so far, no centriolar-like structure has been detected at any stage of the cell cycle and, in particular, at the spindle poles during mitosis. This hindered considerably the identification and the characterization of plant MTOCs [12]. However, recent progress has been made where new components of plant MTOCs have been discovered, and where microtubule-nucleation sites have been functionally characterized.

## Molecular composition of higher plant MTOCs

Since microtubule-nucleating factors are thought to be partly conserved between animal and plant kingdoms, antibodies which specifically stain the pericentriolar material of centrosomes have been used to identify components of higher plant MTOCs. Until now, only four of these antibodies were found to cross-react with plant proteins. The human auto-immune serum, 5051, stains the spindle poles from pre-prophase to late anaphase, as well as the surface of both interphase and telophase nuclei in higher plant cells [13,14]. However, the specificity of this labelling has been questioned [15]. Recently, the 5051 antigen was identified in mammalian cells as pericentrin, a highly conserved centrosome protein involved in microtubule organization [16]. Evidence for the existence of pericentrin, or a related protein in plant cells, is still lacking.

A perinuclear labelling by the mammalian anti-centrosome monoclonal antibody, mAb 6C6, was also reported [17]. mAb 6C6 stains the interphase nuclear surface of higher plants in a 'dotty' pattern throughout interphase, until the nuclear envelope breaks down, and reappears in late telophase, as the two daughter nuclei reconstitute. Double labelling with mAb 6C6 and tubulin antibodies revealed that the mAb 6C6 spots at the nuclear surface correspond to discrete foci, where plant microtubules converge. The mAb 6C6 antigen can be solubilized from both centrosomes and isolated maize nuclei by urea treatment. A 100 kDa maize polypeptide is detected by immunoblotting [18]. These results suggest that the mAb 6C6 antigen is a component of plant MTOCs. Unfortunately, its molecular identity in both mammalian and plant cells remains unknown.

An important step toward the elucidation of the properties of MTOCs was the discovery of soluble multiprotein complexes as a precursor to centrosomes in mammalian cells, and in *Drosophila*. These complexes, named gamma-somes, contain $\gamma$-tubulin (as well as $\alpha$- and $\beta$-tubulins), actin, heat-shock protein (hsp70), the protein synthesis elongation factor EF-1$\alpha$ and MAPs [19–22]. Interesting recent data about two of these gamma-some components have been obtained in higher plant cells.

First, using monoclonal antibodies raised against sea urchin centrosomes, S. Hasezawa and T. Nagata [23] detected a 49 kDa protein related to EF-1$\alpha$ [24] which co-localized with cortical microtubules, the PPB and the perinuclear region during interphase, as well as with the spindle/phragmoplast microtubules during mitosis. The association of the 49 kDa protein along microtubules has also been observed in sea urchins, where the protein recognized by the monoclonal antibodies is localized at the centrosome, as well as along the mitotic spindle microtubules [25]. Whether the 49 kDa protein is involved in the plant microtubule-nucleation process, or whether it is a microtubule-interacting protein which may modulate plant microtubule dynamics, remains to be elucidated.

The second gamma-like component identified in plant cells is $\gamma$-tubulin, a highly conserved centrosome protein implicated in microtubule nucleation [26]. $\gamma$-Tubulin is associated with all plant microtubule arrays in a cell cycle, and in a developmentally dependent manner [27–30]. These include cortical microtubules at the $G_1$ stage, the perinuclear region and the PPB at the $G_2$ stage, the polar caps of the late prophase spindle, kinetochore fibres during metaphase and anaphase, the poles in late anaphase, and the phragmoplasts in telophase. Such a widespread localization resembles the distribution of the EF-1$\alpha$ related protein described above. However, the EF-1$\alpha$ related protein is detected along the whole length of the microtubule, whereas $\gamma$-tubulin is associated with microtubules in a punctuated manner. So far, $\gamma$-tubulin has been cloned in *Arabidopsis* [28] and in maize [31]. Sequence comparisons revealed that plant $\gamma$-tubulin shares a high amino-acid similarity (approximatively 70%) with $\gamma$-tubulin of animals and fungi, favouring the idea that $\gamma$-tubulin has the same function as a microtubule nucleator in all eukaryotic cells, including higher plant cells [32]. However, the multiplicity of sites where the $\gamma$-tubulin is detected in higher plant cells is very perplexing (see below for discussion).

## The plant nuclear surface acts as a MTOC

Structural studies using anti-tubulin labelling revealed the existence of nuclear-associated microtubules in higher plant cells. Hence, microtubule nucleation sites could be located at the nuclear surface [33–36]. Preferential incorporation of exogenously labelled tubulin at the nuclear surface of lysed *Haemanthus* endosperm cells during both interphase and telophase strongly reinforced this hypothesis [37]. Recently, an *in vitro* functional assay was developed, in which microtubule assembly is strictly dependent on isolated plant nuclei [18,38]. In the presence of a tubulin concentration insufficient for spontaneous microtubule assembly, isolated plant nuclei could nucleate microtubules *in vitro* (Fig. 2). The staining of the isolated nuclei by mAb 6C6 was abolished after incubation with 2 M urea, and the disappearance of the staining was correlated with the loss of the ability of these isolated nuclei to nucleate microtubules [18]. Furthermore, using this *in vitro* nucleation assay, isolated plant nuclei and purified mammalian centrosomes were similarly affected by external factors, such as plant MAPs [39]. The interphase nuclear surface is the only site in higher plant cells for which an ability to nucleate microtubules has been functionally characterized so far.

Although the plant nuclear surface, the mammalian centrosomes, and the spindle pole bodies of fungi share the property of nucleating microtubules, they are

---

**Fig. 2.**          **Microtubule nucleation around maize nuclei**

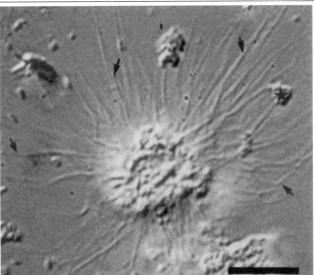

*Isolated nuclei were incubated with purified neurotubulin (14mM) and GTP (1mM) for 20min at 308C. The plant nuclear surface nucleates microtubule assembly (arrows), indicating an MTOC-like activity. Bar: 10mm. Reproduced by permission from Plant Cell [18].*

structurally very different. MTOCs appear to be dispersed in higher plant cells around the nucleus, whereas in mammalian and fungal cell MTOCs, they are focused on discrete organelles [40]. Such a difference may influence microtubule nucleation and microtubule dynamics, since diffusion of tubulin subunits can limit the density of growing microtubules [41]. Assuming that the density of microtubule nucleating sites, and, consequently, the density of microtubules, is lower at the plant nuclear surface than at the centrosome, such a difference may have an effect on dynamics of either nucleus- or centrosome-nucleated microtubules. *In vitro,* microtubules nucleated by isolated plant nuclei are shorter than those nucleated at the centrosomes [39]. These observations may suggest that physical, as well as biochemical, characteristics of MTOCs, in combination with the activities of cell cycle regulated cytoplasmic factors, can regulate microtubule length.

## Does γ-tubulin reveal microtubule-nucleation sites in higher plant cells?

As described above, the plant nuclear surface acts as an MTOC. However, important questions remain. Should we consider the plant nuclear surface as the unique plant MTOC? Or, does each plant microtubule array possess its own nucleation sites? The origin of the cortical microtubules, and of the PPB in interphase, has long been highly debated. It has been hypothesized that cortical microtubules are nucleated either at the nuclear surface [13,14,36], or at the cell cortex [42,43]. Alternative hypotheses to these two opposite models have been proposed [5,10,23,44].

The recent report of the cell cycle dependent association of γ-tubulin with all plant microtubule arrays raised new insights and questions in this debate. We know from studies in mammals and fungi that γ-tubulin is essential for microtubule nucleation [45,46], and that it is associated with both centriolar [45,47–49] and acentriolar [50,51] MTOCs. Furthermore, it was recently demonstrated that, *in vitro,* γ-tubulin specifically binds to the minus-end of microtubules, suggesting that γ-tubulin might link microtubules to MTOCs [52]. Therefore, it is likely that γ-tubulin staining reveals MTOC activity in higher plant cells. Whether each location where γ-tubulin is detected corresponds to a microtubule nucleation site remains to be shown. Three, not mutually exclusive, hypotheses can be considered: (i) each γ-tubulin spot corresponds to a microtubule-nucleation site, suggesting that higher plant cells contain multiple MTOCs located at the nuclear surface and within the cortical microtubule arrays during interphase [42,43]; (ii) some γ-tubulin dots could correspond to γ-tubulin molecules, located at the minus-end of microtubules, which have been detached from a unique MTOC, the nuclear surface, and then translocated to the cell cortex. (However, data obtained from neural cells do not favour this hypothesis, since γ-tubulin has not been detected at the minus-ends of axonal microtubules [53], which are first nucleated at the centrosome in the cell body, and then are detached and translocated into the axon [54]); and (iii) some γ-tubulin spots might correspond to γ-tubulin molecules which have detached from both the MTOCs and the minus-end of microtubules. These γ-tubulin molecules (associated with other components in a multiprotein complex?) could either bind to the walls of microtubules at the minus-end proximal region to stabilize them [55], or be redistributed to other microtubule-nucleation sites in a cell cycle dependent manner [3]. This model is consistent with recent reports, which indicate that in centrosome-containing mammalian cells, γ-tubulin is not restricted to the pericentriolar material, but is also detected in the spindle [56] and in the midbody [57]. The determination of the function(s) of γ-tubulin, and of the other aforementioned centrosome-related proteins, should give an insight into the elucidation of microtubule nucleation during the cell cycle in higher plants. Furthermore, the role of MAPs in the nucleation of the cortical microtubule arrays can be considered [39,58].

One may also hypothesize that, in addition to the activity of centrosome-like proteins such as γ-tubulin, other mechanisms may participate in plant microtubule nucleation. In mitosis, the formation of the microtubule spindle may result from both spontaneous self-reorganization/rearrangement of microtubules, and from lateral interactions of microtubules [59,60]. Given the present state of our knowledge, no single model can be favoured.

## Cell-cycle regulation of plant microtubule-nucleating activity

In mammalian cells, microtubule nucleation at the $G_2/M$ transition may be regulated by both the accumulation of microtubule-nucleating factors at the centrosome, and the phosphorylation of centrosomal components [61,62]. In higher plant cells, microtubules associated with the nuclear surface are more numerous in $G_2$ cells than in $G_1$ cells. Interestingly, the MPM-2 antibody, which specifically recognizes a mitotic phosphoepitope of centrosomal proteins [63], strongly stains the periphery of the plant nucleus at the interphase/prophase transition [64]. Furthermore, the intensity of the γ-tubulin-labelling around the nucleus increases as the cell progresses to mitosis [27]. However, nuclei isolated from either $G_1$ or $G_2$ plant cells possess a similar capacity to nucleate microtubules *in vitro* [39]. All these results suggest that the regulation of microtubule nucleation, and redistribution during the cell cycle of higher plant cells, is a complex phenomenon which may involve at least three families of proteins: (i) microtubule-nucleating factors, which may accumulate at the MTOCs at specific stages of the cell cycle; (ii) cell cycle dependent kinases and phosphatases, which may specifically phosphorylate or dephosphorylate some components of the MTOCs; and (iii) microtubule-interacting proteins, such as MAPs or microtubule-severing factors, which may affect microtubule dynamics and/or microtubule integrity.

# What is the role of MAPs in microtubule organization during the cell cycle of higher plants?

## General background

In eukaryotic cells, progression through the cell cycle requires changes in microtubule organization (see Fig. 1). As cells progress into mitosis, the interphase microtubule network is eliminated and gradually replaced by the bipolar spindle, which will ensure the accurate segregation of the chromosomes between the two daughter cells. All these events are accompanied by changes in the dynamic properties of microtubules, mitotic microtubules being much more dynamic than interphase microtubules in both animal cells [65–67] and plant cells [68]. In addition, dynamic behaviour of kinetochore microtubules is regulated differently during metaphase than during anaphase [69]. Clearly, dynamic instability allows rapid and extensive rearrangements in the microtubule arrays of animal cells both *in vitro* and *in vivo*. Hence, it will be challenging to investigate its role in the generation of new microtubule populations at the cell-cycle-phase transition in higher plant cells.

Recently, much effort has been devoted to an understanding of the molecular mechanisms which regulate the dynamic properties of microtubules. Notably, MAPs, together with other factors, play a critical role in controlling microtubule stability and function. These MAPs are also called 'structural' MAPs, because they bind to microtubules without a requirement for nucleotide involvement, and can be co-purified with tubulin through several cycles of microtubule polymerization/depolymerization [4,70]. The best characterized MAPs have been purified from neural cells, partly because the microtubular proteins are very abundant in nervous tissue (about 20% of the soluble protein). Neuro-MAPs, such as MAP2 and tau, affect the dynamic instability of microtubules by decreasing the frequency of catastrophe, and increasing the frequency of rescue events *in vitro* [71–73]. Thus, each of these MAPs modulates, but does not abolish, the dynamic behaviour of microtubules. Consequently, these MAPs induce an increase of microtubule polymerization and stability, both *in vitro* and *in vivo* [74].

In dividing cells, the most abundant and well characterized MAP is MAP4 [75]. *In vitro*, MAP4 stabilizes microtubules, mainly by increasing rescue frequency at the plus-end of microtubules. When phosphorylated, MAP4 loses its microtubule-stabilizing activity, leading to a decrease in its affinity for microtubules, and an increase of the microtubule dynamics [76]. *In vivo*, MAP4 becomes phosphorylated during mitosis, but still remains associated with the mitotic spindle [77]. In addition to MAP4, the phosphorylation of several other MAPs, such as P220 [78], XMAP230 [79], and E-MAP-115 [80] are cell cycle regulated. Phosphorylation at the $G_2/M$ phase transition seems to be a prerequisite for increasing the dynamic properties of interphase microtubules, which leads to the assembly of the mitotic spindle. In other words, one might expect that MAPs play a crucial role in the maintenance of the integrity of the interphase microtubule network. MAP phosphorylation could lead to the disruption of the microtubule interphase array to allow the assembly of the mitotic spindle.

In higher plant cells, the role of MAPs in microtubule dynamics and organization remains entirely unknown. As mentioned in the introduction to this chapter, the reorganization of microtubules during the interphase–mitosis transition phase proceeds differently from that of animal cells (Fig. 1), and one wonders what role is played by plant MAPs, since these appear to be key regulators in such processes.

In addition to the spatial arrangement of microtubules by assembly/ disassembly, translocation or reorientation of existing microtubules may take place. Several studies using injection of fluorescent tubulin into interphase plant cells suggest that changes which occur in the cortical microtubule array may involve microtubule translocation [81–83]. Stable microtubules could then be relocated into preferred arrangements. Asada and Shibaoka [84] reported a microtubule-translocating activity in plant cell extracts, which occurs in the presence of ATP or GTP. This mechanism could involve motor-MAPs, such as kinesin- or dynein-related proteins, both of which have been identified in plant cells [85–87].

Of special interest, is the study of a novel class of microtubule-related proteins, called microtubule-severing proteins. Indeed, these proteins may disassemble microtubule arrays at the $G_2/M$ phase transition [88–90]. Although microtubule-severing proteins have not yet been reported in plant cells, such proteins could play a role in the disruption of microtubule cortical arrays when cells progress to mitosis. On the other hand, the higher plant EF-1α homologue causes the stabilization and formation of microtubule bundles *in vitro* [91], whereas it has a severing activity in animal cells [92,93]. Even if these data appear contradictory, they provide stimulating debate, and may indicate important different features between plant and animal cytoskeletal proteins.

## Higher plant MAPs

Compared to the animal MAPs mentioned so far, the characterization of structural plant MAPs remains poorly documented, mainly because of the difficulties in purifying sufficient quantities of proteins [94,95].

Recently, several putative plant MAPs have been described. Primarily, higher plant MAPs have been characterized by their ability to bind to taxol-stabilized brain microtubules [96]. These authors reported the purification of a 76 kDa microtubule-interacting protein in carrots, which increases neurotubulin stability against cold-induced depolymerization, and promotes neurotubular bundling. In interphase, it co-localizes with cortical microtubules. From purified fractions of putative MAPs, Chang-Jie and Sonobe [97] isolated a 65 kDa polypeptide in tobacco, which co-localizes with all plant microtubule arrays throughout the cell cycle. *In vitro,* this polypeptide exhibits microtubule-bundling activity, leading to the formation of cross-bridges between microtubules. More recently, two immunologically related putative MAPs of 100 kDa (P100) and 50 kDa (P50), respectively, were characterized in maize [98,99]. Both of these proteins share common epitopes with the neuro-MAP tau. While P100 is expressed in cultured and differentiated maize cells, P50 is expressed in coleoptiles upon phytochrome-induced cell elongation, and is co-localized with transverse cortical microtubules [99].

    None of these aforementioned microtubule-interacting proteins can clearly be defined as structural MAPs, and their effect on microtubule dynamics is not yet known. So far, only fractions of putative MAPs isolated from maize cultured cells initiate and promote MAP-free tubulin assembly under conditions where microtubule self-assembly is not possible [95,100]. These properties, characteristic of MAPs, are similar to those reported for non-fractionated heat-stable neural MAPs [72], or purified tau and MAP2 [71,73]. However, how the plant MAP-fraction affects the rate at which tubulin is added to the microtubule and, more generally, how it modulates microtubule dynamic instability, remains to be investigated.

## Do plant MAPs regulate microtubule assembly in a cell cycle dependent manner?

As mentioned above, the interest in identifying plant structural MAPs is especially because of their possible role in the assembly of microtubule arrays during the cell cycle. The idea that MAPs may play a role in microtubule nucleation, microtubule assembly, and/or microtubule distribution in higher plant cells is especially attractive, because these cells lack a defined MTOC (see above), and possess a highly complex and dynamic microtubular network [68,81–83].

    In order to identify MAPs involved in the regulation of plant microtubule dynamics, one approach is to isolate MAP fractions at different stages of the cell cycle. The effects of these MAPs on microtubule nucleation and growth can then be investigated using a nucleus-mediated microtubule-nucleation assay [18]. Compared to a tubulin-assembly-free system, the use of an MTOC allows discrimination between microtubule nucleation and microtubule elongation. We reported that nuclei were isolated from tobacco BY-2 cells synchronized in $G_1$, and MAPs were purified from BY-2 cells synchronized at different stages of the cell cycle [39]. MAPs were isolated by their ability to bind taxol-stabilized neurotubules [95]. The effects of tobacco MAP fractions on microtubule nucleation at mammalian centrosomes were analyzed in parallel. Below the critical tubulin concentration for spontaneous assembly, both interphase (mostly $G_1$) and mitotic (metaphase–anaphase) MAP fractions promoted microtubule assembly around tobacco nuclei, as well as at purified mammalian centrosomes. However, differences appeared in the activity of these fractions on microtubule assembly. While the mean nucleated-microtubule length increased with the amount of interphase MAPs added to tubulin

**Fig. 3.    Effects of interphase tobacco MAPs on microtubule assembly in presence of tobacco nuclei and mammalian centrosomes *in vitro***

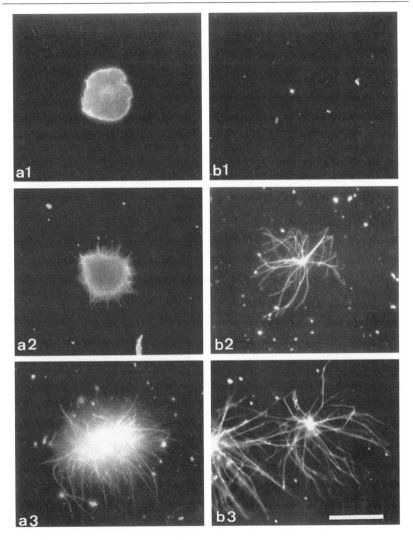

*Tobacco nuclei (a1–3) and mammalian centrosomes (b1–3) were incubated with 10 μM purified tubulin in the presence of various amounts of interphase tobacco MAPs for 20 min at 30 °C. Samples were then fixed and immunolabelled with anti-tubulin antibodies. The ratio (w/w) of interphase tobacco MAPs to tubulin was: a1, b1, 0 (no MAPs); a2, b2, 1:10; a3,b3, 1:7. The length of nucleus- and centrosome-nucleated microtubules increases with the amount of MAPs added to tubulin. Bar: 10 μm. Reproduced by permission from Eur. J Cell Biol. [39].*

(Fig. 3), the mean nucleated-microtubule length did not increase with the quantity of mitotic MAPs added to tubulin (Fig.4).

There are several hypotheses which can be proposed to explain the different microtubule length distribution we obtained. None are mutually exclusive. First, MAPs which could promote microtubule elongation during interphase may be either lacking, or inactivated by phosphorylation during mitosis, and thus microtubule elongation is inhibited. Such an activity has been proposed for the *Xenopus* XMAP230 [79]. Secondly, the short length of microtubules nucleated in the presence of mitotic MAPs may reflect higher dynamics. Indeed, if microtubules grow quickly and shrink rapidly, then according to the model of dynamic instability, their mean length at steady-state should be short. This would

**Fig. 4.**          **Effects of mitotic tobacco MAPs on frequency distribution of centrosome-nucleated microtubules length**

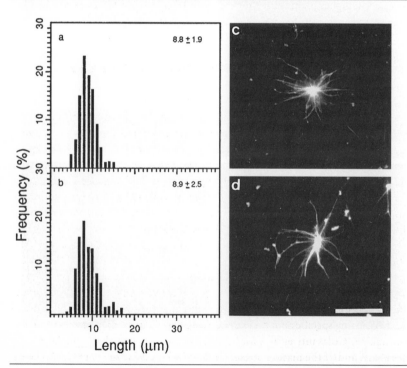

Assays were done for 20 min at 30 °C with purified tubulin (10 μM), mammalian centrosomes, and mitotic MAPs. Mitotic MAP/tubulin ratios (w/w) were: 1:40 (a, c) and 1:20 (b, d). For each experiment, 250 microtubules were measured. Whatever the quantity of mitotic MAPs added to tubulin, the microtubule length is always short and constant. Bar: 10 μm. Reproduced by permission from Eur. J Cell Biol. [39].

imply that mitotic MAPs could regulate the dynamic behaviour of microtubules *in vivo*. By elegant photobleaching experiments, Hush et al. [68] demonstrated that the turnover of plant spindle microtubules is comparable to that reported for animal cells, and is more rapid than the turnover of interphase microtubules. The expression and/or modification of specific MAPs at certain stages of the cell cycle could regulate the stability of microtubules, and, consequently, regulate the size of the mitotic spindle. Phosphorylation of MAPs could be involved in such mechanisms, since it affects binding of MAPs to microtubules, and increases microtubule dynamic instability ([76,79,80], see below). The presence of phospho-proteins associated with the plant mitotic spindle has been extensively documented using staining with the monoclonal MPM-2 antibody [64,101–103]. In endosperm cells of the higher plant *Haemanthus*, MPM-2 appears to be located along microtubule arrays from prophase to telophase, whereas it gives a diffuse staining within the cytoplasm during interphase [64]. These authors suggest a reversible phosphorylation/dephosphorylation of microtubules during interphase–mitosis transition in higher plant cells. Thus, it seems likely that MAPs carry the MPM-2 epitopes, and are targets for mitotic-phase-activated kinases.

## Are plant MAPs substrates of cell cycle dependent kinases?

The requirement of active $p34^{cdc2}$ kinase to generate M-phase-specific events has been extensively studied in recent years. However, few data are available concerning the spatial targeting of this kinase. The first evidence for a role of $p34^{cdc2}$ kinase on microtubule dynamics was described by Verde et al. [104,105], who demonstrated that treatment of *Xenopus* oocyte extracts with $p34^{cdc2}$ induces a decrease in microtubule stability similar to that observed during the interphase–mitosis transition. Since MAPs control microtubule dynamics, they could be the targets of these kinases. Reinforcing this hypothesis, the phosphorylation of a 220 kDa MAP from *Xenopus* [78] and MAP4 [76] by Cdc2 kinase abolishes their microtubule-stabilizing activity.

Several groups have confirmed the presence of a functional $p34^{cdc2}$ kinase homologue in higher plant cells ([106-110]; see J. De Veylder et al., chapter 1). Many recent experiments to examine its intracellular location have been performed [107,111]. Using specific antibodies directed against a functional $p34^{cdc2}$ homologue from maize, Colasanti et al. [107] reported the staining of the nuclei during interphase, and of the mature microtubule PPB in cells during late G2/M (see V. Sundaresan et al., chapter 3). This specific localization suggests that $p34^{cdc2}$ kinase could be involved in the disruption of the PPB, rather than nucleation of new microtubules. Furthermore, the use of inhibitors of protein kinases affect microtubule PPB breakdown in tobacco TBY-2 cells [112], whereas inhibitors of protein phosphatases block its formation [113]. No specific association of $p34^{cdc2}$

with either the mitotic spindle or the spindle poles have been reported, whereas p13[suc1], which associates with and inactivates p34[cdc2], is co-localized with the mitotic spindle when injected in living cells [114]. In animal and yeast cells, p34[cdc2] is associated with the spindle poles in M phase. These differences may be owing to the lack of defined structural MTOCs, such as centrioles in these areas in higher plant cells.

In conclusion, all the indirect information we have today suggests strongly that phosphorylation of MAPs by cell cycle regulated protein kinases is a key regulatory step in the change of microtubule dynamics during interphase–mitosis transitions in higher plant cells. This phosphorylation is only one step of a cascade of signals and regulatory events which drive these processes. In particular, it would be extremely interesting to determine whether the association of plant cdc2 homologues with microtubules is directly mediated by MAPs, or through inter-actions with particular cyclins, as reported for MAP4 [76]. In the near future, we may learn more about Cdk/cyclin protein complexes in higher plant cells (see other chapters in this book), and gain a greater understanding to further investigate changes in microtubule reorganization during the plant cell cycle.

## Summary and concluding remarks

In this review, we have mainly focused attention on the mechanisms which control microtubule dynamics and organization during the cell cycle in higher plant cells. We have seen that the characterization of MTOC components and microtubule-interacting proteins has progressed during the last few years, and given new insights into microtubule-nucleating sites. We have also seen that original approaches have recently been developed using *in vitro* assays. These assays combine neurotubulin, isolated nuclei and putative MAPs isolated from synchronized tobacco cells at specific stages of the cell cycle. They allowed us to study both the nucleating capacity of the only plant MTOC characterized so far, the nuclear surface, and the MAPs involved in microtubule assembly. Together with these *in vitro* assays, recent experiments on living cells, such as microinjection of fluorescent tubulin, suggest that plant microtubule nucleation and dynamics may be regulated in a cell cycle dependent manner.

Much effort has been devoted to an understanding of the molecular mechanisms which control the activity of plant MTOC proteins and MAPs during the cell cycle. Current evidence favours the idea that these proteins could be targets for activated Cdks in living cells (described in several chapters of this book). The study of the effects of phosphorylation on the activity of these proteins with respect to microtubule assembly should be focused upon, in particular during $G_2$ (formation of the PPB), and during the $G_2/M$ transition when the mitotic spindle is forming.

In parallel with the identification of microtubule-nucleating proteins that could be substrates for plant Cdks, the next challenge will be to study how the phosphorylation of MAPs by Cdks regulate the change in microtubule dynamic instability during the cell cycle. The assembly of short and highly dynamic microtubules during mitosis *in vivo*, and the presence of phosphoproteins associated with the mitotic spindle, suggest that phosphorylation of MAPs by Cdks could be involved in the regulation of the catastrophe frequency events, as has been suggested for animal cells.

Since no specific polar MTOCs have been characterized so far in higher plant cells, the question of microtubule nucleation and assembly remains, as to where and how these occur within the two half mitotic spindles. In addition, we can ask whether plant MTOC components are duplicated, as is the case for centrosomes or spindle pole bodies. If so, are they redistributed to ensure spindle pole formation, or do poles originate from reorganization of perinuclear microtubules?

The co-localization of γ-tubulin with all plant microtubule arrays, as well as its distribution at the MTOC(s), remains enigmatic, since no functional data are currently available. In mammalian cells, γ-tubulin has been recently found in both the core structure of centrioles, and in the pericentriolar material [115], suggesting that γ-tubulin not only plays an essential role in microtubule nucleation, but could also have a structural role in centrosome duplication. Corollaries of these main functions in higher plant cells remain unknown.

Owing to their specific features, higher plant cells represent a model of particular interest which provides original approaches for studying cell cycle regulation of microtubule nucleation and dynamics.

## References

1.  Mitchison, T.J. and Kirschner, M.W. (1984) Nature (London) **312**, 237–242
2.  Gelfand, V.I. and Bershadsky, A.D. (1991) Annu. Rev. Cell Biol. **7**, 93–116
3.  Lambert, A.M. and Lloyd, C.W. (1994) in Microtubules (Hyams, J.S. and Lloyd, C.W., eds.), pp. 325–341, Wiley-Liss, New York
4.  Mandelkow, E. and Mandelkow, E.M. (1995) Curr. Opin. Cell Biol. **7**, 72–81
5.  Lambert, A.M. (1993) Curr. Opin. Cell Biol. **5**, 116–122
6.  Staiger, C. and Doonan, J. (1993) Curr. Opin. Cell Biol. **5**, 226–231
7.  Cyr, R. (1994) Annu. Rev. Cell Biol. **10**, 153–180
8.  Goddart, R.H., Wick, S.M., Silflow, C.D. and Snustad, D.P. (1994) Plant Cell Physiol. **104**, 1–6
9.  Shibaoka, H. and Nagai, R. (1994) Curr. Opin. Cell Biol. **6**, 10–15
10. Cyr, R.J. and Palevitz, B.A. (1995) Curr. Opin. Cell Biol. **7**, 65–71
11. Brinkley, B.R. (1985) Annu. Rev. Cell Biol. **1**, 197–224
12. Lambert, A.M. (1995) Bot. Acta **108**, 535–537
13. Clayton, L., Black, C.M. and Lloyd, C.W. (1985) J. Cell Biol. **101**, 319–324
14. Wick, S.M. (1985) Cytobios **43**, 285–294
15. Harper, J.D.I., Mitchison, J.M., Williamson, R.E. and John, P.C.L. (1989) Cell Biol. Int. Rep. **13**, 471–483
16. Doxsey, S.J., Stein, P., Evans, L., Calarco, P.D. and Kirschner, M. (1994) Cell **76**, 639–650
17. Chevrier, V., Komesli, S., Schmit, A.C., Vantard, M., Lambert, A.M. and Job, D. (1992) J. Cell Sci. **101**, 823–835
18. Stoppin, V., Vantard, M., Schmit, A.C. and Lambert, A.M. (1994) Plant Cell **6**, 1099–1106
19. Marchesi, V.T. and Ngo, N. (1993) Proc. Natl. Acad. Sci. U.S.A. **90**, 3028–3032
20. Raff, J.W., Kellog, D.R. and Alberts, B. (1993) J. Cell Biol. **121**, 823–835
21. Stearns, T. and Kirschner, M. (1994) Cell **76**, 623–637

22. Felix, M.A., Antony, C., Wright, M. and Maro, B. (1994) J. Cell Biol. **124**, 19–31
23. Hasezawa, S. and Nagata, T. (1993) Protoplasma **176**, 64–74
24. Kumagai, F., Hasezawa, S., Takahashi, Y. and Nagata, T. (1995). Bot. Acta **108**, 467–473
25. Ohta, K., Toriyama, M., Endo, S. and Sakai, H. (1988) Cell Motil. Cytoskeleton **10**, 496–505
26. Oakley, B.R. (1992) Trends Cell Biol. **2**, 1–6
27. Liu, B., Marc, J., Joshi, H.C. and Palevitz, B.A. (1993) J. Cell Sci. **104**, 1217–1228
28. Liu, B., Joshi, H.C., Wilson, T.J., Silflow, C.D., Palevitz, B.A. and Snustad, D.P. (1994) Plant Cell **6**, 303–314
29. McDonald, A.R., Liu, B., Joshi, H.C. and Palevitz, B.A. (1993) Planta **191**, 357–361
30. Palevitz, B.A., Liu, B. and Joshi, H.C. (1994) Sex. Plant Reprod. **7**, 209–214
31. Lopez, I., Kahn, S., Sevik, M., Cande, W.Z. and Hussey, P.J. (1995) Plant Physiol. **107**, 309–310
32. Joshi, H.C. (1994) Curr. Opin. Cell Biol. **6**, 55–62
33. Lambert, A.M. (1980) Chromosoma **76**, 295–308
34. De Mey, J., Lambert, A.M., Bajer, A.S., Moeremans, M. and De Brabander, M. (1982) Proc. Natl. Acad. Sci. U.S.A. **79**, 1898–1902
35. Wick, S.M. (1985) Cell Biol. Int. Rep. **9**, 357–371
36. Flanders, D.J., Rawlins, D.J., Shaw, P.J. and Lloyd, C.W. (1990) J. Cell Biol. **110**, 1111–1122
37. Vantard, M., Levilliers, N., Hill, A.M., Adoutte, A. and Lambert, A.M. (1990) Proc. Natl. Acad. Sci. U.S.A. **87**, 8825–8829
38. Mizuno, K. (1993) Protoplasma **173**, 77–85
39. Stoppin, V., Lambert, A.M. and Vantard, M. (1996) Eur. J. Cell Biol. **1**, 11–23
40. Mazia, D. (1987) Int. Rev. Cyt. **100**, 49–92
41. Dogterom, M. and Leibler, S. (1993) Phys. Rev. Lett. **70**, 1347–1350
42. Gunning, B.E.S., Hardham, A.R. and Hughes, J.E. (1978) Planta **143**, 161–179
43. Panteris, E., Apostolakos, P. and Galatis, B. (1995) Protoplasma **188**, 78–84
44. Nagata, T., Kumagai, F. and Hasezawa, S. (1994) Planta **193**, 567–572
45. Oakley, B.R., Oakley, C.E., Yoon, Y. and Jung, M.K. (1990) Cell **61**, 1289–1301
46. Joshi, H.C., Palacios, M.J., McNamara, L. and Cleveland, D.W. (1992) Nature (London) **356**, 80–83
47. Horio, T., Uzawa, S., Jung, M.K., Oakley, B.R., Tanaka, K. and Yanagida, M. (1991) J. Cell Sci. **99**, 693–700
48. Stearns, T., Evans, L. and Kirschner, M. (1991) Cell **65**, 825–836
49. Zheng, Y., Jung, M.K. and Oakley, B.R. (1991) Cell **65**, 817–823
50. Gueth-Hallonet, C., Anthony, C., Aghion, J., Santa-Maria, A., Lajoie-Mazenc, I., Wright, M. and Maro, B. (1993) J. Cell Sci. **105**, 157–166
51. Palacios, M.J., Joshi, H.C., Simerly, C. and Schatten, G. (1993) J. Cell Sci. **104**, 383–389
52. Li, Q. and Joshi, H.C. (1995) J. Cell Biol. **131**, 207–214
53. Baas, P.W. and Joshi, H.C. (1992) J. Cell Biol. **119**, 171–178
54. Joshi, H.C. and Baas, P.W. (1993) J. Cell Biol. **121**, 1191–1196
55. Liu, B., Joshi, H.C. and Palevitz, B.A. (1995) Cell Motil. Cytoskeleton **31**, 113–129
56. Lajoie-Mazenc, I., Tollon, Y., Detraves, C., Julian, M., Moisand, A., Gueth-Hallonet, C., Debec, A., Salles-Passador, I., Puget, A., Mazarguil, H., Raynaud-Messina, B. and Wright, M. (1994) J. Cell Sci. **107**, 2825–2837
57. Julian, M., Tollon, Y., Lajoie-Mazenc, I., Moisand, A., Mazarguil, H., Puget, A. and Wright, M. (1993) J. Cell Sci. **105**, 145–156
58. Cyr, R.J. (1991) in The Cytoskeletal Basis of Plant Growth and Form (Lloyd, C.W., ed.), pp. 57–67, Academic Press, New York
59. Bajer, A.S. and Molè-Bajer, J. (1986) J. Cell Biol. **102**, 263–283
60. Smirnova, E.A. and Bajer, A.S. (1992) Cell Motil. Cytoskeleton **23**, 1–7
61. Centoze, V.E. and Borisy, G.G. (1990) J. Cell Sci. **95**, 405–411
62. Masuda, H., Sevik, M. and Cande, Z.W. (1992) J. Cell Biol **117**, 1055–1066
63. Davis, F.M., Tsao, T.Y., Fowler, S.K. and Rao, P.N. (1983) Proc. Natl. Acad. Sci. U.S.A. **80**, 2926–2930
64. Smirnova, E.A. and Bajer, A.S. (1995) Cell Motil. Cytoskeleton **31**, 34–44
65. Salmon, E.D., Leslie, R.J., Saxton, W.M., Karow, M.L. and McIntosh, J.R. (1984) J. Cell Biol. **99**, 2165–2174

66. Saxton, W.M., Stemple, D.L., Leslie, R.J., Salmon, E.D., Zavortink, M. and McIntosh, J.R. (1984) J. Cell Biol. **99**, 2175–2186
67. Belmont, L.D., Hyman, A.A., Sawin, K.E. and Mitchison, T.J. (1990) Cell, **62**, 579–589
68. Hush, J.M., Wadsworth, P., Callaham, D.A. and Hepler, P.K. (1994) J. Cell Sci. **107**, 775–784
69. Zhai, Y., Kronebush, P.J. and Borisy, G.G. (1995) J. Cell Biol. **131**, 721–734
70. Hirokawa, N. (1994) Curr. Opin. Cell Biol. **6**, 74–81
71. Bré, M.H. and Karsenti, E. (1990) Cell Motil. Cytoskeleton **15**, 88–98
72. Pryer, N.K., Walker, R.A., Pretrie-Skeen, V., Bourns, B.D., Soboeiro, M.F. and Salmon, E.D. (1992) J. Cell Sci. **103**, 965–976
73. Brandt, R. and Lee, G. (1993) J. Neurochem. **61**, 997–1004
74. Takemura, R., Okabe, S., Umeyama, T., Kanai, Y., Cowan, N.J. and Hirokawa, N. (1992) J. Cell Sci. **103**, 953–964
75. Bulinski, C. (1994) in Microtubules (Hyams, J.S. and Lloyd, C.W., eds.), pp. 167–182, Wiley-Liss, New York
76. Ookata, K., Hisanaga, S., Bulinski, J.C., Murofushi, H., Aizawa, H., Itoh, T.J., Hotani, H., Okumura, E., Tachibana, K. and Kishimoto, T. (1995) J. Cell Biol. **128**, 849–862
77. Vandré, D.D., Centonze, V.E., Peloquin, J., Tombes, R.M. and Borisy, G.G. (1991) J. Cell Sci **98**, 577–588
78. Shiina, N., Gotoh, Y. and Nishida, E. (1992) EMBO J. **11**, 4723–4731
79. Andersen, S.S., Buendia, B., Dominguez, J.E., Sawyer, A. and Karsenti, E. (1994) J. Cell Biol. **127**, 1289–1299
80. Masson, D. and Kreiss, T.E. (1995) J. Cell Biol. **131**, 1015–1024
81. Wasteneys, G.O., Gunning, D.E.S. and Hepler, P.K. (1993) Cell Motil. Cytoskeleton **24**, 205–213
82. Yuan, M., Shaw, P.J., Warn, R.M. and Lloyd, C. (1994) Proc. Natl. Acad. Sci. U.S.A. **91**, 6050–6053
83. Yuan, M., Warn, R.M., Shaw, P.J. and Lloyd, C. (1995) Plant J. **7**, 17–23
84. Asada, T. and Shibaoka, H. (1994) J. Cell Sci. **107**, 2249–2257
85. Cai, G., Bartalesi, A., Delcasino, C., Moscatelli, A., Tiezzi, A. and Cresti, M. (1993) Planta **191**, 496–506
86. Mitsui, H., Yamagushi-Shinosaki, K., Shinosaki, K., Nishikawa, K. and Takahashi, H. (1993) Mol. Gen. Genet. **238**, 362–368
87. Moscatelli. A., Delcasino, C., Lozzi, L., Cai, G., Scall, M., Tiezzi, A. and Cresti, M. (1995) J. Cell Sci. **108**, 1117–1125
88. Vale, R.D. (1991) Cell **64**, 827–839
89. Shiina, N., Moriguchi, T., Ohta, K., Gotoh, Y. and Nishida, E. (1992) EMBO J. **11**, 3877–3984
90. McNally, F.J. and Vale, R.D. (1993) Cell **75**, 419–429
91. Durso, N.A. and Cyr, R.J. (1994) Plant Cell **6**, 893–905
92. Shiina, N., Gotoh, Y., Kobumura, N., Iwamatsu, A. and Nishida, E. (1994) Science **266**, 282–285
93. Shiina, N., Gotoh, Y. and Nishida, E. (1995) Trends Cell Biol. **5**, 283–286
94. Bokros, C., Hugdahl, J., Hanesworth, V., Murthy, J. and Morejohn, L. (1993) Biochemistry **32**, 3437–3447
95. Schellenbaum, P., Vantard, M., Peter, C., Fellous, A. and Lambert, A.M. (1993) Plant J. **3**, 253–260
96. Cyr, R.J. and Palevitz, B.A. (1989) Planta **177**, 245–260
97. Chang-Jie, J. and Sonobe, S. (1993) J. Cell Sci. **105**, 891–901
98. Vantard, M., Peter, C., Fellous, A., Schellenbaum, P. and Lambert, A.M. (1994) Eur. J. Biochem. **220**, 847–853
99. Nick, P., Lambert, A.M. and Vantard, M. (1995) Plant J. **8**, 835–844
100. Vantard, M., Schellenbaum, P., Fellous, A. and Lambert, A.M. (1991) Biochemistry **30**, 9334–9340
101. Traas, J.A., Beven, A.F., Doonan, J.H., Cordewener, J. and Shaw, P.J. (1992) Plant J. **2**, 723–732
102. Binarova, P., Rennie, P. and Fowke, L. (1994) Protoplasma **180**, 106–117
103. Young, T., Hyams, J.S. and Lloyd, C.W. (1994) Plant J. **5**, 279–284
104. Verde, F., Labbé, J., Dorée, M. and Karsenti, E. (1990) Nature (London) **343**, 233–238

105. Verde, F., Dogterom, M., Stelzer, E., Karsenti, E. and Leibler, S. (1992) J. Cell Biol. **118**, 1097–1108
106. Colasanti, J., Tyers, M. and Sundaresan, V. (1991) Proc. Natl. Acad. Sci. U.S.A. **88**, 3377–3381
107. Colasanti, J., Cho, S.O., Wick, S. and Sundaresan, V. (1993) Plant Cell **5**, 1101–1111
108. Ferreira, P.C., Hemerly, A.S., Villarroel, R., Van Montagu, M. and Inzé, D. (1991) Plant Cell **3**, 531–540
109. Hirt, H., Pay, A., Gyorgyey, J., Bako, L., Nemeth, K., Bögre, L., Schweyen, R.J., Heberle-Bors, E. and Dudits, D. (1991) Proc. Natl. Acad. Sci. U.S.A. **88**, 1636–1640
110. Hemerly, A.S., Ferreira, P., Engler, J.A., Van Montagu, M., Engler, G. and Inzé, D. (1993) Plant Cell **5**, 1711–1723
111. Mineyuki, Y., Yamashita, M. and Nagahama, Y. (1991) Protoplasma **162**, 182–186
112. Katsuta, J. and Shibaoka, H. (1992) J. Cell Sci. **103**, 397–405
113. Hasezawa, S. and Nagata, T. (1992) Bot. Acta **105**, 63–69
114. Hepler, P.K., Sek, F.J. and John, P.C. (1994) Proc. Natl. Acad. Sci. U.S.A. **91**, 2176–2180
115. Fuller, S.D., Gowen, B.E., Reinsch, S., Sawyer, A., Buendia, B., Wepf, R. and Karsenti, E. (1995) Curr. Biol. **5**, 1384–1393

# 16

# Cell cycle mutants in higher plants: a phenotypical overview

## Jan Traas* and Patrick Laufs

INRA, Laboratoire de Biologie Cellulaire, Route de Saint Cyr, 78026
Versailles Cedex, France

## Introduction

From the preceding chapters, it is clear that significant advances have been made in our understanding of mitosis in plants. Indeed, recent work has allowed the identification of different cell cycle regulators, such as cyclins, CDKs and different phosphatases, confirming how extremely well conserved the basic mechanism is in all eukaryotes [1]. Now that we have identified some of the essential regulators, we can start to ask questions about the spatial and temporal regulation of mitosis during plant development. It is unlikely that the upstream regulation has been as well conserved as the basic machinery, and therefore we cannot base our research entirely on a further search for plant homologues of yeast and animal genes. An essential tool in the elucidation of the cell-cycle phosphorylation cascade has been the use of mutational genetic dissection. Although it could yield important extra information, such a genetic approach has never been used to analyse developmental control of plant cell division. This is not surprising. Mutant screens for cell division control in multicellular organisms are hard to design, because, in a broad sense, all mutations affecting growth, including those perturbing metabolic pathways, will also have their effects on mitosis. What we need, therefore, is a set of characteristics that will allow us to identify relevant mutations affecting directly the spatial regulation and mechanisms of higher plant cell cycle control. As a first step towards such a list of criteria, we describe a range of mutants isolated by different laboratories, including our own, in terms of cell division patterns. Most of these mutants were isolated in *Arabidopsis thaliana*, but some relevant mutants from other species were also considered. There are probably many other phenotypes that have been described in the past. However, since many of these have not been studied in terms of cell division control, they are difficult to trace using standard literature search procedures. Therefore, we have mainly restricted ourselves to the recent literature. In this chapter, we will present the first classification of these mutants, and discuss their potential importance for our understanding of the developmental control of cell division. This review only

*To whom correspondence should be addressed

concerns sporophytic development, and mutants perturbed during the gameto-phytic phase will not be considered here. In addition, we have not included embryolethals (for reviews, see [2,3]), as no detailed analyses of cell division activity in the majority of these mutants are available yet. This does not mean that plants with gametophytic and embryonic lethal phenotypes are irrelevant for our understanding of plant cell division (see [4], and references therein).

Plant development has evolved under the constraint imposed by the cell walls that tightly cement together all cells, and fix their positions within tissues. Most of the mitotic activity is restricted to relatively small cell populations, i.e. the meristems, and the young growing tissues surrounding them. This is accompanied by mechanisms which regulate division plane alignment, which control the correct positioning of daughter cells after mitosis. Although both the spatial control of mitotic activity, and division plane alignment, link morphogenesis and cell division, they are two distinct processes (see V. Sundaresan et al., chapter 3). We will consider them separately in this chapter.

# The spatial regulation of mitotic activity

## Mitotic activity during development

In plants, most of the mitotic activity is restricted to the meristems and the young growing tissues surrounding them. Meristems play a central role in plant development ([5,6]; see D. Francis, chapter 9; J.H. Doonan, chapter 10). The pri-mary meristems have to initiate the formation of organs and determine their basic structure, whereas another type of meristematic cell, the cambial cell, is responsible for secondary growth and increasing the width of stems and branches. In addition, the meristems have to maintain themselves in order to guarantee a constant supply of new cells. This functional complexity is reflected in meristem structure. For instance, careful histological analysis has shown that the apical meristems are organized into layers that correspond to the basic structure of the more mature organs. Thus, the L1, L2 and L3 layers in the shoot apical meristem will preferen-tially give rise to the epidermis, mesophyll and vascular tissues, respectively. Superimposed on this partition in layers, are zones which are supposed to be func-tionally different. Again, in the shoot apex, we can distinguish the central zone and the peripheral zones, which are thought to be involved in meristem maintenance and organ initiation, respectively. This subdivision in distinct zones coincides with important differences in mitotic activity (see D. Francis, chapter 9). In roots, the cells of the quiescent centre divide at a low rate compared to their neighbours. In maize roots, cells of the quiescent centre divide up to 10 times slower than those in the cortex (reviewed in [6]). Likewise, the cells in the peripheral part of the shoot apex divide more rapidly than those at the summit of the meristem, although here the differences are less dramatic (e.g. a 3-fold difference found in *Pisum*). In roots of many species, the cell cycle tends to be controlled at the $G_1/S$ transition, as it is mainly the $G_1$ phase that is prolonged [6]. In shoots, this is less evident, and, depending on the species studied, either $G_1$ or all interphase states can vary in length. The factors that determine these differences in division rate, and those that control the mitotic state of the cells in the meristems, are completely unknown.

Cell division does not occur only within the meristem. In the shoot apex, cells in the young growing organs retain mitotic activity and, even later, groups of specialized cells can still divide, such as in the 'meristemoids' that will form stomatal complexes. In growing organs, however, the cells will eventually lose their capacity to divide, depending on their age. The fully developed first leaves of *Arabidopsis* contain about 135 000 cells [7]. Given the number of meristematic cells that participate in the formation of the first leaf primordium, which can be reasonably estimated at about 25 [8,9], 12–13 rounds of division should be sufficient to generate the entire leaf, after which cell division gradually stops, from the top to the bottom of the leaf.

## Cell division patterns and cell cycle gene expression

It is interesting to compare cell division patterns with the gene expression patterns of the cell cycle regulators, in order to see to what extent they overlap. Although the results are still scarce, different sets of experiments confirm that three distinct expression patterns exist, depending on the gene studied. In *Arabidopsis*, the functional *cdc2* homologue is expressed in all dividing and in all growing (but not necessarily mitotic) cells. Similar patterns are found for a range of genes involved in cell division (for example, [10]; also see L. De Veylder et al., chapter 1). In contrast, genes encoding mitotic cyclins, and others, such as those encoding histones, are only expressed preferentially in cycling cells during specific stages of the cell cycle ([10]; also see D. Dudits et al., chapter 2; N. Chaubet and C. Gigot, chapter 13). Recent expression studies on the $G_1$-type or D-cyclins show that there is a third type of pattern, as these genes are only expressed in cells that are actively dividing (see J.A.H. Murray et al., chapter 5; J.H. Doonan, chapter 10). From these studies, we can conclude that, from a cell cycle point of view, there are at least three positions which the cells can occupy: (i) not competent to divide, (ii) competent to divide, but not dividing, and (iii) dividing, the last, of course, being subdivided into the $G_1/S/G_2$ and M phases. Each position (or state) is characterized by a specific combination of cell cycle expression patterns (see J.H. Doonan, chapter 10). Hence, it is likely that, during development, mitotic activity is at least partially regulated at the mRNA level by the activation and inactivation of specific sets of genes that form the basic cell-cycle mechanism. This leaves us with the question: what are the upstream regulators that organize cell cycle gene expression and cell division in three dimensions during development?

## Mutants perturbed in the spatial regulation of cell division

One of the possible approaches to answer this question is to look for mutants perturbed in the spatial control of cell division. As already mentioned, it did not seem useful to start extensive screens for mutants, as it was unclear what the most interesting phenotypes would be. Therefore, we limited ourselves to existing mutants available in our laboratory, or phenotypes described in the literature. This first analysis confirmed our expectation that many mutants show abnormal cell division patterns. For many of these cases, however, this is accompanied by severely perturbed growth outside the meristems and even necrosis. These mutants will not be considered here, although, strictly speaking, these genes are necessary for the co-ordination of mitotic activity. With regard to the remaining mutants, it is often difficult, at this stage, to attribute a specific role to the genes, especially since only a minority have been identified. For a classification, we chose to distinguish between mitosis inside and outside the meristems. Although this might seem arbitrary, it can be argued that mitotic activity has different functions in the two zones. Outside the meristems, as soon as the primordia are generated, mitosis accompanies growth and probably serves, as put in a recent review by Jacobs [1], not only to "to subdivide space", but to also increase the amount of DNA to sustain expansion. This is not necessarily the case in meristems, however, where the regulation of mitotic activity could be necessary for establishing basic plant architecture, i.e. to create functionally different units. Therefore, it appears logical to consider mitotic activity inside and outside the meristems separately.

## Genes regulating cell proliferation in the meristems

A whole range of genes seem to control cell division in the meristems. As it is not possible to discuss each of them within the context of this chapter, we will limit ourselves to three subgroups, which have been partially characterized.

The first set of genes appears to be necessary for maintaining the meristematic state, and, as a consequence, for maintaining populations of actively dividing cells at the apex of roots and shoots (Table 1). For the shoot apex, at least, one family of homoeobox genes has been identified by mutation which appears to be necessary for installing and maintaining meristematic activity. In maize, the best characterized member of this family is *KNOTTED (KN1)* [11,12]. This gene was originally identified by a transposon insertion into an intron, causing ectopic expression in leaf tissue. Interestingly, this resulted in a dominant phenotype characterized by extra cell divisions in and around the vascular tissues. *In situ* hybridization experiments later showed that this gene was only expressed in vegetative meristems in wild-type plants. The gene has been introduced into tobacco, under control of the 35S promoter, and, again, extra cell divisions were observed in and around vascular tissues [13]. Strong overexpression caused the formation of ectopic meristems at the same location. Although these results suggest that *KN1* is involved in meristem function, its exact role in the meristem is not known, as mutations reducing its activity have not been isolated. More recently, another member of the *KN1* family was identified by mutation: the *SHOOT-MERISTEMLESS (STM)* gene [14,15]. Inactivation of this gene resulted in a complete absence of meristematic and mitotic activity at the shoot apex. These results suggest not only that the *KN1*-like genes function in the initiation of meristems, but that they can also activate the mitotic machinery in non-meristematic cells, as shown by the leaf phenotypes in overexpressers. Interestingly, these supernumerary divisions only occurred in vascular tissues which, at least in *Arabidopsis*, have been shown to contain non-dividing, but competent, cells. This would imply that the protein product is unable to induce cell division in non-competent cells, such as mesophyll and epidermis. Recently, Cheng et al. [16] identified two loci that, when mutated, block outgrowth of the root meristem. The genes were therefore termed ROOTMERISTEMLESS (RML). This name, however, does not accurately define the phenotype, as mutants are able to form embryonic root meristems. Development is blocked only postembryonically, as root cells are unable to divide further. The mutants were able to form lateral roots at the collet, but, again, these laterals only developed the basic root meristem, which then failed to develop further. This shows that root formation is a two-step process, whereby the *RML* genes in the second step appear to be necessary specifically to start mitotic activity. In tobacco, a comparable phenotype was identified, which was termed *rootless* [17]. This mutation confers a dominant phenotype, however, and although the root is unable to develop, the aerial parts of the plant remain unaffected, and the plant is able to develop after grafting on to a wild-type specimen. Therefore, like *rml*, the *rootless* mutant is unable to induce mitosis in the meristematic cells, which nevertheless elongate and differentiate normally. As a consequence, the root meristem disappears. In a number of plants, the vascular cells at the root tip start to divide, and eventually will give rise to a callus. This phenotype is associated with perturbed concentrations of not only auxins, but also cytokinin and abscisic acid. In addition, it was shown to be tolerant to the auxin, indoleacetic acid (IAA).

The second group of genes controlling meristematic activity functions in an opposite way to the *KNOTTED* and *RML* type. Mutations in these genes cause meristems to increase in size dramatically, which can be accompanied by either an increased or a reduced production of organs. Some of the mutants are almost incapable of forming primordia at all, suggesting that the transition from the meristematic to the non-meristematic state is perturbed, as in the T37 mutant isolated in our laboratory (C. Jonak and J. Traas, unpublished data, Fig. 1). Many of the mutants in this group show fasciation. Examples of these mutants in *Arabidopsis* are *clavata* (*clv*1–3) [18,19], *fasciata* (*fas*1–2) [20] and *mgoun* (*mgo*) (P.Laufs et al., unpublished data, Fig. 1), but similar phenotypes are found in many species (e.g. see [21]). Although the division patterns are perturbed in these mutants, it is difficult to define their exact function in division control. In very general terms, they seem to restrict the size of the meristem and, therefore, the number of mitotically active cells in the growing plant (see Table 1). In the past, hormones, such as 2,4-dichlorophe-noxyacetic acid (2,4-D), and hormone inhibitors, such as 2,3,5-tri-iodobenzoic acid (TIBA) can induce fasciation when sprayed on the growing plants [e.g. 21]. These results are not always easy to interpret and, at best, they suggest that some of the

**Table I      Mutants perturbed in the spatial regulation of cell division**

**I.          Genes regulating cell division in the meristems**

(a) Genes necessary for inducing and/or maintaining mitotic activity in the meristem

| Name | Species | Phenotype of mutant |
|---|---|---|
| SHOOTMERISTEMLESS | A. thaliana | No shoot apical meristem |
| ROOTMERISTEMLESS I | A. thaliana | No functional root apical meristem |
| ROOTMERISTEMLESS 2 | A. thaliana | No functional root apical meristem |
| WUSCHEL (43) | A. thaliana | Reduced mitotic and meristematic activity |
| HANDSHAKE (*) | A. thaliana | Reduced meristem size |
| REVOLUTA (26) | A. thaliana | Reduced meristematic and mitotic activity |
| FOREVERYOUNG (44) | A. thaliana | Reduction/arrest of mitotic activity in shoot |
| PINFORMED | A. thaliana | No division in floral meristem primordium |
| PINOID | A. thaliana | No division in floral meristem primordium |
| KNOTTED | Z. mays | Dominant, ectopic expression of the gene, ectopic cell division |
| ROOTLESS | N. tabacum | Dominant, arrest of cell division in the root meristem |

(b) Genes necessary for restricting proliferation of the meristem

| Name | Species | Phenotype of mutant |
|---|---|---|
| CLAVATA I | A. thaliana | Increased meristem size, fasciation |
| CLAVATA 2 | A. thaliana | Increased meristem size, fasciation |
| CLAVATA 3 | A. thaliana | Increased meristem size, fasciation |
| FASCIATA I | A. thaliana | Increased meristem size, fasciation |
| FASCIATA 2 | A. thaliana | Increased meristem size, fasciation |
| HANABA TARANU (2) | A. thaliana | Increased meristem size, fasciation |
| MGOUN | A. thaliana | Increased meristem size, fasciation |
| T37 | A. thaliana | Increased meristem size, finger-like leaves |
| ALTERED MER.PROGRAMME (45) | A. thaliana | Cytokinin-resistant, increased proliferation/rate of leaf prod |
| DEETIOLATED | A. thaliana | Activation of mitotic activity in the dark |
| CONSTITUTIVE PHOTO-MORPHOGENIC | A.thaliana | Activation of mitotic activity in the dark |

**Table 1 (contd.)  Mutants perturbed in the spatial regulation of cell division**

**2.**  **Genes necessary for the regulation of mitotic activity outside the meristems**

(a) Genes necessary to maintain correct cell numbers outside the meristems

| Name | Species | Phenotype of mutant |
| --- | --- | --- |
| PALE CRESS | A. thaliana | 50% reduction of mesophyll cell numbers |
| CRISTAL | A. thaliana | 90% reduction of mesophyll cell numbers |
| T37 | A. thaliana | Reduced cell numbers in leaves, reduced leaf size |
| PIN-FORMED | A. thaliana | No cell division outside shoot apical meristem |
| PINOID | A. thaliana | No cell division outside shoot apical meristem |

(b) Genes necessary to restrict mitotic activity outside the meristems

| Name | Species | Phenotype of mutant |
| --- | --- | --- |
| REVOLUTA | A. thaliana | Extra cell divisions in leaves and stems |
| FAT (46) | N. sylvestris | Extra cell divisions in leaves |
| MGOUN | A. thaliana | Ectopic cell divisions on stems |
| SUPERROOT | A. thaliana | Formation of multiple root meristems on hypocotyl |

For details, see text. The references describing mutants not discussed in the text are shown in parentheses. Asterisk between parentheses: Orbovic, J. Traas and Höfte, unpublished data.

**Fig. I**     **Phenotypes of two mutants with abnormal cell proliferation in and outside the meristems**

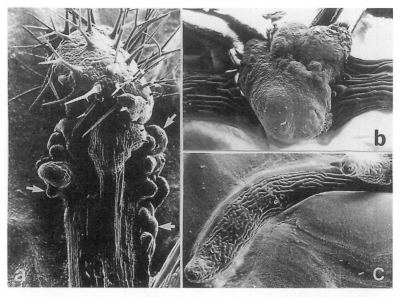

*(a) Scanning electron microscopy (SEM) of mgoun in Arabidopsis, showing a fasciated inflorescence with ectopic meristematic tissue along the stem (arrows). (b) Shoot apical meristem of T37, 3 weeks old. The meristematic cells proliferate, but are initially unable to form leaves. This results in a callus-like meristem. (c) Leaf of a T37 mutant. In most cases, the mutants are able to form finger-like leaves. Many of these leaves remain small in size, with reduced cell numbers. The size of the individual cells is normal. Note that all epidermal cell types are present.*

results are not always easy to interpret and, at best, they suggest that some of the genes are necessary to maintain correct hormone levels in the apical meristem.

Cell proliferation in the meristems is dependent on environmental factors. One of the best characterized factors is light, which is an absolute prerequisite for activation of the shoot and (to a lesser extent) the root apical meristems after germination. In the dark, cell proliferation is inhibited by a set of genes, called *DEETIOLATED (DET)* and *CONSTITUTIVE PHOTOMORPHENIC (COP)* respectively, which were defined by mutation (for reviews, see [22–24]). When these genes are inactivated, the shoot and root apical meristems are active in the dark, forming leaves and lateral roots. Mutants do not have a phenotype in the light. It is not clear how these genes interact with the meristem genes, or the cell cycle regulators. Hemerly et al. [28a] showed that, in the dark, *cdc2a* is not expressed in the shoot apex. Consequently, the *DET* and *COP* genes at least must be necessary to keep the meristematic cells in a 'non-competent' state in the dark. Their targets and their exact function, however, are unknown.

## Genes involved in cell division control outside the meristems

After a group of cells has left the shoot apical meristem, they will go through a limited number of divisions, before mitotic activity ceases. This means that, initially, mitotic activity has to be maintained, in spite of important changes in the developmental programme. It also means that cell division has to be arrested at a specific moment in the development of the growing organs. In terms of the expression of cell-cycle genes, as explained above, this concurs with a transition from active mitotic (*cdc2a* and cyclin expression in *Arabidopsis*) to non-competent to divide (no *cdc2a* and cyclin detectable). Mutants perturbed in this pattern can again be subdivided into several subsets (Table 1). A first group of genes appears to be necessary to inhibit cell divisions outside of the meristem. Phenotypes can vary from supernumerary cell divisions in the leaves or other organs, to the presence of ectopic meristems. Several of these genes appear to be involved in meristem function, and therefore also belong to class I (Table 1) discussed above. *Mgoun*, for example, showed ectopic cell division and the formation of meristems along the stem (P. Laufs et al., unpublished data). The *SUPERROOT (SUR)* gene [25] is necessary to inhibit extra cell divisions and root meristem formation in the root and the hypocotyl (Table 2). In the *sur* mutant, extra root meristems are induced along the central cylinder in the hypocotyl. This mutant has increased levels of free and conjugated auxin. Levels of other hormones do not seem to be affected. In addition, a phenocopy can be obtained when wild-type plants are grown on high auxin concentrations. This shows that hormones are able to trigger meristematic and mitotic activity (see J.A.H. Murray et al., chapter 5). The ectopic mitoses are not randomly distributed; only cells from the pericycle seem to respond. It is interesting to note that these cells also express *cdc2a*, in contrast to the surrounding tissues (see L. De Veylder et al., chapter 1). Therefore, it seems that the mutation affects mainly cells that are competent to divide, comparable to the way in which *KNOTTED* overexpression does in the leaf.

A second set of genes has the opposite function. These are necessary to keep populations mitotically active, until sufficient cells are generated. When these genes are inactivated, the number of cells in the differentiated organs is reduced. Such mutants are most easily identified by screening for abnormal leaf shape. Most of the mutations affect all tissue types within the leaves ([26], J. Traas et al., unpublished data, Fig. 1). There are some interesting exceptions, however. In particular, the *pale cress (pac)* mutants have reduced numbers of mesophyll cells [27]. More dramatic phenotypes were isolated, such as in the *cristal* mutant, which showed a 90% reduction in the number of mesophyll cells [27a]. So far, none of these cell-number mutants have been analysed in sufficient detail to allow further conclusions to be drawn, but we may expect to find different mutants that are perturbed in 'counting' the amount of cell divisions that are needed to complete organ formation. How this counting is regulated, and how tightly cell numbers and leaf size are coupled, is unknown (see D. Francis, chapter 9). Recent results using antisense *cdc2a* constructs in tobacco [28] suggest that cell numbers in the leaves could directly depend on the expression levels of the gene. The results with *cdc2* antisense also show that cell numbers can be reduced 5–10-fold without affecting leaf size, and that cell division and growth can, at least partially, be uncoupled. On the other hand, we have isolated mutants that show reduced cell numbers and reduced leaf size, although cell size is normal (Fig. 1). Clearly, the analysis of organ size and cell number mutants will shed more light on the relationship between mitotic activity and growth.

Several mutants affecting cell divisions outside the inflorescence meristems have also been identified. In particular, the *pin* [29] and *pinoid* [30] mutants in *Arabidopsis* show a reduced capacity to generate mature flowers, flower organs and cauline leaves. Many mutants are unable to generate any organs at all, although scanning electron microscopy (SEM) suggests that primordia are initiated. We can expect therefore that one of the functions of these genes is to maintain mitotic activity once the primordia are formed. The biochemical and physiological basis of the phenotype is still completely unknown. However, a reduced level of polar auxin transport could possibly be involved.

## Cell division plane alignment

A second aspect which characterizes dividing cells in the developing plant is the process of division plane determination. The precise alignment of root cells in files, and the anticlinal division observed in the L1 layers in the shoot meristem, both appear to indicate that the orientation of the division plane and, hence, the correct positioning of daughter cells, is a prerequisite for ordered development in higher plants. In this respect, a major role has been attributed to the cytoskeleton (e.g. see [31]). Microscopical analyses have revealed not only a cytoskeletal ring at the future division site containing microtubules, but also other elements. This structure, the preprophase band (PPB), appears at $G_2$, and disappears at late prophase (see D. Dudits et al., chapter 2, and V. Sundaresan et al., chapter 3). There is a strong correlation between this structure, and the capacity of plant cells to align their cell walls. It has been proposed that the PPB not only predicts, but also fixes, the plane of division, and it is generally assumed that the presence of a PPB is essential for normal morphogenesis.

A number of questions, however, have remained unanswered. Although plant cells have a predetermined future division site, it is not known how this is achieved, or why they preferentially exhibit specific division planes. Moreover, the function of the PPB has only been established at the level of a correlation; there is no strict proof that this structure actually determines the division site. In more general terms, we do not know to what extent division plane alignment is essential

## Fig. 2    Pattern formation in the root meristem of wild-type and
### *hyp2* roots (*N. plumbaginifolia*)

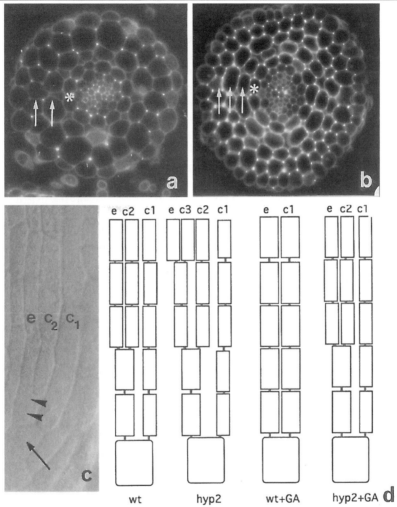

(a,b) Cross-sections through a mature part of the root of the wild-type (a) and hyp2 (b). Arrows indicate the cortical-cell layers. Asterisks indicate endodermis. Note that hyp2 has one extra layer. (c) Longitudinal, optical section through a wild-type root. At the bases of the endodermis (e) and the inner cortex (c2) cell files, a single 'stem' cell can be observed (arrowheads). In a similar way, there is one cell (arrow) at the bases of the e/c2 layer and the c1 layer. (d) Schematic representation of the cell division patterns giving rise to the outer (c1) and inner (c2) cortex layers and the endodermis (e). In hyp2, there is one extra longitudinal cell division giving rise to a c3 layer. When wt and hyp2 are treated with gibberellic acid (GA), the amount of longitudinal cell divisions is reduced, and one cortical layer disappears. Figs. (a), (b) and (c) reproduced with permission from J. Traas et al., Plant J. **7**, 785–796 for plant morphogenesis at all. Finally, from a cellular point of view, it is not known how the reorganization of the cytoskeleton during PPB formation is regulated. Different lines of research have indicated a direct relationship with the phosphorylation cascade, in particular the plant *cdc2* homologue, which is tightly associated with the cytoskeleton during the $G_2/M$ transition [32].

In order to answer these questions, we initiated a search for mutants. Our expectation was that such plants would be severely affected in their development. But, since it was again not clear what all the possible phenotypes were, we restricted ourselves to existing mutants. In particular, we were looking for mutants that would develop as callus-like structures from the embryo stage onwards, since previous work suggested that PPB formation was perturbed in calli [33]. We also, therefore, looked at a number of embryolethals.

The results, so far, allow us to identify about 10 phenotypes, which appear to be of interest (Table 2). Their phenotypes are variable, ranging from only minor modifications in cell arrangements to the formation of callus-like organs. Relatively minor modifications are observed in mutants like *gib-1* [34] in tomato, and *hyp2* in *Nicotiana plumbaginifolia* [35]. Both mutants show abnormal cell alignment in the root meristems, causing the formation of extra cell layers. In the *hyp2* mutant, it was possible to define this phenotype, in terms of abnormal division plane alignment in the root meristem. Two modifications were observed: first, extra anticlinal divisions were observed in all cell layers from pericycle to epidermis, and secondly, we observed one extra round of periclinal, longitudinal division in the cells that normally give rise to the endodermis (Fig. 2). This resulted in the formation of one extra cortical cell layer. This phenotype was in agreement with abnormal extra lateral expansion in the mutant. Since the gibberellin-deficient *gib-1* mutant in tomato also has extra cell layers [34], we tested the effect of gibberellic acid on the division patterns. Gibberellic acid reduced the extra lateral cell expansion, and, at the same time, inhibited the formation of the extra cell layer. Interestingly, the hormone also reduced the number of periclinal divisions in the wild type (Fig. 2). Experiments using an inhibitor of gibberellic acid, paclobutrazole, increased lateral cell expansion in both mutant and wild type and, simultaneously, increased the number of cell layers. This led us to conclude that the hormone and the *HYP2* gene both act independently on cell expansion and pattern formation in the root of *N. plumbaginifolia*. These results also supported the hypothesis instigated by different authors, that cell-expansion characteristics can directly influence the decision of cells to divide in a particular plane. A mutant with more severe alterations in cell patterns is *tangled*, of maize. Mutations in the *TAN* gene alter the shape and arrangement of cells in all tissues, although differentiation patterns and morphogenesis are not severely affected. Smith *et al.* [36] further investigated this phenotype, reporting particular changes in the orientation of the division planes. Instead of the regular patterns of transverse and longitudinal divisions observed during the growth of normal leaves, highly variable orientations of new cross walls were observed. A more quantitative analysis revealed that, in fact, only longitudinal divisions were affected, since the proportion of transverse planes was unchanged in the mutant. The *TAN* gene therefore appears to be necessary for the programming of longitudinal divisions only.

| Table 2 | Mutants with perturbed division plane alignment | |
|---|---|---|
| **Name** | **Species** | **Phenotype** |
| HYPOCOTYL 2 | N. plumbaginifolia | Increased longitudinal division planes |
| MERISTEM 8 (**) | N. plumbaginifolia | Random division planes in root meristem |
| GIBBERELLIN 1 | L. esculentum | Increased amount of longitudinal divisions (?) |
| TANGLED | Z. mays | Perturbed longitudinal division planes |
| GNOM | A. thaliana | Abnormal first division plane |
| FASS | A. thaliana | Random division planes, no PPBs |
| TONNEAU | A. thaliana | Random division planes, no PPBs |
| LAM | N. sylvestris | Perturbed periclinal divisions in L3 during leaf formation |
| FAT | N. sylvestris | For abnormal periclinal divisions in mesophyll |

For details, see text. The meristem 8 mutant (**): J. Traas and M. Caboche, unpublished data.

Mutations like *tan* and *hyp2* appear to affect the upstream regulation of division plane alignment, and not the process itself. A number of mutations in *Arabidopsis* affect cell division patterns from the first embryo stages, onwards. Some of these genes are therefore likely to perform basic functions in the mechanism of division plane alignment itself. An example is the *GNOM* gene [37]. It has been proposed, based on the early phenotype of *gnom* mutants, that this gene is essential for setting up the initial asymmetric division in the zygote, giving rise to the suspensor and the embryo proper. Perturbed activity of this gene results in a symmetric division and, ultimately, in a severely perturbed phenotype, lacking shoot and root meristems [37]. The gene, which was recently cloned, shows some homology with the gene encoding the SEC7 protein in yeast [38]. This suggests that a basic cellular process is affected in this mutant, and that the gene does not play a specific role in the early stages of embryo formation, However, this conclusion is still open for discussion and, clearly, further functional analysis is necessary to determine the exact role of the *GNOM* gene.

Another example of a mutation affecting cell division plane alignment throughout development is *fass*. Mutants grow as short, misshapen barrel-shaped plants ([39,40], Fig. 3), which nevertheless are able to form roots, hypocotyls, cotyledons, leaves and flowers, all in their correct relative positions. This phenotype is associated with abnormal cell shape and division plane alignment [39,40]. Cellular

**Fig. 3**          **Comparison of wild-type and *ton* seedlings in *Arabidopsis***

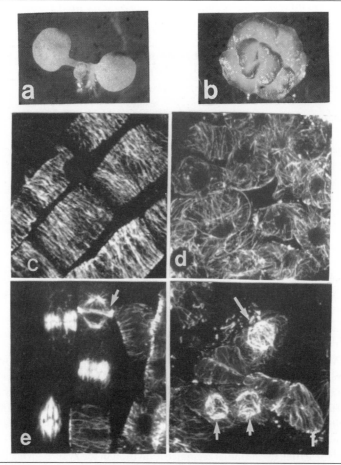

*(a) Wild-type seedling, (b) ton seedling, showing thick, round leaves and perturbed phyllotaxis. (c–f) cytoskeleton in root cells of wild-type and mutant plants. (c) Interphase cortical arrays in meristematic root cells. Microtubules are transversely oriented, perpendicular to the axis of cell expansion. (d) Interphase microtubular array of a ton mutant. Microtubules are randomly oriented. (e) Mitotic cells in wild-type root cells, showing microtubules in the preprophase band (PPB) (arrow), phragmoplast and spindle. (f) Cells of a ton root. Three cells are in prophase, and show microtubules forming the spindle around the nucleus before nuclear envelope breakdown. Cytoplasmic microtubules are still present, but no PPB is observed. Figs. reproduced with permission from J. Traas et al., (1995) Nature (London) 375, 676–677.*

form is highly irregular in all tissues, and cells appear to be unable to orient their axes. This is associated with an inability to align division planes, and even cells in the embryo divide in apparently random patterns. As a consequence, the regular cell files, usually observed in the root and hypocotyl, are completely absent. In our laboratory, six mutants with a comparable phenotype were isolated [40]. Genetic analysis showed that they belonged to two different complementation groups, initially termed *ton* 1 and *ton* 2 (from *tonneau*, the French for *fass*), the latter represented by five alleles. Further complementation tests showed that *ton2* was identical to *fass* (C. Bellini and P. Nacry, personal communication).

To further characterize the role of these genes in plant development and division plane alignment, we compared the microtubular organization in wild-type and mutant plants (Fig. 3). Wild-type root meristems showed the typical interphase arrays of parallel microtubules arranged perpendicular to the root axis. PPBs were observed in many interphase and prophase cells, as well as spindles and phragmoplasts. In differentiated cells, various orientations of microtubules were observed, ranging from helical (in the root, hypocotyl and petioles), to longitudinal (root hairs), or random (mesophyll). This organization corresponds to observations made for many other species. In *fass* and *ton* mutants, the microtubules show highly modified arrangements of the cortical arrays [40]. During interphase, the microtubules are randomly oriented in the root meristem, and even in differentiated cells the cortical array is disorganized. Only one cell type, root hairs, does not seem to be affected. A second major difference from the wild type is that PPBs are never observed, although spindles and phragmoplasts form normally. The conclusion is, therefore, that the *FASS* and *TON* genes are necessary for the organization of the cortical microtubules in transverse and helical arrays, and for the formation of PPBs. Further work, now in progress, is necessary to characterize the exact role of these two genes in cytoskeletal organization, but, at present, the mutants already allow us to draw conclusions on the importance of PPB formation in plant morphogenesis. It is generally assumed that both PPB formation and ordered cortical arrays are essential to set up the basic differentiation patterns of the developing organs. The *ton* and *fass* mutants, however, show that this is not the case, and they support the hypothesis that polarized cell expansion and division plane alignment are not required for spatial positioning of organs and tissue types. Although, strictly speaking, it is unclear whether all characteristics of the phenotypes are owing to the abnormal arrangements of the cytoskeleton, the small size and slow development of the mutants suggest that correct alignment of the cortical microtubules is essential, at least, for rapid and efficient growth.

## Mutations in cell cycle genes?

Most of this chapter is concerned with the upstream regulation of mitotic activity and cell division patterns. At first sight, it appears unlikely that any mutant screen at the embryonic or postembryonic level would allow us to identify mutants perturbed in the basic cell-cycle machinery, as the gametes carrying such mutations would not be viable. On the other hand, one may expect to find leaky mutations, where the protein has retained some of its activity. Moreover, it has been shown that both cyclins and CDKs form multigene families, so that related genes may substitute for the inactivated members of the same family. Finally, there is the possibility that some of the members of these multigene families are not expressed during the haplophase, as was shown for other genes, like those encoding the tubulins and actins. Recent reports from different laboratories support this idea. Springer et al. [41] used gene-trap tagging to identify the MCM2-3-5 homologue, *PROLIFERA*. Although, supposedly, this gene is necessary for DNA synthesis, the *prolifera* mutant is able to develop until the embryonic stage is reached. Female gametophyte development was inhibited, although viable gametes could be obtained. Unexpectedly, the male gametophyte remained unaffected.

Another example is the *KNOLLE* gene. Although it was originally identified as a potential pattern mutant, recent cloning of the gene revealed it to be a member of the syntaxin family, involved in membrane trafficking. Interestingly, this gene shows a cell cycle dependent expression pattern [42]. This expression pattern, in combination with the observation that the mutant exhibits cells with abnormal phragmoplasts, argues strongly for a specific role for the gene during cell division. Other examples of viable mutants perturbed in basic cellular processes have already been mentioned, such as *ton*, *fass* and probably also *gnom*. In addition, Hemerly et al. [28] showed using antisense constructs that plants with greatly reduced levels of *cdc2* can give viable plants. This implies that, probably, many of the basic cell-cycle regulators will be accessible for mutation (and it may be that many of these mutants are already available in the stock centres). The constitution of large collections of insertional mutants in model plants, such as A*ntirrhinum*, maize, *Petunia* and *Arabidopsis*, will also allow us to search directly for inserts in known cell-cycle genes, so that we may expect to have such mutants in the near future.

## Concluding remarks

In this inventory, we have tried to show that the developmental control of cell division is amenable for genetic dissection. From the phenotypes and the genes that were cloned so far, it appears that it will even be possible to identify mutations in basic cell-cycle functions, such as DNA synthesis, PPB formation and cytokinesis. Theoretically, we should therefore also have access to basic regulators that possibly have not been identified previously in other eukaryotes. Considering the observation that four such genes were identified at the embryonic level, it might be worthwhile to analyse in more detail the many embryo lethal mutants described by different groups.

For the immediate future, it is hard to propose a particular mutant screen that will allow us to identify other upstream regulators of mitotic activity. It appears, however, that leaves are very convenient for screening purposes, as leaf size, cell size and cell numbers can be determined relatively easily. Nevertheless, other criteria, such as root size, root cell numbers, or stem length and cell numbers, might also be useful. Since no thorough search for these mutants has been performed so far, it is likely that such screens will provide a range of additional mutants, as we are probably still far from saturation. Moreover, the existing mutants can be used to look for other genes in suppressor screens, for instance. In parallel, we can also start to isolate the genes that were identified by mutation, and continue their functional analysis. This should involve a careful analysis of the expression of cell-cycle genes in mutant backgrounds. In several phenotypes, we can expect to find abnormal patterns, as in mutants showing ectopic cell divisions.

Not all the mutants described in this review will be useful for furthering our understanding of cell cycle control. However, we can anticipate that a rigorous genetic analysis will provide a solid basis for our future research on the role of cell division in plant development.

## References

1.  Jacobs, T.W. (1995) Annu. Rev. Plant Mol. Biol. Pl. Physiol. **46**, 317–339
2.  Bowman, J., ed., (1993) *Arabidopsis*, An Atlas of Morphology and Development, Springer Verlag, New York
3.  Meinke, D. (1994) in *Arabidopsis* (Meyerowitz E.M. and Somerville C.R. eds.), pp. 253–296, Cold Spring Harbor Laboratory Press, Cold Spring Harbor
4.  Staiger, C. and Cande, W.Z. (1990) Dev. Biol. **138**, 231–242
5.  Steeves, T.A. and Sussex, I.M. (1991) in Patterns in Plant Development, Cambridge University Press, Cambridge
6.  Lyndon, R.F. (1990) in Topics in Plant Physiology 3: Plant Development, Black and Chapman, Boston
7.  Pyke, K.A., Marrison, J.L. and Leech, R.M. (1991) J. Exp. Bot. **244**, 1407–1416
8.  Irish, V.F. and Sussex, I. (1992) Development **115**, 745–753
9.  Telfer, A. and Poethig, R.S. (1994), in *Arabidopsis* (Meyerowitz E.M. and Somerville C.R. eds.), pp. 379–401, Cold Spring Harbor Laboratory Press, Cold Spring Harbor
10. Fobert, P.R., Coen, E.S., Murphy, G. and Doonan, J. (1994) EMBO J. **13**, 616–624
11. Vollbrecht, E., Veit, B., Sinha, N. and Hake, S. (1991) Nature (London) **350**, 241–243
12. Jackson, D., Veit, B. and Hake, S. (1994) Development **120**, 405–413
13. Sinha, N.R., Williams, R.E. and Hake, S. (1993) Genes Dev. **7**, 787–795
14. Barton, K.M., and Poethig, R. S. (1993) Development **119**, 823–831
15. Long, J., Moon, E., Medford, J. and Barton, K. (1996) Nature (London) **379**, 66–69
16. Cheng, J., Seeley, K.A. and Sung, Z.R. (1995) Plant Physiol. **107**, 365–376
17. Pelese, F., Megnegneau, B., Sotta, B., Sossountzov, L., Caboche, M. and Miginiac, E. (1989) Plant Physiol. **89**, 86–92
18. Clark, S.E., Running, M.P. and Meyerowitz, E.M. (1993) Development **119**, 397–418
19. Clark, S.E., Running, M. and Meyerowitz, E.M. (1995) Development **121**, 2057–2067
20. Leyser, H. and Furner, I.J. (1992) Development **116**, 397–403
21. Gorter, C. (1965) in Encyclopaedia of Plant Physiology (Ruhland, W. ed.), pp. 330–351, Springer Verlag, New York
22. Deng, Z., Caspar, T. and Quail, P. (1991) Genes Dev. **5**, 1172–1182
23. Chory, J.C., Peto, C., Feinbaum, R., Pratt, L. and Ausubel, F. (1989) Cell **58**, 991–999
24. Chory, J. and Susek, R.E. (1994) in *Arabidopsis* (Meyerowitz, E. and Somerville, C. eds.), pp. 579–614, Cold Spring Harbor Laboratory Press, Cold Spring Harbor
25. Boerjan, W., Cervera, M.T., Delarue, M., Beeckman, T., Dewitte, W., Bellini, C., Caboche, M., Van Onckelen, H., Van Montagu, M. and Inzé, D. (1995) Plant Cell **7**, 1405–1419
26. Talbert, P.F., Adler, H.T., Parks, D.W. and Comai, L. (1995) Development **121**, 2723–2735

27.    Rieter, R.S., Coomber, S.A., Bourett, T.M., Barley, G.E. and Scolnik, P.A. (1994) Plant Cell **6**, 1253–1264
27a.   Delarve, M. Santoni, V., Cabochi, M., Bellini, C. (1997) Planta **202**, 51–61
28.    Bennett, S.R.M., Hemerly, A., Almeida-Engler, J., Bergounioux, C., Van Montagu, M., Engler, G., Inzé, D. and Ferreira, P. (1995) EMBO J. **14**, 3925–3936
28a.   Hemerly, A., Ferreira, P., Almeida-Engler, J., Van Montagu, M., Engler, G. and Inzé, D. (1993) Plant Cell **5**, 1711–1723
29.    Okada, K., Ueda, J., Komaki, M., Bell, C. and Shimura, Y. (1991) Plant Cell **3**, 667–684
30.    Bennett, S.R.M., Alvarez, J., Bossinger, G. and Smyth, D. (1995) Plant J. **8**, 505–520
31.    Lloyd, C.W. (ed.), in The Cytoskeletal Basis of Plant Growth and Form, Academic Press, London
32.    Colasanti, J., Cho, S., Wick, S. and Sundaresan, V. (1993) Plant Cell **5**, 1101–1111
33.    Traas, J.A., Renaudin, J.P. and Teysendier de la Seve, B. (1990) Plant Sci. **68**, 249–256
34.    Barlow, P.W. (1993) in Molecular and Cell Biology of the Plant Cell Cycle (Ormrod, J. and Francis, D., eds.), pp. 179–199, Kluwer Academic Publishers, Dordrecht
35.    Traas, J.A., Laufs, P., Jullien, M. and Caboche, M. (1995) Plant J. **7**, 785–796
36.    Smith, L., Hake, S. and Sylvester, A. (1996) Development **122**, 481–489
37.    Mayer, U., Buttner, G. and Jurgens, G. (1993) Development, **117**, 149–162
38.    Shevel, D.E., Leu, W.M., Gillmor, C., Zia, G., Feldmann, K. and Chua, N.H. (1994) Cell **77**, 1051–1062
39.    Torres-Ruiz, R. and Jurgens, G. (1994) Development **120**, 2967–2978
40.    Traas, J., Bellini, C., Nacry, P., Kronenberger, J., Bouchez, D. and Caboche, M. (1995) Nature (London) **375**, 676–677
41.    Springer, P., McCombie, W., Sundaresan, V. and Martienssen, R. (1995) Science **268**, 877–878
42.    Lukowitz, W., Mayer, U. and Jurgens, G. (1995) Cell **84**, 61–71
43.    Jurgens, G. (1994) in *Arabidopsis* (Meyerowitz, E.M. and Somerville, C.R. eds.), pp. 297–312, Cold Spring Harbor Laboratory Press, Cold Spring Harbor
44.    Medford, J., Behringer, F., Callos, J. and Feldmann, K. (1992) Plant Cell **4**, 631–643
45.    Chaudbury, A.S., Letham, S., Craig, S. and Dennis, E.S. (1993) Plant J. **4**, 907–916
46.    McHale, N. (1993) Plant Cell **5**, 1029–1038

# Index